Gas Processing

Gas Processing
Environmental aspects and methods

James G. Speight

Butterworth-Heinemann Ltd
Linacre House, Jordan Hill, Oxford OX2 8DP

PART OF REED INTERNATIONAL BOOKS

OXFORD   LONDON   BOSTON
MUNICH   NEW DELHI   SINGAPORE   SYDNEY
TOKYO   TORONTO   WELLINGTON

First published 1993

© Butterworth-Heinemann 1993

British Library Cataloguing in Publication Data
Speight, James G.
  Gas Processing: Environmental Aspects and Methods
  I. Title
  665.7

ISBN 0 7506 1132 4

Library of Congress Cataloguing-in-Publication Data
Speight, J. G.
  Gas processing: environmental aspects and methods/James G. Speight.
  p. cm.
  Includes bibliographical references and index.
  ISBN 0 7506 1132 4
  1. Gas manufacture and works.   2. Natural gas.   I. Title.
  TP751.S66   1993
  665.7–dc20   92–43354   CIP

Typeset by Vision Typesetting, Manchester

Printed and bound in Great Britain

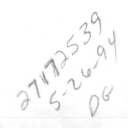

Contents

Part Two: Gas cleaning 173

9 Processing – general concepts 175

10 General process classification 210

11 Processing equipment 231

Preface

The last two decades have seen not only perturbations of energy supply systems but also changes in attitudes of governments and voters alike towards environmental issues that are unlikely to be settled within the next two decades. With respect to the former, disruptions of oil supply and quantum leaps (as well as quantum decreases!) in oil prices have, to say the least, sent shudders through industry and to the consumers that are still being felt and discussed. With respect to the latter, environmental issues will be with us as long as we cut down trees, manufacture consumer goods, and burn fossil fuels for energy. And it is this latter issue that is the subject of this text.

Optimists live for the moment when the environment will be pristine and clean again whereas realists assume that this will never be the case and look to return to a stage of minimal pollution that will allow life to proceed without the threat of extinction.

At the same time, it must be recognized that supplies of crude petroleum in North America are dwindling, and, unless new fields are found, serious shortages will ensue. Indeed, current estimates also put a finite lifetime on both OPEC and non-OPEC petroleum. It is, therefore, quite evident that other fuel sources must be used. In the general context of the present text, these sources are coal and natural gas. These resources represent tremendous potential since they also encourage fulfilment of the goal of moving towards a degree of self-sufficiency. Thus, the use of these energy resources as combustible fuels and the resulting effect on the environment are major issues that need to be addressed. The use of coal as a combustible fuel has been known for centuries and the use of natural gas as a supplementary fuel at industrial plants is a reality. But so also is the generation of gaseous products that are not indigenous to the atmosphere, at least in any great quantities, and the effect of these products on the surrounding flora and fauna.

On this issue, it is necessary to consider the gradual rise in temperature of the earth's atmosphere that has occurred over the last few decades. This so-called greenhouse effect arises because of the increased concentration of carbon dioxide, a gas that is indigenous to the atmosphere but certainly not in the quantities that are being produced by a whole host of fossil fuel conversion plants.

The emissions of carbon dioxide into the atmosphere are known to come from the combustion of fossil fuels – particularly the heavier fuels such as coal. It is believed that there is a strong need to move to the higher hydrogen/carbon fossil fuels to combat carbon dioxide production. Thus, there is a need not only to promote the use of natural gas as a fuel for combustion, which produces less carbon dioxide per unit of fuel, but also to understand the chemistry and engineering of the combustion of fossil fuels and the means by which the gaseous pollutants are produced. Following on from this, it may then be possible to design suitable new methods by which their emission to the atmosphere is not only reduced but completely mitigated. It is the purpose of this book to outline the current methods and known technologies that will aid in the development of processes by which this might be accomplished.

Gas processing, although generally understandable using chemical and/or physical principles, is often confusing because of the frequent changes in terminology and, often, the lack of cross-referencing. For the purposes of this book, and in an attempt to alleviate some of the confusion that arises from uncertainties in the terminology, gas cleaning (gas processing) can be conveniently subdivided into two parts: Part One deals with the origins of process gases and also contains chapters dealing with recovery, properties and composition. Part Two deals with the chemistry and engineering aspects of the methods, and principles involved, by which the gas streams produced during industrial operations might be cleaned from their noxious constituents.

Thus, although gas processing employs different process types there is always an overlap between the various concepts. Therefore, in any text such as this, there is a need, for the convenience of the reader, for repetition and this is achieved by subdividing the subject categories. When necessary, cross-referencing is employed so that the reader will not miss any particular aspect of the processing operations.

The sections relating to testing procedures contain references to the relevant standard test method. In some cases, reference is given to older methods as well as to the current methods. Even though some of the older test methods are no longer in use, in so far as they have been replaced by newer methods, it is considered useful to refer to these methods as they have played an important role in the evolution of the newer methods. Indeed, some of the older methods are still preferred by many laboratories, hence their inclusion here.

Finally, attempts have been made to render each chapter a stand-alone segment of the book. Whilst every effort is made to ensure adequate cross-referencing, sufficient information is included in each chapter to give the reader the necessary background.

In summary, this book provides a ready, easy-to-use reference source to compare the scientific and technological aspects of gas-processing operations and the means by which the environment might be protected.

Author's note

For the sake of simplicity, illustrations in the book are usually line drawings. Further, no attempt has been made to illustrate the myriad of valves, heat exchangers and the like that may occur within a processing sequence.

In addition, the various units are presented as those familiar to North American readers and as SI units. As an example, training in the engineering disciplines is more likely to employ the Fahrenheit temperature scale whereas many of the other scientific disciplines employ the Celsius (Centigrade) temperature scale. For the sake of clarity, the text contains temperatures in both scales but it should be noted that exact conversion is not always possible. Accordingly, the interconversion of these scales in this text is often to the nearest 5° where such licence (especially in describing processing temperatures) would not cause serious errors or misconceptions. In all other cases, conversion is as close as possible.

Part One

Origin, properties and composition of gases

1
Historical and modern perspectives

Introduction

The use of fossil fuel resources has increased by several orders of magnitude since the early decades of this century. The expansion of the industrialized system and the increased use of automobiles, to mention only two examples, have been the major driving forces behind the expansion of fossil fuel usage. But, in concert with this increased usage, there has also followed the onset of detrimental side effects – an excellent example of the medicine almost killing the patient!

The emissions resulting from the use of the various fossil fuels have had deleterious effects on the environment and promise further detriment unless adequate curbs are taken to control not only the nature but also the amount of gaseous products being released into the atmosphere.

Fossil fuels are a necessary part of our modern world, hence the need for stringent controls over the amounts and types of emissions from the use of these fuels. The necessity to clean up process gases is real. Moreover, to intimate that the only fossil fuel needing to be cleaned is natural gas would be a gross error. Gaseous products and by-products are produced in petroleum refineries, gas from coal is an old technology, and gases are produced in a variety of other industries (Austin, 1984; Probstein and Hicks, 1990; Speight, 1990, 1991). These gaseous products all contain quantities of noxious materials that are a severe detriment to the environment. Not to admit to this would be a serious omission of fact. All fossil fuels produce noxious products that need to be removed before use or discharge to the atmosphere.

This book focuses on clean-up procedures that might be applied to all process gases. Of course, the major focus is on natural gas as well as gases from the coal and petroleum refinery industries. Whilst these industries might be the major source of gaseous pollutants, there are other industries that also produce noxious emissions and many of the concepts and processes outlined in this text can be applied to gas clean-up in a variety of industries.

But first, since fossil fuels are a major source of potential pollutants, it is necessary to understand, by way of introduction, the various types of fossil fuels that generate noxious emissions and the evolution of their use at present. Fossil fuels are, in general, found in solid, liquid and gas (or vapour) phases.

Coal is a very abundant fossil fuel and forms a major part of the earth's fossil fuel resources (Berkowitz, 1979; Hessley et al., 1986; Hessley, 1990), the amount available being subject to the method of estimation and to the

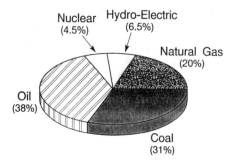

Figure 1.1 Distribution of world energy supplies

definition (see Chapter 2) of the resources (EIA, 1988, 1989, 1991a,b, 1992). For the purposes of this text, in terms of measurable reserves coal is assumed to constitute about 3% of the fossil fuel supplies of the world (Figure 1.1). But, in the relative terms of energy content, it is generally recognized that coal (68%) and natural gas (13%) are the major energy-containing fossil fuels, with petroleum (19%) making up the remainder (EIA 1989, 1991a,b, 1992).

Coal is, perhaps, the most familiar of the fossil fuels, not necessarily because of its usage throughout previous centuries (Galloway, 1882) but more because of its extremely common usage during the nineteenth century; it became responsible, in large part, for the industrial revolution. Coal is extremely important in the present context because of the associated, and very necessary, clean-up of the emissions when it is burned.

Coal occurs in various forms defined by rank or type (Table 1.1). Not only is coal a solid hydrocarbonaceous material with the potential to produce considerable quantities of carbon dioxide as a result of combustion, but also many coals contain considerable quantities of sulphur (Table 1.2), the concentration of which varies (Table 1.3) but which, nevertheless, opens up both the possibility and the reality of sulphur dioxide production (Manowitz and Lipfert, 1990).

Another solid fossil fuel, oil shale (Scouten, 1990), has received much less attention but it also has the capability of producing noxious emissions. Oil shale is composed of an organic material (kerogen) which occurs in a matrix of shaley strata and has the potential to produce pollutant emissions by virtue of its composition (Table 1.4). However, it will receive no attention here and is included for definition only. Indeed, the oil shale industry is virtually non-existent, but even if it did exist it would also suffer from the necessity to discharge aqueous and solid pollutants into the environment (Probstein and Hicks, 1990). This aspect of fossil fuel processing is not dealt with in this text.

Liquid fossil fuels are usually given one or more of the following names: oil, crude oil, petroleum, heavy oil. The near-solid bitumens (which are also referred to as natural asphalts but, in the truest sense of the terminology, incorrectly so since asphalt is a product of refinery processing and may be an altered, rather than natural, material ; bitumen occurs in various locations throughout the world (Speight, 1990, 1991)) are often classified as a heavy oil although various subdivisions of this classification are possible (Tables 1.5 and 1.6).

The gaseous component that often occurs in reservoirs of petroleum is

Table 1.1 Analytical data for coals of the United States

No.	Coal rank Class	Coal rank Group	State	County	M	VM	GC	A	S	Btu	Rank FC	Rank Btu
										(Coal analysis bed moisture basis)		
1	I	1	PA	Schuylkill	4.5	1.7	84.1	9.7	0.77	12 745	99.2	14 280
2	I	2	PA	Lackawanna	2.5	6.2	79.4	11.9	0.60	12 925	94.1	14 880
3	I	3	VA	Montgomery	2.0	10.6	67.2	20.2	0.62	11 925	88.7	15 340
4	II	1	WV	McDowell	1.0	16.6	77.3	5.1	0.74	14 715	82.8	15 600
5	II	1	PA	Cambria	1.3	17.5	70.9	10.3	1.68	13 800	81.3	15 595
6	II	2	PA	Somerset	1.5	20.8	67.5	10.2	1.68	13 720	77.5	15 485
7	II	2	PA	Indiana	1.5	23.4	64.9	10.2	2.20	13 800	74.5	15 580
8	II	3	PA	Westmoreland	1.5	30.7	56.6	11.2	1.82	13 325	65.8	15 230
9	II	3	KY	Pike	2.5	36.7	57.5	3.3	0.70	14 480	61.3	15 040
10	II	3	OH	Belmont	3.6	40.0	47.3	9.1	4.00	12 850	55.4	14 380
11	II	4	IL	Williamson	5.8	36.2	46.3	11.7	2.70	11 910	57.3	13 710
12	II	4	UT	Emery	5.2	38.2	50.2	6.4	0.90	12 600	57.3	13 560
13	II	5	IL	Vermilion	12.2	38.8	40.0	9.0	3.20	11 340	51.8	12 630
14	III	1	MT	Musselshell	14.1	32.2	46.7	7.0	0.43	11 140	59.0	12 075
15	III	2	WY	Sheridan	25.0	30.5	40.8	3.7	0.30	9 345	57.5	9 745
16	III	3	WY	Campbell	31.0	31.4	32.8	4.8	0.55	8 320	51.5	8 790
17	IV	1	ND	Mercer	37.0	26.6	32.2	4.2	0.40	7 255	55.2	7 610

Note: For definition of coal rank class, see Hessley (1990).
Date on coal (bed moisture basis):
M = equilibrium moisture, %; VM = volatile matter, %; FC = fixed carbon, %; A = ash, %; S = sulphur, %; Btu = high heating value, Btu per lb; rank
Btu = moist mineral matter free Btu per lb; all calculations by Parr formulae.
Source: Douglas M. Considine (ed.) (1977) *Energy Technology Handbook*, McGraw-Hill, New York, pp. 1–20.

Table 1.2 Representative sulphur contents of coals of the United States

Region	No. samples	Organic S(%)	Pyritic S(%)	Total S(%)
N. Appalachian	227	1.00	2.07	3.01
S. Appalachian	35	0.67	0.37	1.04
E. Midwest	95	1.63	2.29	3.92
W. Midwest	44	1.67	3.58	5.25
Western	44	0.45	0.23	0.68
Alabama	10	0.64	0.69	1.33

Source: Gilleland and Swisher (1986).

Table 1.3 Distribution of sulphur within the coal groups of the United States

	Sulphur content range (%)			
Rank	0–0.7	0.8–1.0	1.1–3.0	3.1+
	Percentage of total in each range			
Anthracite	96.5	0.6	2.9	
Bituminous	14.3	15.2	26.2	44.3
Subbituminous	66.0	33.6	0.4	
Lignite	77.0	13.7	9.3	
US average	46	19	15	20

Source: Gilleland and Swisher (1986).

Table 1.4 Analytical data for various kerogen concentrates

	Green River (Rifle, CO)	Aleksinac (Yugoslavia)	Irati (Brazil)	Pumpherston (Scotland)
Organics (wt %):				
C	77.39	71.87	78.83	75.70
H	10.26	8.73	9.47	10.37
N	3.10	3.21	3.92	4.18
O+S (diff.)	9.25	16.19	7.78	9.75
Atomic H/C	1.59	1.46	1.44	1.64

Table 1.5 Classification of petroleum by density-gravity

Type of crude	Characteristics
1 Conventional or 'light' crude oil	Density-gravity range less than 934 kg/g^3 ($>20°$ API)
2 'Heavy' crude oil	Density-gravity range from 1000 kg/m^3 to more than 934 kg/m^3 (10$°$ API to $<20°$ API)
	Maximum viscosity of 10 000 mPa s (cp)
3 'Extra-heavy' crude oil; may also include atmospheric residue (b.p. $>340°$C; $>650°$F)	Density-gravity greater than 1000 kg/m^3 ($<10°$ API)
	Maximum viscosity of 10 000 mPa s (cp)
4 Tar sand bitumen or natural asphalt; may also include vacuum residue (b.p. $>510°$C; $>950°$F)	Viscosity greater than 10 000 mPa s (cp) Density-gravity greater than 1000 kg/m^3 ($<10°$ API)

Table 1.6 Subclassification of petroleum and various derivatives

Natural materials	Manufactured materials	Derived materials
Petroleum	Wax	Oils
Heavy oil	Residuum[b]	Resins
Mineral wax	Asphalt[c]	Asphaltenes
Bitumen (native asphalt)	Tar	Carbenes
Asphaltite	Pitch	Carboids
Asphaltoid	Coke	
Migrabitumen	Synthetic crude oil	
Bituminous rock[a]		
Bituminous sand		
Kerogen		
Natural gas		

[a] Tar sand is a misnomer; tar is a product of coal processing. Oil sands is also a misnomer but is equivalent to usage of 'oil shale'. Bituminous sand is more correct; bitumen is a naturally occurring asphalt.
[b] The non-volatile portion of petroleum, often further defined as 'atmospheric' (b.p. $> 350\,°C$; $> 660\,°F$) or vacuum (b.p. $> 565\,°C$; $> 1050\,°F$).
[c] A product of a refinery operation, usually made from a residuum.

natural gas. It is usually processed in a petroleum refinery but there are sources of natural gas which occur without the associated presence of petroleum (Chapter 3). Even though the gas may, to all intents and purposes, be characterized as methane, there are those constituents of it which present the potential for pollution and must be removed.

The rate of primary energy use (Figure 1.2) (OPEC, 1991), as energy from fossil fuel resources is often termed, is subject to many factors, all of which mirror the evolution of technology (Gibbons *et al.*, 1989). This, in turn, is related to population growth.

For example, after the Second World War, a phenomenal rise occurred in the use of petroleum and its various products, mostly through the development of the internal combustion engine (the automobile) and the accompanying expansion of industrial operations. At the same time, the growth in the use of electricity during the last four decades is related to the rise in the use of household consumer products such as refrigerators, cookers, etc.

So it is predictable that fossil fuels will be the primary source of energy for the next few decades, well into the next century, and therefore the message is clear: until other energy sources supplant coal, natural gas and petroleum the challenge is to develop technological concepts that will provide the maximum recovery of energy from these fossil fuel resources (Fulkerson *et al.*, 1990). Thus, it is absolutely essential that energy from such resources be obtained cheaply and efficiently (Dryden, 1975; Yeager and Baruch, 1987), and with minimal detriment to the environment.

To complete the energy resource scenario and to put fossil energy resources into perspective, there are non-fossil fuel energy resources (Figure 1.3) which are derived from the sun and generically referred to as non-fossil resources, or renewable resources, or/and geophysical energy resources. These resources include direct solar energy, hydroelectric energy, and elements of minerals from which nuclear energy may be derived (Hafele, 1990; Weinberg and

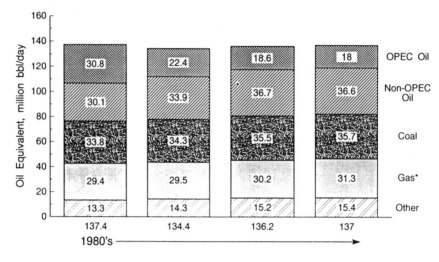

' Includes Natural Gas Liquids

Figure 1.2 World primary energy consumption

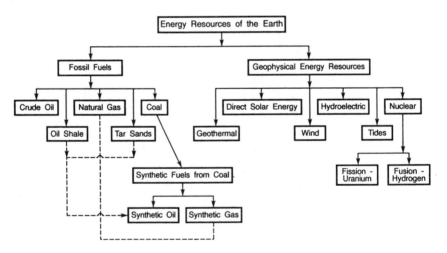

Figure 1.3 Energy resources of the earth

Williams, 1990), but they are not the subject of this book. Nuclear energy, which can be as hazardous as any fossil fuel in terms of destruction of the environment, deserves some comment in so far as an appreciation of its use may contribute to the general energy scenario and aid in putting fossil fuel resources into a more complete context.

Energy from a nuclear source may be the result of fission, which uses uranium ore or refined uranium-235 as its basic energy source. The ore must be concentrated into fissionable isotopes through nuclear processing. Eventually, the spent material must be replaced, but the resources of uranium ore are also finite. Hence there is a need to explore the potential for generating more of the fissionable isotope. However, the use of such materials has become a major controversy which threatens to be, at least for the time being, the death-knell of the nuclear industry.

Nuclear fusion, unlike nuclear fission which involves the use of fissionable material, involves the fusion of hydrogen nuclei. The concept has been suggested as having the potential to provide a clean and virtually unlimited supply of energy. However, it is generally felt (Halpern, 1980) that the technology is at the stage where 'several' decades appear to be needed for laboratory-scale experiments to be performed which are critical to the development of controlled energy from fusion. Indeed, the claims of fusion 'at the bench' under ambient conditions and utilizing conventional glassware has set the idea back on its heels and the onset of a fusion industry is now clouded by uncertainties.

Beyond this, there is much doubt about the time when we can look forward to the generation of energy from 'unconventional' sources. The timescale for the generation of electrical energy from such sources is generally unknown or, at best, open to much speculation and criticism. Hence, there is the continued use of the so-called 'conventional' fuel resources derived from the remains of ancient plants and animals. These same resources have been instrumental in the phenomenal expansion of the industrialized world. And these same resources can have serious adverse effects on the flora and fauna (including man) of the world.

Historical perspectives

In a combination of historical and geological perspectives, and remembering the phenomenal growth of technology, man's rapid emergence and thrusting appearance in the twentieth century has been intimately related to his discovery and use of fossil fuels for energy. From the first use of fire as a source of warmth to the overwhelming use of coal as the fuel of the industrial revolution, man has made his mark in using fossil fuels as energy sources.

In order to put this development into the perspective of geological time, which spans some 4.5 billion (4.5×10^9) years on earth (Table 1.7), it is necessary to remember that the earth was, for aeons, virtually uninhabited. For example, the Pre-Cambrian period spans approximately 3.8 billion (3.8×10^9) years or 85% of this timescale. This geological period is very distinct in so far as Pre-Cambrian rocks, in general, do not contain fossil fuels. In short, there were few, if any, or only simplified forms of life for much of the earth's history. Man is a comparative newcomer to the earth, the johnny-come-lately of the earth's fauna.

In addition, records show that the timescale for fossil fuel use is an even smaller fraction of the total age of the earth. Of course, this does not recognize that a considerable portion of the earth's history was employed in the production of these fuels! Indeed, the recorded use of petroleum, natural gas and the non-volatile derivative of petroleum, asphalt, has been known for only about 6000 years (Table 1.8) (Speight, 1991). For example, asphalt was being used thousands of years before Christ by the Sumerians, who knew how to apply the asphalt (often referred to in ancient records as pitch, slime or bitumen) as an insulating material for buildings (Agricola, 1556; Abraham,

Table 1.7 Geological timescales

System and period	Series and epoch	Distinctive records of life	Began (millions of years ago)
Cenozoic era:			
Quarternary	Recent (last 11 000 years)		
	Pleistocene	Early man	2+
	Pliocene	Large carnivores	10
	Miocene	Whales, apes, grazing forms	27
Tertiary	Oligocene	Large browsing mammals	38
	Eocene	Rise of flowering plants	55
	Palaeocene	First placental mammals	65–70
Mesozoic era;			
Cretaceous		Extinction of dinosaurs; appearance of floras with modern aspects	130
Jurassic		Dinosaurs' zenith, primitive birds, first small mammals	180
Triassic		Appearance of dinosaurs	225
Palaeozoic era:			
Permian		Conifers abundant, reptiles developed	260
Carboniferous			
Upper (Pennsylvanian)		First reptiles, great coal forests	300
Lower (Mississippian)		Sharks abundant	340
Devonian		Amphibians appeared, fishes abundant	405
Silurian		Earliest land plants and animals	435
Ordovician		First primitive fishes	480
Cambrian		Large faunas of marine invertebrates	550–570
Pre-Cambrian time:			
		Plants and animals with soft tissues, few fossils	Samples of isotopic dates
No known basis for systematic division			1500
			1900
			3200
			3490

1945; Speight, 1991). There are references to oil seepages, the product being frequently referred to as 'pitch', in the ancient world of the Greeks and Persians (Herodotus, 447 BC) and Alexander the Great is reputed to have discovered crude oil on the banks of the River Oxus (Abraham, 1945).

There are also records of the use of mixtures of pitch and sulphur as a weapon of war during the Battle of Palatea, Greece, in the year 429 BC (Forbes, 1959). There are references to the use of a liquid material, 'naft' (presumably the volatile fraction of petroleum which we now call naphtha and which is used as a solvent or as a precursor to gasoline), as an incendiary material during various battles of the pre-Christian era. This is the famed 'Greek fire', a precursor and distant cousin to napalm, of which so much is known and the effects recognized by those who have experienced its use as a weapon, as well as by historians.

Just as petroleum and its derivatives were used in antiquity, natural gas was

Table 1.8 Documented uses of petroleum and asphalt

3800 BC	First documented use of asphalt for caulking reed boats
3500 BC	Asphalt used as cement for jewellery and for ornamental applications
3000 BC	Documented use of asphalt as a construction cement by Sumerians; also believed to be used as a road material; asphalt used to seal bathing pool or water tank at Mohenjo-Daro
2500 BC	Documented use of asphalt and other petroleum liquids (oils) in the embalming process; asphalt believed to be widely used for caulking boats
1500 BC	Documented use of asphalt for medicinal purposes and (when mixed with beer) as a sedative for the stomach; continued reference to use of asphalt liquids (oil) as illuminant in lamps
1000 BC	Documented use of asphalt as a waterproofing agent by lake dwellers in Switzerland
500 BC	Documented use of asphalt mixed with sulphur as an incendiary device in Greek wars; also use of asphalt liquid (oil) in warfare
350 BC	Documented occurrence of flammable oils in wells in Persia
300 BC	Documented use of asphalt and liquid asphalt as incendiary device (Greek fire) in warfare
250 BC–AD 250	Documented occurrences of asphalt and oil seepages in several areas of the Fertile Crescent (Mesopotamia); repeated documentation of the use of liquid asphalt (oil) as an illuminant in lamps
AD 750	First documented use in Italy of asphalt as a colour in paintings
AD 950–1000	Report of destructive distillation of asphalt to produce an oil; reference to oil as nafta (naphtha)
AD 1100	Documented use of asphalt for covering (lacquering) metalwork
AD 1200	Continued use of asphalt and naphthas as an incendiary device in warfare; use of naphtha as an illuminant and incendiary material
AD 1500–1600	Documentation of asphalt deposits in the Americas; first attempted documentation of the relationship of asphalts and naphtha (petroleum)
AD 1600–1800	Asphalt used for a variety of tasks; relationship of asphalt to coal and wood tar studied; asphalt studied, used for paving; continued documentation of the use of naphtha as an illuminant and the production of naphtha from asphalt; importance of naphtha as fuel realized
AD 1859	Discovery of petroleum in North America; birth of modern petroleum science and refining

also known but, in general terms, its use is less well documented. There is fragmentary evidence in old texts for the use of natural gas but this is usually inferred since the gas is not named specifically. Its initial use seems to have been more for religious purposes rather than as a fuel.

In fact, gas wells were an important aspect of religious life to ancient Persians because of the importance of fire in their religion; exhibitions of 'burning pillars of fire' must have been awe inspiring to say the least, since, as the old name *varishnak* implies, they needed no food (Forbes, 1958). As another example, in the book of Daniel (Chapter 3; Christian Bible) note is made of the 'eternal fires' that existed in the time of King Nebuchadnezzar. These fires, which have been quoted as varying between 9 cubits (13.5 ft, 4 m) and 40 cubits (60 ft, 18.3 m) high, have since been reported as being due to the seepage of a self-igniting natural gas located close to where the Tigris and Euphrates rivers meet and which ignites periodically.

There is also a very distinct possibility that the 'voices of the gods', as referenced in old texts, were due to natural gas forcing its way through fissures in the earth's surface (Scheil and Gauthier, 1909; Schroder, 1920). There is a passage in the records of the Assyrian King Tukulti Ninurta (*ca.* 885 BC) in which the 'voices of the gods' are heard arising from rocks near to Hit, the place of the bitumen deposits so well known as a building mastic. Similar sounds also attributed to be the 'voices of the gods' are noted to have occurred in the region around Kirkuk.

In classical times these wells often flared and must have been a very impressive sight, to say the least (Lockhart, 1939; Forbes, 1958, 1964). They are depicted as burning near Apollo's shrine, on the coins of Appollonia, and near the present Selenizza (Albania) (Forbes, 1958). Plutarch mentions that Alexander the Great saw burning gas wells near to Ecbatana which he (Plutarch) described as a 'gulf of fire streaming from an inexhaustible source'!

Historical records also indicate that the use of natural gas (for other than religious purposes) dates back to about AD 250 when it was used as a fuel in China. The gas was obtained from shallow wells and was distributed through a piping system constructed from hollow bamboo stems. Gas wells were also known in Europe in the Middle Ages and were reputed to eject oil, such as the phenomena observed at the site near to the town of Mineo in Sicily (Forbes, 1958). Natural gas was used on a small scale for heating and lighting in northern Italy during the early seventeenth century.

From this, it might be conjectured that natural gas found some use from the seventeenth century to the present day, while recognizing that gas from coal would be a strong competitor.

In a more modern context, there is the record in 1775 of a 'burning spring' near Charleston, West Virginia, as well as on land owned by George Washington (Lincoln, 1785). The possibility exists that the fire may have been caused by the ignition of naphtha seepages, but there is also the very strong, probably the only, possibility that the fire was caused by a gas seepage that was ignited by a flaming torch held by one of the investigators as he brought it near to the place where the gas came to the surface. Furthermore, the first record of a natural gas well being drilled (to a depth of 27 ft, 8.2 m) in the United States occurred near a 'burning spring' at Fredonia, New York, in 1821.

In the years following this discovery, natural gas usage was restricted to local environs since the available technology for storage and transportation (bamboo pipes notwithstanding!) was not well developed and, at that time, natural gas had little or no commercial value. In fact, in the 1930s, when petroleum refining was commencing an expansion in technology that still continues, gas was not considered to be a major fuel source and was only produced as an unwanted by-product of crude oil production.

The principal gaseous fuel source at that time (i.e. the 1930s) was the gas produced by the surface gasification of coal. In fact, each town of any size had a plant for the gasification of coal (hence the use of the term 'town gas'). Most of the natural gas produced at the petroleum fields was vented to the air or burned in a flare stack – only a small amount was pipelined to industrial areas for commercial use. It was only in the years after the Second World War that natural gas became a popular fuel commodity leading to the recognition that it has at the present time.

Coal has probably been known and used for an equal length of time but the records are somewhat less than complete. There are frequent references to coal in the Christian Bible (Cruden, 1930) but, all in all, the recorded use of coal in antiquity is very sketchy. However, there are excellent examples of coal mining in Britain from the year AD 1200 which marked, perhaps, the first documented use of mined coal in England (Galloway, 1882).

On another historical note, it is recorded that in 1257 a very singular event occurred which threatened the very existence of coal use and its future as a fuel. Eleanor, Queen of Henry III of England, was obliged to leave the town of Nottingham where she had been staying during the absence of the King on an expedition into Wales. The Queen was unable to remain in Nottingham because of the troublesome smoke from the coal being used for heating and cooking! Over the subsequent decades, a variety of proclamations were issued by Henry and by his son, Edward I, which threatened the population with the loss of various liberties, even life, if the consumption of coal was not seriously decreased and, in some cases, halted (Galloway, 1882).

One wonders whether these worthy kings of England realized the environmental consequences of burning coal or whether they were more interested in the use of wood, from royal forests of course, and the resulting income therefrom. Whilst the royal positions on the pollution problem, if that really be the issue, may have solved part of the problem at the time, it did not have any lasting effect; coal burning has continued in England from that time!

In fact, in the late 1500s, an increasing shortage of wood in Europe resulted in the search for another form of combustible energy and coal became an exploitable resource in Britain, France, Germany and Belgium. In the mid-to-late 1700s, the use of coal increased dramatically in Britain with the successful development of coke smelters and the ensuing use of coal to produce steam power. By the 1800s, coal was supplying most of Britain's energy requirements; the use of gas from coal ('town gas') for lighting was also established.

In contrast, in the United States, wood was more plentiful and the population density was much lower than in Europe; any coal required for energy was imported from Britain and Nova Scotia. But after the Revolutionary War, coal entered the picture as an increasing source of energy; as an example, the state of Virginia supplied coal to New York city. However, attempts to open the market to accept coal as a fuel were generally ineffective in the United States and progress was slow, extremely slow. It was not until the period from 1850 to 1885 that coal use in the United States increased, spurred by the emerging railroad industry both for the manufacture of steel rails and as a fuel for locomotives. At last, coal seemed to be undergoing a transformation as a fuel on both sides of the Atlantic Ocean.

Then petroleum entered the picture again, but this time in a never-to-be-forgotten manner in 1859, with the advent of the modern petroleum era resulting from the rush of liquid from the Drake well in Titusville, Pennsylvania.

Modern perspectives

The increased use and popularity of fossil fuels is due, no doubt, to the relative ease of accessibility which has remained unchanged over the centuries. Petroleum is now an occasional exception because of various physical and political reasons.

The relatively simple means by which fossil fuels can be utilized have also been a major factor in determining their popularity. In addition, fossil fuels are interchangeable on a purely physical basis in so far as one form can be readily converted to another:

Gas ↔ Liquid ↔ Solid

Thus, the ease with which fossil fuels were available to the world did indeed play a major role in their increased use. The conversion of these natural products to fuel products and to chemicals, as evidenced, first, by the coal chemicals industry of the nineteenth century and, second, by the petrochemical industry of the twentieth century, also served to increase their popularity.

Projections that the era of fossil fuels (gas, petroleum and coal) will be almost over when the cumulative production of fossil resources reaches 85% of their initial total reserves (Hubbert, 1973) may or may not have some merit when looking back with some two decades or more of experience (some two decades or more of 20–20 hindsight!) and are alarming and worthy of some attention. The scarcity (relative to a few decades ago) of petroleum is real but it seems that the remaining reserves of petroleum, coal and natural gas make it likely that there will be an adequate supply of energy for several decades to come (Martin, 1985; MacDonald, 1990; Banks, 1992).

But energy in what form and at what cost (Bending et al., 1987; Hertzmark, 1987; Meyers, 1987; Sathaye et al., 1987)? There are those people who advertise regularly in the news media for the use of clean energy, such as electricity, as a means of ameliorating the environmental problems that go with the use of 'unclean' forms of energy, but who never mention, or appear to give thought to, the source of electricity – usually coal!

Therefore, technologies to ameliorate the effects of fossil fuel combustion on acid rain deposition, urban air pollution and global warming must be pursued vigorously. This is a challenge that must not be ignored, for the effects of acid rain in soil and water, for example, leave no doubt about the need to control its causes (Mohnen, 1988). Indeed, recognition of the need to address these issues is the driving force behind recent energy strategies as well as a variety of research and development programmes (Stigliani and Shaw, 1990; United States Department of Energy, 1990; United States General Accounting Office, 1990).

As new technology is developed, emissions may be reduced by repowering, in which ageing equipment is replaced by more advanced and efficient substitutes (Hyland, 1991). Such repowering might, for example, involve an exchange in which an ageing unit is replaced by a newer combustion chamber, such as the atmospheric fluidized-bed combustor (AFBC) or the pressurized fluidized-bed combustor (PFBC) (Chapter 8). In the PFBC, pressure is maintained in the boiler, often at an order of magnitude greater than in the AFBC, and additional efficiency is achieved by judicious use of the hot gases in the combustion chamber (combined cycle).

Both these combustors burn coal with limestone or dolomite in a fluidized bed which allows, with recent modifications to the system, the limestone sorbent to take up some of the sulphur that would normally be emitted as sulphur dioxide. In addition, combustion can be achieved at a lower temperature than in a conventional combustor thereby reducing the formation of nitrogen oxide(s).

An important repowering approach attracting great interest is the integrated coal–gasification combined cycle (IGCC) system (Chapter 8). The major innovation introduced with the IGCC technology is the conversion of coal into synthesis gas, a mixture of mainly hydrogen (H_2) and carbon monoxide (CO) with lesser quantities of methane (CH_4), carbon dioxide (CO_2) and hydrogen sulphide (H_2S). Up to 99% of the hydrogen sulphide can be removed by commercially available processes (Chapters 9, 10 and 12) before the gas is burned. The synthesis gas then powers a combined cycle in which the hot gases are burned in a combustion chamber to power a gas turbine and the exhaust gases from the turbine generate steam to drive a steam turbine.

The solution to the issue of acid rain deposition lies in the control of sulphur oxides (usually sulphur dioxide, SO_2) emissions as well as of nitrogen oxide (NO_x) emissions. These gases react with the water in the atmosphere and the result is an acid:

$$SO_2 + H_2O = H_2SO_3$$
$$2SO_2 + O_2 = 2SO_3$$
$$SO_3 + H_2O = H_2SO_4$$
$$2NO + H_2O = HNO_2$$
$$2NO + O_2 = 2NO_2$$
$$NO_2 + H_2O = HNO_3$$

Indeed, the combustion of coal can account for the large majority of the sulphur oxides and nitrogen oxides released to the atmosphere. Whichever technologies succeed in reducing the amounts of these gases in the atmosphere should also succeed in reducing the amounts of urban smog, those notorious brown and grey clouds that are easily recognizable, at some considerable distance from urban areas, not only by their appearance but also by their odour.

The oxides of carbon (carbon monoxide, CO, and carbon dioxide, CO_2) are also of importance in so far as the fossil fuels, all carbon-based materials, produce either or both of these gases during use. And both gases have the potential to harm the environment.

A reduction in the emissions of these gases, particularly carbon dioxide which is the final combustion product of fossil fuels, has very little chance of being achieved without a traumatic switch to non-fossil energy sources (Figure 1.4). However, such emissions can be moderated by trapping and recovering the carbon dioxide at the time of fossil fuel usage. On the other hand, it has been widely suggested that coal be replaced by natural gas which produces mainly carbon dioxide. Any oxides of nitrogen and sulphur come from impurities in the gas and can be removed first.

In fact, it has been estimated that the combustion of natural gas, compared with coal, produces about 70% more energy for each unit of carbon dioxide produced. Other factors, such as the efficiency with which the gas can be handled, lack of ash by-product and virtually no sulphur, allow more energy to be diverted to the clean-up processes.

Current awareness of these issues at a variety of levels of government has resulted, in the United States, in the institution of the Clean Coal Program. This Program has greatly facilitated the development of pollution abatement techniques, and it has led to successful partnerships between government and industry (United States Department of Energy, 1991). Although the technologies for using natural gas efficiently are improving (Figure 1.5), there are significant limitations to substituting gas for coal.

In the first place, natural gas is far less abundant than coal and the inadvertent release of natural gas into the atmosphere (methane, CH_4, is a

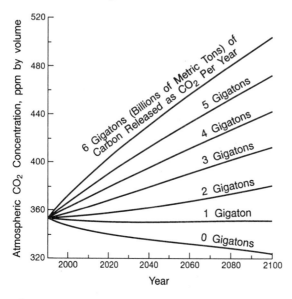

Figure 1.4 Projected atmospheric concentrations of carbon dioxide which are required for stabilization of the atmosphere

	Plant Cost (Dollars/Kilowatt of Capacity)	Percent Reduction in SO_2 Emitted	NO_x Emitted (Milligrams per Million Joules of Electricity)	Estimated Carbon Emitted as CO_2 (Kilograms of Carbon per CO_2 Kilowatt Hour)	Efficiency (Percent of Stored Energy Converted to Electricity)
Conventional (Steam Turbine)					
Gas	760	*	180	.14	36
Coal (With Scrubber)	1,600	90	300	.25	34
Combined Cycle (Steam and Gas)					
Gas	520	*	15	.10	47
Coal (With Gasification)	1,700	99	25	.20	42
Pressurized Fluidized-Bed Combustion					
Coal (Combined Cycle)	1,200	90	60	.19	42
Steam-Injected Gas Turbine					
Gas	410	*	15	.12	40
Coal (With Gasification)	1,300	99	25	.24	36
Intercooled Steam-Injected Gas Turbine					
Gas	400	*	10	.10	47
Coal (With Gasification)	1,030	99	20	.20	42
Advanced Fuel Cells					
Gas	600-800	*	5-20	.09-.10	50-55
Coal (With Gasification)	1,000-1,500	99	10-35	.17-.19	45-52

* Indicates Only Trace Emission of SO_2

Figure 1.5 Comparison of installation costs for advanced fuel technologies

'greenhouse' gas; Graedel and Crutzen, 1989; Hileman, 1992) may tend to offset any of the advantages of its use. Indeed, the release of other gases resulting from fossil fuel usage into the atmosphere has been projected to cause environmental perturbations (Table 1.9) (Graedel and Crutzen, 1989) that may be difficult to ignore at the present rates of release.

It is also necessary to note here, on the positive side for natural gas, that methane produces less carbon dioxide per unit of energy than coal (Michaels, 1991). In fact, coal produces more carbon dioxide per unit of energy than any other fossil fuel.

Thus, there is the potential that new laws, such as the Clean Air Act amendments in the United States, might be a positive factor and supportive of the use of natural gas. However, this will not occur without some changes in the nature of the gas-processing and refining operations which are also dependent upon market demand, perhaps even driving the market (Cannon, 1990; Dalton, 1991; Haun et al., 1991; Scherr et al., 1991; Larson, 1992).

But it is not just the production of carbon dioxide from coal, and other fossil fuels, that needs to be decreased. The production of other pollutants such as sulphur dioxide (SO_2) and the oxides of nitrogen (NO_x, where $x = 1$ or 2) also needs attention.

Recognition of the production of these atmospheric pollutants in considerable quantities every year has led to the institution of national emission standards for all such pollutants. Using sulphur dioxide as an example (Table 1.10) (Kyte, 1991), the various standards are not only very specific but will become more stringent with the passage of time (IEA Coal Research, 1991). Atmospheric pollution is being taken very seriously and there is also the threat, or promise, of heavy fines and/or gaol terms for any pollution-minded miscreants who seek to flaunt the laws!

Be that as it may, there is a trend towards the increased use of natural gas and coal that will require a more stringent approach to issues related to environmental protection than we have ever known at any time in the past. The need to protect the environment is strong. One example which attests to this fact is the passage of amendments to the Clean Air Act in the United States (United States Congress, 1990; Stensvaag, 1991).

Thus, as the alternatives in energy vacillate from coal to oil to gas and back again, to be followed, presumably, by the eras of nuclear fuels and solar energy, there is an even greater need to ensure that emissions are clean and that the utility of the fossil fuels does not overcome the need to protect the environment (Clark, 1989).

It is for this reason that this book is written – so that those already in the fossil energy field and those engineers and scientists about to be recruited into it are made aware of the issues, given a means by which the emissions can be controlled, and provided with 'food for thought' that, hopefully, will result in the conception and development of innovative technologies for future use, particularly those technologies that trap the pollutants and, where possible, convert them to usable materials.

Table 1.9 Gases which cause environmental perturbations

Gas	Greenhouse effect	Stratospheric ozone depletion	Acid deposition	Smog	Corrosion	Decreased visibility	Decreased self-cleaning of atmosphere
Carbon monoxide (CO)	+						+
Carbon dioxide (CO_2)	+	+/−					
Methane (CH_4)	+	+/−					+/−
NO_x: nitric oxide (NO) and nitrogen dioxide (NO_2)		+/−	+	+		+	
Nitrous oxide (N_2O)	+	+/−					−
Sulphur dioxide (SO_2)	−		+		+	+	
Chlorofluoro-carbons	+	+					
Ozone (O_3)	+			+			−

Gas	Major anthropogenic sources	Anthropogenic/total emissions per year (millions of tons)	Average residence time in atmosphere	Average concentration 100 years ago (ppb)	Approximate current concentration (ppb)	Projected concentration in year 2030 (ppb)
Carbon monoxide (CO)	Fossil fuel combustion, biomass burning	700/2000	Months	?, N. Hem. 40 to 80, S. Hem. (clean atmospheres)	100 to 200, N. Hem. 40 to 80, S. Hem. (clean atmospheres)	Probably increasing
Carbon dioxide (CO_2)	Fossil fuel combustion, deforestation	5500/∼5500	100 years	290000	350000	400000 to 550000
Methane (CH_4)	Rice fields, cattle, landfills, fossil fuel production	300 to 400/550	10 years	900	1700	2200 to 2500

Gas	Source		Lifetime			
NO$_x$ gases	Fossil fuel combustion, biomass burning	20 to 30/ 30 to 50	Days	0.001 to? (clean to industrial)	0.001 to 50 (clean to industrial)	0.001 to 50 (clean to industrial)
Nitrous oxide (N$_2$O)	Nitrogenous fertilizers, deforestation, biomass burning	6/25	170 years	285	310	330 to 350
Sulphur dioxide (SO$_2$)	Fossil fuel combustion, ore smelting	100 to 130/ 150 to 200	Days to weeks	0.03 (clean to industrial)	0.03 to 50 (clean to industrial)	0.03 to 50 (clean to industrial)
Chlorofluoro-carbons	Aerosol sprays, refrigerants, foams	~1/1	60 to 100 years	0	About 3 (chlorine atoms)	2.4 to 6 (chlorine atoms)

Source: Graedel and Crutzen (1989).

Table 1.10 Current national emission standards for sulphur (mg SO_2/m^3) (IEA, 1991)

Country	New plants	Existing plants
Austria	200–400	200–400
Belgium[c]	400–2000(250)[d]	
Canada[a]	740	
Denmark[c]	400–2000	
Finland	400–660	660
France[c]	400–2000	
FRG[c]	400–2000	400–2500
Greece[c]	400–2000	
Ireland[c]	400–2000	
Italy[c]	400–2000	400–2000
Japan	[b]	[b]
Luxembourg[c]	400–2000	
Netherlands[c]	200–700	400–700
Poland	540	2700–4000 (1800–3000)
Portugal[c]	400–2000	
Spain[c]	2400–9000	2400–9000
Sweden	290	290–570
Switzerland	400–2000	400–2000
Taiwan	2145–4000	2145–4000
Turkey	400–2000	
UK[c]	400–2000	
USA	740–1480[e]	[e]

[a] Guidelines.
[b] Set on a plant-by-plant basis according to nationally defined formulae.
[c] EC countries.
[d] From 1995.
[e] Clean Air Act Amendments, 1990.
Source: Kyte (1991).

References

Abraham, H. (1945) *Asphalts and Allied Substances*, Van Nostrand, New York
Agricola, Georgius (Bauer, Georg) (1556) *De Re Metallica*, Froben, Basle
Austin, G.T. (1984) *Shreve's Chemical Process Industries*, McGraw-Hill, New York
Banks, F.E. (1992) *Opec Bulletin*, **XXIII**, (2), 20
Bending, R.C., Cattell, R.K. and Eden, R.J. (1987) *Annual Review of Energy*, **12**, 185
Berkowitz, N. (1979) *An Introduction to Coal Technology*, Academic Press, New York
Cannon, R.E. (1990) *Oil and Gas Journal*, **88**, (28), 47
Clark, W.C. (1989) *Scientific American*, **261**, (3), 46
Cruden, A. (1930) *Complete Concordance to the Bible*, Lutterworth Press, London
Dalton, S.M. (1991) *Desulphurisation 2: Technologies and Strategies for Reducing Sulphur Emissions* (W.S. Kyte Chairman-editor), Symposium Series No. 123, Institute of Chemical Engineers, Rugby, Warwickshire
Dryden, I.G.C. (1975) *The Efficient Use of Energy*, IPC Business Press, Guildford
EIA (1988) *International Energy Outlook: Projections to 2000*. Report No. DOE/EIA-0484(87), United States Department of Energy, Energy Information Administration, Washington, DC
EIA (1989) *Annual Energy Outlook: Long Term Projections*. Report No. DOE/EIA-0383(89), United States Department of Energy, Energy Information Administration, Washington, DC
EIA (1991a) *Annual Energy Review 1990*. Report No. DOE/EIA-0384(90), United

States Department of Energy, Energy Information Administration, Washington, DC

EIA (1991b) *Annual Outlook for Oil and Gas 1991*. Report No. DOE/EIA-0517(91), United States Department of Energy, Energy Information Administration, Washington, DC

EIA (1992) *Annual Energy Outlook: With Projections to 2010*. Report No. DOE/EIA-0383(92), United States Department of Energy, Energy Information Administration, Washington, DC

Forbes, R.J. (1958) *Studies in Early Petroleum Chemistry*, Brill, Leiden

Forbes, R.J. (1959) *More Studies in Early Petroleum History*, Brill, Leiden

Forbes, R.J. (1964) *Studies in Ancient Technology*, Vol. I, Brill, Leiden

Fulkerson, W., Judkins, R.R. and Sanghvi, M.K. (1990) In *Energy for Planet Earth*, Freeman, New York, Chapter 8

Galloway, R.L. (1882) *A History of Coal Mining in Great Britain*, Macmillan, London

Gibbons, J.H., Blair, P.D. and Gwin, H.L. (1989) *Scientific American*, **261**, (3), 136

Gilleland, D.S., and Swisher, J.H. (eds.) (1986) *Acid Rain Control II: The promise of New Technology*, Southern Illinois University Press, Carbondale, ILL

Graedel, T.E. and Crutzen, P.J. (1989) *Scientific American*, **261**, (3), 58

Hafele, W. (1990) In *Energy for Planet Earth*, Freeman, New York, Chapter 9

Halpern, G.M. (1980) Fusion Energy. In *Encyclopedia of Chemical Technology*, Vol. 11, Wiley, New York, p. 590

Haun, R.R., Ellington, E.E. and Otto, K.W. (1991) *Oil and Gas Journal*, **89**, (29), 46

Herodotus (447 BC) (1956) *Historia. The History of Herodotus* (transl. George Rawlinson, ed. M. Kamroff), Tudor, New York

Hertzmark, D.I. (1987) *Annual Review of Energy*, **12**, 23

Hessley, R.K. (1990) *Fuel Science and Technology Handbook* (ed. J.G. Speight), Marcel Dekker, New York

Hessley, R.K., Reasoner, J.W. and Rilewy, J.T. (1986) *Coal Science*, Wiley, New York

Hileman, B. (1992) *Chemical and Engineering News*, **70**, (6), 26

Hubbert, M.K. (1973) *American Association of Petroleum Geologists Bulletin*, **57**, (9), 1843

Hyland, R.P. (1991) *Hydrocarbon Processing*, **70**, (5), 113

IEA Coal Research (1991) *Emissions Standard Data Base*, International Energy Agency Coal Research, London

Kyte. W.S. (1991) *Desulphurisation 2: Technologies and Strategies for Reducing Sulphur Emissions*, Institute of Chemical Engineers, Rugby, Warwickshire

Larson, K. (1992) *Control for the Process Industries*, **V**, (3), 38

Lincoln, B. (1785) *Memoirs of the American Academy of Arts and Sciences*, Vol. I, p. 372

Lockhart, L. (1939) *Journal of the Institute of Petroleum*, **25**, 1

MacDonald, G.J. (1990) *Annual Reviews of Energy*, **15**, 53

Manowitz, B. and Lipfert, F.W. (1990) In *Geochemistry of Sulfur in Fossil Fuels* (ed. W.L. Orr and C.M. White), American Chemical Society, Washington, DC, Chapter 3

Martin, A.J. (1985) In *Prospects for the World Oil Industry* (ed. T. Niblock and R. Lawless), Croom Helm, Beckenham, Kent, Chapter 1

Meyers, S. (1987) *Annual Review of Energy*, **12**, 81

Michaels, P.J. (1991) *Journal of Coal Quality*, **10**, (1), 1

Mohnen, V.A. (1988) *Scientific American*, **259**, (2), 30

OPEC (1991) *Facts and Figure: A Graphical Analysis of World Energy up to 1990*, Organization of Petroleum Exporting Countries, Vienna

Probstein, R.F. and Hicks, R.E. (1990) *Synthetic Fuels*, pH Press, Cambridge, MA

Sathaye, J., Ghirardi, A. and Schipper, L. (1987) *Annual Review of Energy*, **12**, 253

Scheil, V. and Gauthier, A. (1909) *Annales de Tukulti Ninip II*, Paris

Scherr, R.C., Smalley, G.A. Jr and Norman, M.E. (1991) *Oil and Gas Journal*, p. 68 (27 May) and p. 35 (10 June)

Schroder, O. (1920) *Keilschriftetexte aus Assur verscheidenen XIV*, Leipzig

Scouten, C.S. (1990) *Fuel Science and Technology Handbook* (ed. J.G. Speight), Marcel Dekker, New York

Speight, J.G. (1990) *Fuel Science and Technology Handbook*, Marcel Dekker, New York

Speight, J.G. (1991) *The Chemistry and Technology of Petroleum*, 2nd edn, Marcel Dekker, New York

Stensvaag, J-M. (1991) *Clean Air Act Amendments: Law and Practice*, Wiley, New York

Stigliani, W.M. and Shaw, R.W. (1990) *Annual Review of Energy*, **15**, 201

United States Congress (1990) *Public Law 101–549. An Act to Amend the Clean Air Act to Provide for Attainment and Maintenance of Health Protective National Ambient Air Quality Standards, and for Other Purposes*, 15 November

United States Department of Energy (1990) *Gas Research Program: Implementation Plan*, DOE/FE-0187P, United States Department of Energy, Washington, DC, April

United States Department of Energy (1991) *Clean Coal Technology Demonstration Program*, DOE/FE-0219P, United States Department of Energy, Washington, DC, February

United States General Accounting Office (1990) *Energy Policy: Developing Strategies for Energy Policies in the 1990s.* Report to Congressional Committees, GAO/RCED-90-85, United States General Accounting Office, Washington, DC, June

Weinberg, C.J. and Williams, R.H. (1990) In *Energy for Planet Earth*, Freeman, New York, Chapter 10

Yeager, K.E. and Baruch, S.B. (1987) *Annual Review of Energy*, **12**, 471

2

Environmental aspects

Introduction

Industrial operations which produce any product from a natural resource must make every attempt to ensure that natural resources such as air, land and water remain as unpolluted as possible, and the environmental aspects of such operations need to be carefully addressed.

Whilst the focus of this text is on the clean-up of potentially harmful gaseous emissions produced by a variety of industries which can cause serious pollution problems, there is also the need to recognize that gas-processing operations can also cause other environmental damage. For example, the spill of an acid solution or the inadvertent discharge of a wash solution can cause severe damage to flora and fauna as well as to aquatic life in the region. Thus, caution is advised from all environmental aspects and not just that most closely related to the process operation.

Indeed, any of the products from a gas-cleaning plant can contain contaminants. The very nature of the gas-cleaning plant dictates that this is so but there are many options available to assist in the clean-up of the plant's products. However, it is very necessary, in view of the scale of such plants, that controls be placed on the release of materials to the environment to minimize potential damage to the immediate environment; this damage must be avoided.

In order to understand the means by which gaseous products can cause pollution, it is necessary to understand the various cycles that exist in the biosphere and atmosphere. Such cycles (Figures 2.1–2.5) (Clark, 1989; Graedel and Crutzen, 1989; Schneider, 1989; Maurits la Rivière, 1989; Frosch and Gallopoulos, 1989) have dominated the global environmental system for millennia.

For example, respiration and fossil fuel combustion release substantial quantities of carbon dioxide (CO_2) to the atmosphere. The decomposition of organic material produces methane (CH_4) which, with the carbon dioxide, can cause increases in the temperature of the earth. Emissions such as sulphur dioxide (SO_2) and nitrogen oxides (NO_x) are major contributors to acid rain (Manowitz and Lipfert, 1990). Other materials, such as chlorofluorocarbons (CFCs) but not discussed, also make a significant contribution to climatic change (NRC, 1990).

The capacity of the environment to absorb the effluents and other impacts of energy technologies is not unlimited as some would have us believe. The

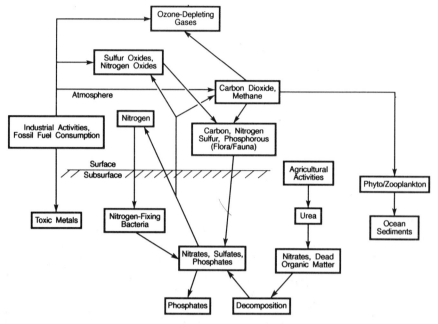

Figure 2.1 Schematic representation of gaseous cycles (Source: Clark, 1989)

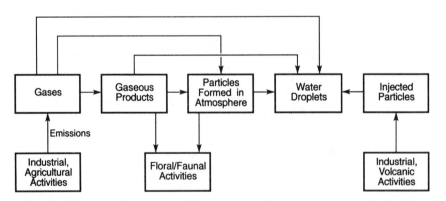

Figure 2.2 Fate of gaseous emissions into the atmosphere (Source: Graedel and Crutzen, 1989)

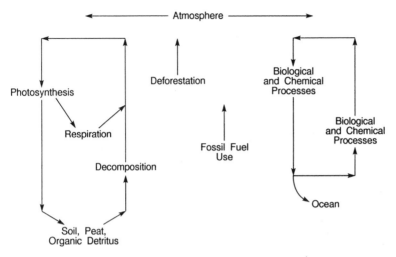

Figure 2.3 Schematic representation of carbon cycles (Source: Schneider, 1989)

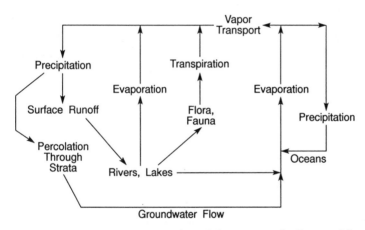

Figure 2.4 Schematic representation of the water cycle (Source: Maurits la Rivière, 1989)

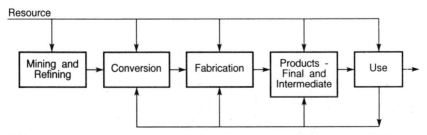

Figure 2.5 The industrial ecosystem (Source: Frosch and Gallopoulos, 1989)

environment should be considered to be an extremely limited resource, and the discharge of chemicals into it should be subject to severe constraints. Indeed, the declining quality of raw materials, especially fossil fuels that give rise to many of the gaseous emissions of interest in this text, dictates that more material must be processed to provide the needed fuels. And the growing magnitude of the effluents from fossil fuel processes has moved beyond the point where the environment has the capability to absorb such (gaseous) effluents without disruption.

For the purposes of this text, it can be assumed (since nuclear materials are not considered here) that the raw materials brought to the plant consist principally of carbonaceous matter (i.e. coal, petroleum and natural gas) to be converted. And these materials must leave as a product, or as a by-product, or as an unsaleable waste stream. However, there are contradictions in fossil fuel conversion in so far as the greatest quantity of material emitted from the plant may have little, or no, environmental impact. On the other hand, the minutest amount of material may be extremely hazardous and have significant impact on the environment.

In addition, and in general terms, the gaseous emissions from fossil fuel conversion facilities may be broadly classed as those originating from four processing steps: pretreatment, conversion and upgrading, as well as those from ancillary processes (Figure 2.6).

The environmental aspects of fossil fuel use have been a major factor in the various processes, and the see-sawing movement of the fossil fuel base between petroleum, natural gas and coal has increased the need for pollutant control for large, fossil-fuelled power plants. These power plants emit pollutants

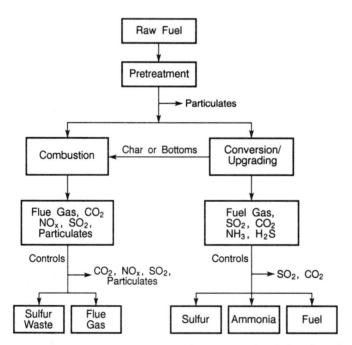

Figure 2.6 General representation of the source of emissions from fossil fuels (Source: Probstein and Hicks, 1990)

which, by atmospheric chemical transformations, may become even more harmful secondary pollutants.

As a summary, in conventional power plants, pulverized coal is burned in a boiler, where the heat vaporizes water in steam tubes. The resulting steam turns the blades of a turbine, and the mechanical energy of the turbine is converted to electricity by a generator. Waste gases produced in the boiler during combustion, among them sulphur dioxide, nitgrogen oxide(s) and carbon dioxide, flow from the boiler to a particulate removal device and then to the stack and the air.

For example, the sulphur dioxide that is produced during the combustion of coal or petroleum in power plants

$$[S]_{fuel} + O_2 = SO_2$$

will react with oxygen and water in the atmosphere to yield the environmentally detrimental sulphuric acid

$$2SO_2 + O_2 = 2SO_3$$
$$SO_3 + H_2O = H_2SO_4$$

or

$$2SO_2 + O_2 + 2H_2O = 2H_2SO_4$$

which is a contributor to acid rain.

Using coal as an example, even though the other two fossil fuels of petroleum and natural gas are not necessarily free from blame and there is a threat to 'outlaw' coal in some countries (C&ST, 1992), the three major types of pollutants emitted by a coal-fired power plant are particulate matter, sulphur dioxide and nitrogen oxides (NO_x).

Particulate matter is produced from mineral matter in the coal that is converted during combustion to finely divided inorganic material referred to as 'fly ash', which can be carried out of the stack with the hot exhaust gases. Furthermore, the practice of burning finely divided coal can contribute to fly ash emissions. Sulphur dioxide is produced by the oxidation of organic sulphur in the coal and is normally cited as the most troublesome of pollutants. From 10 to 50% of the nitrogen inherent in the organic coal structure is converted to nitric oxide during combustion:

$$[N]_{fuel} + O_2 = 2NO$$
$$2NO + O_2 = 2NO_2$$
$$NO + H_2O = HNO_2$$
$$NO_2 + H_2O = HNO_3$$

Although not usually considered to be a major pollutant, the hydrogen chloride that is produced from mineral matter and other inorganic contaminants

$$[Cl]_{coal\ minerals} + [H]_{coal} = HCl$$

is gaining increasing recognition as a pollutant which needs serious attention.

Hydrogen chloride quickly picks up water from the atmosphere to form droplets of hydrochloric acid and, like sulphur dioxide, is a contributor to acid rain. However, hydrogen chloride may exert severe local effects because, unlike sulphur dioxide, it does not need to participate in any further chemical reactions to become an acid. Under atmospheric conditions which favour a buildup of stack emissions in the area of a large power plant, the amount of hydrochloric acid in rainwater could be quite high.

Very little, if anything, can be done during the pretreatment of coal to eliminate nitrogen since the nitrogen is part of the organic coal structure. The situation is less clear in the case of the sources of hydrogen chloride; both organic chlorine and inorganic chloride salts contribute to the formation of hydrogen chloride during combustion. Coal-cleaning processes can reduce the mineral matter content but pretreatment processes do not remove organically bound chlorine, which is more likely to be the precursor to hydrogen chloride in a combustion process.

Pretreatment washing processes are also successful methods for removing inorganic sulphur but they do not affect the organic sulphur content. Thus, even before combustion begins, some of the sulphur can be removed from coal. For instance, commercially available processing methods crush the coal and separate the resulting particles on the basis of density, thereby removing up to about 30% of the sulphur. But whilst pretreatment washing may remove up to 90% of the pyritic sulphur, up to 20% of the combustible coal may also be removed and a balance must be struck between the value of the sulphur removed and the coal lost to the cleaning process.

There are a variety of processes which are designed for sulphur dioxide removal from stack gas (Chapters 9, 10 and 12) but scrubbing processes utilizing limestone ($CaCO_3$) or lime [$Ca(OH)_2$] slurries have received more attention than other stack gas scrubbing processes. Attempts have been made to use dry limestone or dolomite ($CaCO_3.MgCO_3$) within the combustor as an *in situ* method for the removal of sulphur dioxide, thereby eliminating the wet sludge from wet processes. This involves the injection of dry carbonate mineral with the coal followed by recovery of the calcined product along with sulphite and sulphate salts:

$$[S]_{coal} + O_2 = SO_2$$
$$2SO_2 + O_2 = 2SO_3$$
$$2CaCO_3 + SO_2 + SO_3 = CaSO_3 + CaSO_4 + 2CO_2$$

The majority of the stack gas scrubbing processes are designed to remove sulphur dioxide from the gas streams; some processes show potential for the removal of nitrogen oxide(s). However, there is a current line of thinking which pursues the options that enable sulphur oxides and nitrogen oxide(s) to be controlled, at least as far as possible, by modification of the combustion process. Sulphur (as already noted) can be removed by injecting limestone with the coal into a boiler whilst modifications of the combustion chamber, as well as methods of flame temperature regulation and techniques that lower combustion temperatures, such as the injection of steam into the combustion region, are claimed to reduce the emissions of nitrogen oxide(s).

Gaseous emissions

The major gaseous waste streams leaving many plants might be water vapour and carbon dioxide. The former has little effect on the environment whilst the effect of the latter in considered proven but is still open to considerable debate, speculation and often violent scientific and emotional disagreement. Nevertheless, as common sense alone will tell us, it must be assumed that the effects of releasing unlimited amounts of carbon dioxide to the atmosphere will, and most probably does, cause adverse, if not severe, perturbations to the environment. However, it is all a matter of degree. The discharge of liquid water is a different issue, because the potential for dissolved contaminants is real. Water may not be water may not be water!

If it can be assumed that the water used for cooling purposes in a fossil fuel plant is predominantly the type that passes through cooling towers, the other gaseous emissions from such a plant will be carbon dioxide (CO_2), hydrogen sulphide (H_2S), as well as sundry other sulphur compounds of which carbonyl sulphide (COS), carbon disulphide (CS_2) and mercaptans (R-SH) may be cited as examples. Nitrogen compounds, especially the oxides (NO_x, where $x = 1$ or 2), will also be evident, particularly when combustion of the fossil fuel occurs.

For the last few years, there has been much concern about global warming (the 'greenhouse effect' as the phenomenon is often unaffectionately called) (Ember *et al.*, 1986; Keepin, 1986; Kellogg, 1987; Schlesinger and Mitchell, 1987; MacDonald, 1988; Smith, 1988; Hileman, 1989; Douglas, 1990; White, 1990; IEA Coal Research, 1992) whereby increased concentrations of carbon dioxide in the atmosphere, produced by the combustion of fossil fuels, are believed, with much debate and conjecture from both proponents and opponents of the theory, to lead to global long-term climatic changes (Stobaugh and Yergin, 1983; Michaels, 1991). But to maintain the perspective, the possibility that changes in the amount of carbon dioxide in the atmosphere could lead to changes in world climate is not new and has been suggested since at least the end of the last century.

In a simplified illustration of the greenhouse effect (Figure 2.7) (Schneider, 1989) carbon dioxide and water vapour in the atmosphere absorb (or trap) part of the long-wave (infrared) radiation from the earth's surface, while at the same time the atmosphere allows passage of the short-wave (visible) radiation from the sun. An increase in the concentration of 'greenhouse gases' (such as carbon dioxide) in the atmosphere causes an increase in the absorption of the radiation from the earth which, ultimately, causes an increase in the surface temperature. As a very short summary, the differential in the behaviour of the atmosphere towards 'outgoing/incoming' radiation plays a similar role to that

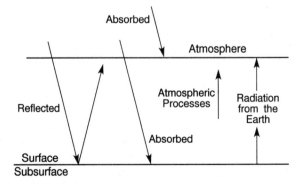

Figure 2.7 Schematic representation of heat trapping ('greenhouse effect') by the earth's atmosphere (Source: Schneider, 1989)

of the glass roof in a small horticultural ecosystem, i.e. a greenhouse. Hence the name 'greenhouse effect'.

However, the term 'greenhouse effect' is deemed by some to be inappropriate because the actual events are much more complex than this simple absorption/radiation example would indicate and many other physical processes and climatic effects must be included in the model. And there are those who consider that the extent of the coupling of the effects is still sufficiently in doubt and that it cannot be stated, or estimated, with any degree of certainty whether or not the surface temperature of the earth will increase, is increasing, or may even be decreasing!

In fact, the only thing that can be stated with any degree of certainty is that opinions will differ and the debate will continue! But instincts alone tell us that the discharge of foreign species, or even the discharge of a large surplus of indigenous species, must alter the delicate systems in an adverse fashion. Perhaps it can be likened to the injection of excessive amounts of steroid materials into the human body, without any short-term beneficial effect. Perhaps the atmosphere is unaffected in the short term but then the long-term effects take over...

It has been argued that the burning of hydrocarbon fuels with the highest hydrogen-to-carbon atomic ratio would minimize the carbon dioxide issue in that the minimum amount of carbon dioxide would be produced per unit of heat generated. And this seems to be true for the natural fuels. Nevertheless, this rationale does not mitigate the need for cleaning gases before they are released to the atmosphere. And there are regulations to this effect.

In the United States, the Clean Air Act of 1970 and the ensuing amendments (United States Congress, 1990; Stensvaag, 1991) to it (as well as the emissions standard for other countries (IEA Coal Research, 1991)) provide the basis for the regulatory constraints imposed on air emissions. The specific regulated pollutants are particulates, sulphur dioxide, photochemical oxidants, hydrocarbons, carbon monoxide, nitrogen oxides and lead (Mintzer, 1990).

In fossil fuel conversion facilities the largest air emissions, apart from carbon dioxide and water vapour, are particulates, sulphur dioxide and nitrogen oxides. Hydrocarbons, ammonia and trace metals may also occur in the effluents and, although the quantities are small, these compounds are still important environmentally. Air emissions from the use of a fossil fuel depend on the efficiency of the control placed on each pollutant as well as on the amounts of the pollutants, or the compounds that form the pollutants, present in the raw material.

As noted above (Figure 2.6), there are several areas where pollutants can be produced during fossil fuel conversion. Indeed, a major generic source of air-polluting emissions from any plant is in the preparation of the raw feedstock, specifically the crushing, screening and storage operations associated with a coal conversion (e.g. utility) plant and the subsequent feeding to the reactors. For coal conversion processes, another major source of particulate matter arises from the ancillary combustion operations where fly ash is produced when coal, or mineral-containing carbonaceous materials, is burned to generate process heat, raise steam, or produce electric power.

Gases leaving a reactor, whether they are product gases from a gasifier or by-product gases from a carbonization reactor or gases from a petroleum-cracking unit, contain components that may be categorized as desirable, neutral or undesirable.

A component is only defined here as desirable or undesirable primarily from a process point of view. A desirable component should be present in the end product, an undesirable one absent. A component may also be undesirable because its presence in a processing stage could be detrimental. For example, hydrogen sulphide is undesirable in gas streams because its level specified for pipeline gas contracts is often orders of magnitude less than what is present in off-gases. In addition to hydrogen sulphide, undesirable components include ammonia, hydrogen chloride, carbon dioxide and particulate matter depending upon the ultimate use of the gas stream (Kohl and Riesenfeld, 1979).

Another effect that must also be considered in gas-cleaning operations is that the gases exiting reactors such as those used for coal gasification and carbonization are hot, with temperatures ranging from 400 to 1500 °C (750 to 2730 °F). Thus, a simple gas-cleaning procedure involves cooling the gas, generally accomplished by indirect heat exchange, to a temperature at which the water and hydrocarbons condense; tars (liquids), often evident as products of coal processing, may also be condensed.

Gas cooling by scrubbing with water is normally employed as one of several means of gas cleaning and purification. The water will remove from the gas stream any soluble gases and particulate matter. Furthermore, even though the acid gases carbon dioxide and hydrogen sulphide by themselves are not very soluble in water, the presence of ammonia (a weak alkali and soluble in water) in the gas stream or in the water will accomplish the solubilization of the carbon dioxide and hydrogen sulphide.

The removal of particulates is essential although there is still some debate about the requirements, which are considered by many to be too generic in so far as many standards do not relate, specifically, either to chemical composition or to particulate size. But increasing emphasis is being placed on inhalable particulates less than 15 μm in size, especially on fine particulates less than 2 to 3 μm.

Among the conventional control technologies are cyclones, scrubbers, electrostatic precipitators and fabric filters. Electrostatic precipitators, also suitable for fine particulate removal, operate on the principle of charging the particles with ions and then collecting the ionized particles on a surface from which they are subsequently removed. Electrostatic precipitators are in common use for fly ash removal from power plant glue gases.

All of the conventional devices have certain advantages and disadvantages and efficiency varies (Figure 2.8). Fabric filtration, which is used for the removal of extremely fine particulate matter, is accomplished in a 'baghouse' in which are hung a number of filter bags through which the gas stream flows.

Disposal and stack height

The use of tall stacks is based on the simple principle that the higher the stack, the lower the concentration of noxious pollutants at ground level. Power generation boiler stacks vary in height (from 250 to 1200 ft, 76 to 366 m) and design with one tall stack often serving multiple boilers (Carlton-Jones and Schneider, 1968; Alba, 1970; Leite, 1990).

In general, five major factors determine the concentration of a pollutant at ground level: (a) stack height; (b) gas exit velocity from the tip; (c) gas exit temperature; (d) concentration and nature of the pollutant in the exit gas; and

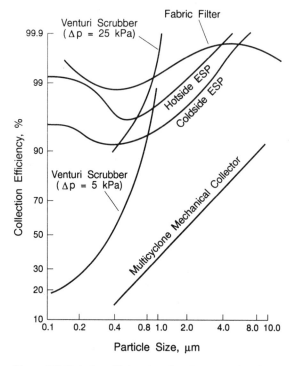

Figure 2.8 Relative efficiencies of various gas-cleaning systems (Source: Probstein and Hicks, 1990)

(e) ambient and wind conditions surrounding the exit tip of the stack (Figures 2.9 and 2.10).

For example, the gas stacks associated with Claus sulphur plants vary in height from 100 ft (30.5 m; non-residential area) to 600 ft (183 m) or higher in areas where there are more stringent pollution abatement standards. Normal current practice is to incinerate the tail gas from a Claus unit, together with fuel gas and about 25% excess air, to about 650°C (1200°F). The incinerator discharges ino the base of a refractory-lined stack (containing a monolithic, castable lining) and the gas may exit the stack at 54–595°C (1000–1100°F).

But it is now recognized that gases being released into the atmosphere at one point can, and usually do, descend to earth at another point in the form of acid rain. The four main stages of chimney plume dispersion can be related to acid rain deposition only relatively short distances away from the stack (Figure 2.11). Thus, stack height is no longer a universally viable means of pollution abatement!

Pollution regulations

Amendments to the United States Clean Air Act of 1967 were made in 1970 and again in 1990 (United States Congress, 1990) and provided for the establishment of national ambient air quality standards for, as an example,

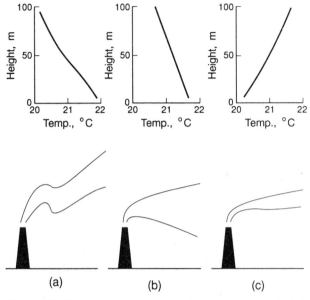

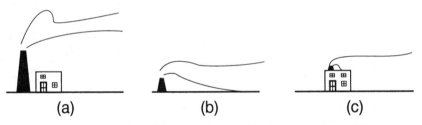

Figure 2.9 General representation of temperature variation with plume height and behaviour

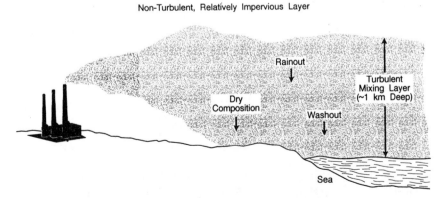

Figure 2.10 Representation of the behaviour of (a) a buoyant plume; (b) a dense gas plume; and (c) a plume affected by a building

Figure 2.11 Representation of the main stages of plume dispersion

sulphur dioxide. These standards do not stand alone and there are many national standards for sulphur emissions (Table 1.10, page 20).

In summary, regulations are becoming stricter as the only hope that exists for ensuring a clean environment.

Consequences for gas-processing options

It has been recognized for some time that gaseous pollutants, especially sulphur dioxide, aggravate existing respiratory disease in humans and contribute to its development (e.g. Houghson, 1966). Sulphur dioxide gas by itself can irritate the upper respiratory tract. It can also be carried deep into the respiratory tract by airborne adsorbents and can cause near-irreversible damage in the lungs. There is also the belief that sulphur dioxide is a contributor to increased respiratory disease death rates. It also contributes to the various types of smog that occur in many industrialized areas of the world. Furthermore, it is harmful to a variety of flora including forage, forest, fibre, and cereal crops as well as many vegetable crops. Vegetation just cannot grow, let alone flourish, in an atmosphere polluted by sulphur dioxide. Indeed, the occurrence of zones of dead vegetation were a common sight at points downwind of sulphide ore smelters!

Thus, it has become very apparent over the last three decades that abatement of air pollution needs to be mandatory now and for the future.

Four main avenues of action are open to decrease the amount of sulphur dioxide emitted from the stacks of power-generating plants: (a) burn low-sulphur fuels; (b) desulphurize available fuels; (c) remove sulphur oxides from flue gases; or (d) generate power by nuclear reactors. Low-sulphur fuels are expensive and not readily available in many areas where the population density is at its greatest. Desulphurization of fuels is also expensive, and the technology for desulphurizing coal is still in the developmental stage. And few nuclear power plants are being built today. Safety and health concerns about nuclear facilities make it unlikely that there will be an increase in building such facilities for some years to come.

Therefore, the abatement solution that appears to be of most immediate application is the removal of sulphur dioxide from utility stack gases.

Fuel desulphurization methods such as the hydrodesulphurization of fuel oil (Speight, 1981), amine treatment of gas streams, and gasification of coal (Speight, 1990) all produce hydrogen sulphide as a by-product of the primary process. Thereafter, the usual practice is to utilize a Claus sulphur recovery unit to convert the hydrogen sulphide to elemental sulphur.

The use of fossil fuels involves at some stage, by deliberate means or accidentally, the generation of gaseous mixtures that can, to say the least, be quite obnoxious in terms of environmental contamination.

Whilst the problem in former heavy industrialized centres may be seemingly less acute than it was decades ago, predominantly because of an increased environmental consciousness, the generation of such noxious materials is still an issue. Industry continues to march forwards but the increased need to maintain a clean, liveable, environment is more evident now than at any time in the past. In the past, a certain amount of pollution was recognized as being almost inevitable, perhaps even fashionable! But now this is not the case. Any industry found guilty of emitting noxious materials can suffer heavy fines. And there is also the possibility of a gaol term for the offending executives! Pollution of the environment will not be tolerated.

Thus, whilst industry marches on using many of the same processes that were in use in the early days of the century, more stringent methods for clean-up are necessary before any product/by-product can be released to the atmosphere. And this is where gas processing becomes an important aspect of industrial life. Furthermore, gas-cleaning processes are now required to be more efficient than ever before.

The beginning

The general prognosis for gas clean-up is not pessimistic and can be looked upon as being quite optimistic, nitrogen and sulphur species notwithstanding. Essentially all of the nitrogen and all of the sulphur in a fuel will be oxidized to nitrogen oxides and sulphur oxides (usually sulphur dioxide). The control of nitrogen oxides is achieved, to date, mainly through modification of the combustion process, and other methods, including various flue gas treating processes, are being developed. Indeed, it is anticipated that flue gas control technologies may achieve a significantly higher control of gaseous emissions than is currently being recorded. In fact, current flue gas desulphurization techniques are capable of removing about 90%.

It is considered likely that much of the environmental impact of fossil fuel usage (including the hazards of coal mining, gaseous emisions, acid precipitation) could be substantially reduced. A considerable investment in retrofitting or replacing existing facilities and equipment might be needed. However, replacing coal with natural gas, which releases less carbon dioxide per unit of energy, is at best a short-term solution. Minimizing the carbon dioxide emissions from coal and from oil (which must also include emissions of hydrogen sulphide) would require a revamping of the current fuel-burning technology. But it is possible, and a conscious goal must be to improve the efficiency with which fossil fuels are transformed and consumed and to shift to alternative fuels, especially from carbon-rich to hydrogen-rich fuels. Such fuels are the biomass fuels.

Ethanol (ethyl alcohol; grain alcohol) is a major biomass fuel today but in the current market conditions it remains significantly more expensive than gasoline. The widespread use of biomass (alcohol) fuels is constrained by the size of the resource base from which they are produced. There is also the argument that methyl alcohol (methanol) produced from natural gas releases as much carbon dioxide as does gasoline. The counter argument is that methanol-fuelled engines have a greater potential for improvement than do gasoline engines (e.g. they can function at higher levels of compression than have heretofore been attempted). Others argue that the energy contained in natural gas might be exploited more efficiently if the fuel were burned directly instead of being converted to methanol. Indeed, natural gas releases about 20% less carbon dioxide per unit of energy than does gasoline. But natural gas must be stored under compression in heavy, bulky tanks, thus limiting the range and payload of vehicles. Moreover, existing distribution systems are designed for liquids and would require fundamental changes to accommodate natural gas.

All of the aforementioned arguments for and against the use of fossil fuels as sources of energy do not belie the obvious: fossil fuels will be with us as energy sources for some time to come. They must be used judiciously in so far as they are sources of pollutants, many of which are gases and must not be released to the atmosphere.

Obviously, much work is needed to accommodate the change to a different fuel source. In the meantime, we use what we have, all the while working to improve efficient usage and to ensure that there is no damage to the environment.

Such is the nature of fossil fuel usage and the expectancy of gas cleaning processes.

References

Alba, C. (1970) *Hydrocarbon Processing*, p. 170 (March)

Carlton-Jones, D. and Schneider, H.B. (1968) *Chemical Engineering*, 14 October, p. 166

Clark, W.C. (1989) *Scientific American*, **261**, (3), 46

C&ST (1992) Puerto Rico considers outlawing coal. *Coal & Synfuels Technology*, 23 March, p. 5

Douglas, J. (1990) *EPRI Journal, Electric Power Research Institute*, June, p. 4

Ember, L.R., Layman, P.L., Lepkowski, W. and Zurer, P. S. (1986) *Chemical and Engineering News*, 24 November, p. 14

Frosch, R. A. and Gallopoulos, N.E. (1989) *Scientific American*, **261**, (3), 144

Graedel, T.E. and Crutzen, P.J. (1989) *Scientific American*, **261**, (3), 58

Hileman, B. (1989) *Chemical and Engineering News*, 13 March, p. 25

Houghson, R.V. (1966) *Chemical Engineering*, 29 August, p. 71

IEA Coal Research (1991) *Emissions Standard Data Base*, International Energy Agency Coal Research, London

IEA Coal Research (1992) *Greenhouse Gases Bulletin*, International Energy Agency Coal Research, London

Keepin, B. (1986) *Annual Review of Energy*, **11**, 357

Kellogg, W.W. (1987) *Climatic Change*, **10**, 113

Kohl, A.L. and Riesenfeld, F.C. (1979) *Gas Purification*, 3rd edn, Gulf Publishing Company, Houston, TX

Leite, O.C. (1990) *Hydrocarbon Processing*, **69**, (11), 64

MacDonald, G.J. (1988) *Journal of Policy Analysis and Management*, **7**, 425

Manowitz, B. and Lipfert, F.W. (1990) In *Geochemistry of Sulfur in Fossil Fuels* (ed. W.L. Orr and C.M. White), American Chemical Society, Washington, DC, Chapter 3

Maurits la Rivière, J.W. (1989) *Scientific American*, **261**, (3), 80

Michaels, P.J. (1991) *Journal of Coal Quality*, **10**, (1), 1

Mintzer, I.M. (1990) *Annual Review of Energy*, **15** 513

NRC (1990) *Confronting Climate Change. A Report of the National Research Council for the United States Department of Energy*. Report No. DOE/EH/89027P-Hl, United States Department of Energy, Washington, DC

Probstein, R.F. and Hicks, R.E. (1990) *Synthetic Fuels*, pH Press, Cambridge, MA

Schlesinger, M.E. and Mitchell, J.F.B. (1987) *Reviews of Geophysics*, **25**, 760

Schneider, S.H. (1989) *Scientific America*, **261**, (3), 70

Smith, I.M. (1988) *CO$_2$ and Climatic Change*. Report No. IEACR/07, International Energy Agency Coal Research, London

Speight, J G. (1981) *The Desulfurization of Heavy Oils and Residua*, Marcel Dekker, New York

Speight, J.G. (1990) *Fuel Science and Technology Handbook*, Marcel Dekker, New York

Stensvaag, J-M. (1991) *Clean Air Act Amendments: Law and Practice*, Wiley, New York

Stobaugh, R. and Yergin, D. (1983) *Energy Future*, Vintage Books/Random House, New York, and references cited therein.

United States Congress (1990) *Public Law 101–549. An Act to Amend the Clean Air Act to Provide for Attainment and Maintenance of Health Protective National Ambient Air Quality Standards, and for Other Purposes*, 15 November.

White, R.M. (1990) *Scientific American*, **263**, (1), 36

3

Classification, definitions and terminology

Introduction

Gas clean-up is the general collection of processes by which the noxious, unwanted, constituents can be removed from a gaseous mixture (Probstein and Hicks, 1990; Speight, 1991). Clean-up processes can be defined as 'external' ('end of the pipe') or 'integrated'. The former terminology refers to those gas-cleaning processes which occur after the formation of the gas and at a point just prior to the preparation of the gas for sale or venting to the atmosphere. Examples are the 'conventional' cleaning processes as might be included in any refinery or coal conversion operation where the gas is treated by an agent to accomplish cleaning or pollutant removal. However, such processes might also be employed to remove a desired product which is recovered at a later stage of the cleaning operation for further use.

"END-OF-PIPE" SYSTEM

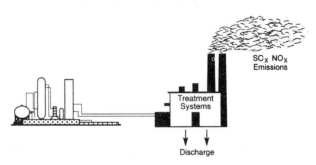

INTEGRATED SYSTEM

Emission Control

Figure 3.1 Schematic representation of an 'external' gas-cleaning operation and an 'integrated' gas-cleaning operation

In the past, there has been a tendency to consider the 'external' processes as being more generally applicable in the petroleum industry (Figure 3.1). The use of gaseous products, not the least of which has been the natural gas constituents that are separated at the outset of refining operations, has been the mainstay of the petrochemical industry; hence the inclusion of gas-processing operations in many refineries. But whilst this may indeed be the case, it is by no means the only case.

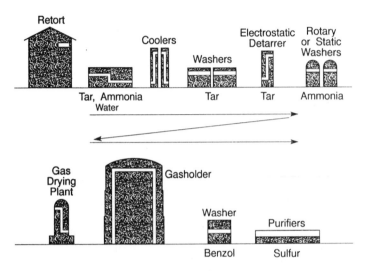

Figure 3.2 General scheme used for coal gas cleaning

Cleaning coal gases to remove distillable tars, liquors and other materials has been practised since the beginning of the commercial coal gas industry (Figure 3.2).

'Integrated' gas-cleaning processes are those processes where an 'agent' is added to, for example, the coal (limestone might be cited as an example) to capture the noxious pollutants at the time of their generation (Figure 3.1).

Gases are produced from a variety of industrial processes (Table 3.1) and, in fact, the composition of one particular gas, say producer gas, can vary depending upon the fuel employed (Table 3.2). In addition, there is the general need for gas treatment to ensure product purity or, more pertinent to the present text, to ensure compliance with a variety of environmental regulations before any such gaseous product is released to the atmosphere.

Therefore, in order to understand the nature of the processes involved in gas processing, it is essential to understand and define the various gas mixtures that might be subject to clean-up from the point of origin and production.

Classification

The gaseous fuels of interest in the present context are the natural hydrocarbon gases and those gases manufactured solely for use as fuels as well as those

Table 3.1 Analyses for various industrial gases

Type	Analysis (% vol)								Specific gravity	Btu/ft³[a]	
	CO_2	O_2	CO	H_2	CH_4	C_2H_6	C_3H_8 and C_4H_{10}	N_2		Gross	Net
Dry natural gas	0.2				99.2			0.6	0.56	1007	906
Propane (LP)						2.6	97.3	0.1	1.55	2558	2358
Refinery oil gas		0.2	1.2	6.1	4.4	72.5	15.0	0.6	1.00	1650	1524
Coke oven gas	2.0	0.3	5.5	51.9	32.3		3.2	4.8	0.40	569	509
Blast furnace gas	11.5		27.5	1.0				60.0	1.02	92	92
Producer gas	8.0	0.1	23.2	17.7	1.0			50.0	0.86	143	133

[a] 1 Btu/ft³ $= 3.8 \times 10^4$ J/m³.

Table 3.2 Composition of producer gases

	CO_2	CO	H_2	CH_4	N_2	Yield $(m^3$ per kg)	MJ/m^3 at s.t.p.	Specific gravity (air = 1)
Coke	5	29	11	0.5	54.5	3.64	5.21	0.9
High volatile matter coal	2.5	30	12	3.0	52.5	3.79	6.48	0.87
High volatile matter coal (Mond gas)	16	12	24	3.0	45.0	4.24	5.73	0.83
Anthracite	6	26	17	1.2	49.8	4.09	5.93	0.85

Table 3.3 Examples of gaseous emissions from different processes

Industry	Air contaminants emitted
Aluminium reduction	Particulates, CO, SO_2, hydrocarbons and fluorides
Cement plants	Particulates, CO, SO_2, NO_2
Coal preparation	Dust, smoke, particulates, sulphur oxides, H_2S
Coke plants	Particulates, SO_2, CO, H_2S, methane, ammonia, H_2, phenols, hydrogen cyanide, N_2, benzene, xylene
Fertilizer industry (chemical)	PH_3, P_2O_5 HF, SiF_4 NH_3, NH_4Cl, SiF_4, HF, NO_2
Kraft pulp mills	Mercaptans, H_2S, H_2, organic sulphides and disulphides
Municipal and industrial incinerators	Particulates, CO, SO_2, ammonia, organic acids, aldehydes, NO_2, hydrocarbons, HCl
Non-ferrous smelters: Copper	SO_2, particulates
Lead	SO_2, CO, particulates
Zinc	Particulates and SO_2, CO
Paint and varnish manufacturing	Acrolein, other aldehydes and fatty acids, phthalic anhydride (sublimed); ketones, fatty acids, formic acids, acetic acid, glycerine, acrolein, other aldehydes, phenols and terpenes; from tall oils, hydrogen sulphide, alkyl sulphide, butyl mercaptan, and thiofene; olefins, branched-chain aromatics and ketone solvents
Rendering plants	SO_2, mercaptans, ammonia
Steel mills	CO, particulates, SO_2, CO_2, NO_2

obtained as by-products of industrial processes (Table 3.3). And, by understanding the means by which the gases are formed as well as their respective compositions, it is possible to classify them accordingly.

For example, natural gas falls within the category of gaseous fuels in the generalized classification scheme for hydrocarbon resources (Figure 3.3). However, as seen from the nomenclature of the different types of gaseous fuels and process emissions other subclassification systems are possible (Figure 3.4).

In addition to the hydrocarbon gases, hydrogen itself is often present as a constituent of many of the industrial gases. It can be employed as part of the

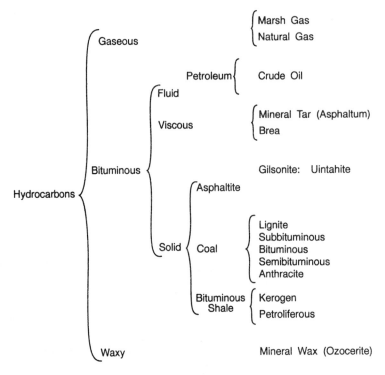

Figure 3.3 General classification for 'hydrocarbon' resources

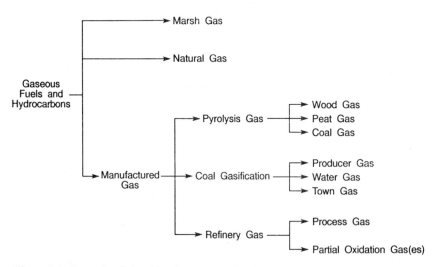

Figure 3.4 General subclassification system for fuel gases and process gases

product as, for example, in the petroleum industry where a hydrogen-rich gas stream (usually 60% or more hydrogen) is a product of the reforming processes and is then used for the various hydroprocesses. On the other hand, hydrogen may be recovered separately for use as a fuel, for synthesis, or for those hydrogenation processes that require the gas in the separated state.

Definitions and terminology

Natural gas

The generic term 'natural gas' applies to gases commonly associated with petroliferous geological formations. As ordinarily found, these gases are combustible, but non-flammable components such as carbon dioxide, nitrogen and helium are often present. Natural gas is generally high in methane and some of the higher paraffins (C_nH_{2n+2}) may be found in small quantities.

Natural gas is an ideal fuel for heating because of its cleanliness, ease of transportation, high heat content, and the high flame temperature.

In more specific terms, natural gas is the naturally occurring gaseous constituents that are found in many petroleum reservoirs (Speight, 1990) and coal seams (Berkowitz, 1979; Speight, 1983; Hessley et al., 1986; Hessley, 1990) and in the decay of organic material. In fact, natural gas is a mixture of combustible hydrocarbon compounds although it does contain non-hydrocarbon compounds (Table 3.4). Not only are there variations in the composition of natural gas with the source (Table 3.5) but also there are variations in the composition of the gas depending upon whether the gas is 'wet' or 'dry' (Table 3.6) (see below for definitions of 'wet' and 'dry').

The gas occurs in the porous rock of the earth's crust either alone or with accumulations of petroleum. In the latter case, the gas forms the gas cap (Figure 3.5) which is the mass of gas trapped between the liquid petroleum and the impervious cap rock of the petroleum reservoir. When the pressure in the reservoir is sufficiently high, the natural gas may be dissolved in the petroleum

Table 3.4 General range of composition of natural gas associated with petroleum

Category	Component	Amount (%)
Paraffinic	Methane (CH_4)	70–98
	Ethane (C_2H_6)	1–10
	Propane (C_3H_8)	Trace – 5
	Butane (C_4H_{10})	Trace – 2
	Pentane (C_5H_{12})	Trace – 1
	Hexane (C_6H_{14})	Trace – 0.5
	Heptane and higher (C_7+)	None–trace
Cyclic	Cyclohexane (C_6H_{12})	Traces
Aromatic	Benzene (C_6H_6), others	Traces
Non-hydrocarbon	Nitrogen (N_2)	Trace – 15
	Carbon dioxide (CO_2)	Trace – 1
	Hydrogen sulphide (H_2S)	Trace occasionally
	Helium (He)	Trace – 5
	Other sulphur and nitrogen compounds	Trace occasionally
	Water (H_2O)	Trace – 5

Table 3.5 Variations of natural gas composition with source

Component	Type of gas field			Natural gas separated from crude oil		
	Dry gas, Los Medanos[a] (mole %)	Sour gas, Jumping Pound[b] (mole %)	Gas condensate, Paloma[a] (mole %)	Ventura[a] 400 lb (mole %)	50 lb (mole %)	Vapour (mole %)
Hydrogen sulphide	0	3.3	0	0	0	0
Carbon dioxide	0	6.7	0.68	0.30	0.68	0.81
Nitrogen and air	0.8	0	0	0	–	2.16
Methane	95.8	84.0	74.55	89.57	81.81	69.08
Ethane	2.9	3.6	8.28	4.65	5.84	5.07
Propane	0.4	1.0	4.74	3.60	6.46	8.76
Isobutane	0.1	0.3	0.89	0.52	0.92	2.14
n-Butane	Trace	0.4	1.93	0.90	2.26	5.20
Isopentane	0		0.75	0.19	0.50	1.42
n-Pentane	0		0.63	0.12	0.48	1.41
Hexane	0	0.7	1.25			
Heptane	0			0.15	1.05	4.13
Octane	0					
Nonane	0		6.30			
	100.0	100.0	100.0	100.0	100.0	100.0

[a] California.
[b] Canada.

Table 3.6 General composition of 'wet' and 'dry' natural gas

Constituents	Composition (vol %)		
Hydrocarbons:	'wet'	←(range)→	'dry'
Methane	84.6		96.0
Ethane	6.4		2.0
Propane	5.3		0.6
Insobutane	1.2		0.18
n-Butane	1.4		0.12
Isopentane	0.4		0.14
n-Pentane	0.2		0.06
Hexanes	0.4		0.10
Heptanes	0.1		0.08
Non-hydrocarbons:			
Carbon dioxide		0–5	
Helium		0–0.5	
Hydrogen sulphide		0–5	
Nitrogen		0–10	
Argon		0–0.05	
Radon, krypton, xenon		Traces	

and released upon penetration of the reservoir as a result of drilling operations. There are also reservoirs in which the gaseous constituents are the only occupants.

The predominant component of natural gas is methane but other hydrocarbons such as ethane, propane and butane may also be present. Carbon dioxide and, on occasion, gaseous sulphur compounds such as hydrogen sulphide also occur in natural gas. Trace amounts of rare gases such as helium may also occur and some natural gas reservoirs are a source of these rare gases.

Natural gas should not be confused with the gaseous products of the destructive distillation or carbonization of wood and coal (Table 3.1; Figures 3.3 and 3.4); such gaseous products are manufactured gases.

The olefin hydrocarbons (C_nH_{2n}), carbon monoxide (CO) and hydrogen (H_2) are not usually present in natural gas. Having made this statement, no doubt exceptions will be noted by the reader!

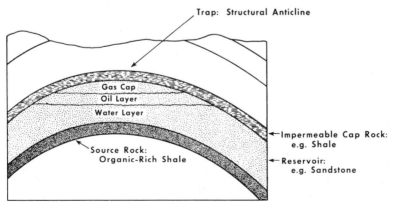

Figure 3.5 Illustration of a petroleum reservoir showing the gas cap

The reactivity of olefins over geological time through the agency of either their natural reactivity or the catalytic effects of minerals in the various formations almost ensures the absence of such materials in natural gas and petroleum. The absence of carbon monoxide and hydrogen from natural gas may be due to similar factors or, perhaps, to a series of virtually constant chemical maturation reactions that prevailed during the formation of natural gas and petroleum from a wide variety of precursors. Local and regional variations in the maturation conditions do not seem to have played a role in allowing many exceptions to this general rule.

Just as petroleum and coal can vary in composition, so too can natural gas. In fact, whilst there are differences in natural gas composition between different reservoirs, two wells in the same field may yield gaseous products that are different in composition and analyses. Indeed, any one particular natural gas field could require different production, processing and handling protocols from another field.

Thus, there is no single composition of components which might be termed 'typical' natural gas. Methane and ethane constitute the bulk of the combustible components, and carbon dioxide and nitrogen, the inert compounds. The net heating value of natural gas served by a utility company is often 1000 to 1100 Btu/ft^3 (3.7 to 4.2×10^7 J/m^3).

In more general terms, the term 'natural gas' does not signify a constant composition of matter. The proportion of non-hydrocarbon constituents can vary over a wide range (Table 3.6; Speight, 1990) and can be classified as two types of chemical materials: (a) diluents such as nitrogen, carbon dioxide and water vapour; and (b) contaminants such as hydrogen sulphide and/or other sulphur compounds.

The diluents are non-combustible gases that reduce the heating value of the gas and are, on occasion, used as 'fillers' when it is necessary to reduce the heat content of the gas. On the other hand, the contaminants are detrimental to production and transportation equipment in addition to being obnoxious pollutants. Thus, the primary reason for gas processing is to prepare the gas for sale to the consumer by removing the unwanted constituents.

The major diluents/contaminants of natural gas are: (a) acid gases, predominantly hydrogen sulphide and carbon dioxide, although the latter does often occur to a much lesser extent; (b) water which includes all entrained free water or water in condensed form; (c) liquids in the gas such as higher-boiling hydrocarbons as well as pump lubricating oil, scrubber oil and, on occasion, methanol; and (d) any solid matter that may be present such as fine silica (sand) and scaling from the transportation/distribution pipe (Curry, 1981).

As a result of the variances in the composition of natural gas from different locations, there are several general definitions that have been applied to the different products. Thus, natural gas can be: (a) 'lean' gas in which methane is the major constituent; (b) 'wet' gas which contains considerable amounts of the higher-molecular-weight hydrocarbons; (c) 'sour' gas which contains hydrogen sulphide; (d) 'sweet' gas which contains very little, if any, hydrogen sulphide; (e) 'residue' gas which is natural gas from which the higher-molecular-weight hydrocarbons have been extracted; and (f) 'casinghead' gas which is derived from petroleum but is separated at the separation facility at the well-head.

To define the terms 'dry' and 'wet' further in quantitative measures, the term 'dry' natural gas indicates that there is less than 0.1 gallon (1 US gallon = 3.79×10^{-3} m^3) of gasoline vapour (higher-molecular-weight paraffins) per 1000 ft^3 (1 ft^3 = 0.028 m^3). The term 'wet' natural gas indicates that there are such paraffins present in the gas, in fact more than 0.1 gal/1000 ft^3.

There are many gas fields (often referred to as 'dry' fields) in which liquids do not occur with the gas and the only processing required is dehydration or, perhaps, heating value adjustment. Other fields (often referred to as 'condensate' fields) are those fields in which a liquid product or condensate is produced with the gas. In these fields, liquids often condense out of the gas (retrograde condensation) as the pressure is reduced (Katz *et al.*, 1959). Liquid also condenses out of the gas in the formation as the pore pressure drops and may not completely vaporize before well abandonment pressure is reached. Cycling plants can be installed to prevent this loss of product, often referred to as 'natural gasoline'.

The produced gas is processed to remove the higher-molecular-weight hydrocarbons and any liquid products. Any residue gas, rather than being sold, is injected to maintain reservoir pressure. When the reservoir has been swept of the higher-molecular-weight materials so that retrograde condensation can no longer occur, the field will then be taken to full production.

As already noted, gas is also produced with crude oil and such gas is usually rich in recoverable hydrocarbon liquids. Thus, it is more than likely that construction of a gas-processing plant in conjunction with the petroleum recovery operations may be economically justifiable even at relatively low gas production rates.

A typical gas-processing plant produces residue gas and a variety of products such as ethane, liquefied petroleum gas (LPG) (Table 3.7) and natural gasoline (Table 3.8). Originally, the gas-processing plants were used to remove the gasoline components to be used as a blending stock for motor gasoline (hence the term 'gasoline' plant was often inappropriately applied to the gas-processing plant). Other fuel needs then caused a shift of focus to the liquefied petroleum gas (propane, butanes and/or mixtures thereof) as well as the gasoline constituents.

More recently the extraction of ethane from natural gas streams for use as a

Table 3.7 Composition of liquefied petroleum gases

	Propane	*Butane*	*Mixture*
Composition: C$_2$ (vol %)	3.3	<0.1	1.7
C$_3$ (vol %)	92.5	13.5	53.0
iso-C$_4$ (vol %)	3.2	35.7	19.4
n-C$_4$ (vol %)	1.0	49.5	25.3
iso-C$_5$ (vol %)	nil	0.8	0.4
n-C$_5$ (vol %)	nil	0.4	0.2
C$_6$ (vol %)	nil	0.1	nil
H$_2$S (ppm)	<1	<1	<1
Mercaptans (ppm)	2.4	1.8	2.1
Specific gravity, 15.5/15.5°C	0.5135	0.5681	0.5408
Reid vapour pressure (psig)	234	96	205
Calorific value (Btu/lb)			
Gross	21 500	21 200	21 350
Net	19 900	19 700	19 800

Table 3.8 Composition of natural gasolines from natural gas

Reid vapour pressure	Ventura gasoline plant			Ten-section gasoline plant
	38 psia	60 psia	100 psia	22 psia
Ethane	Trace	0.5	0.7	0
Propane	1.1	16.0	43.8	0
Isobutane	19.0	16.0	10.7	0.2
n-Butane	41.0	34.7	23.0	22.7
Isopentane	13.2	11.2	7.4	24.1
n-Pentane	11.3	9.5	6.3	21.0
Hexane	6.8	5.7	3.8	12.6
Heptane	5.3	4.4	2.9	13.7
Octane	1.2	1.0	0.7	4.1
Nonane	1.1	1.0	0.7	1.2
Decane	Trace	Trace	Trace	0.4
	100.0	100.0	100.0	100.0

petrochemical feedstock has become an important aspect of gas-processing operations. There has also been much interest shown in the reactions of methane, other than combustion/oxidation reactions, to the extent that methane chemistry (often referred to as 'C_1' chemistry) is a major research area (Fahey, 1986; Mackie, 1991; Wolf, 1992). The work has led to a variety of processes, and emerging process concepts, that have added to the capabilities of the gas-processing and petrochemical industries (Speight, 1990, 1991).

After processing and purification, natural gas is a homogeneous mixture that is odourless, but odour-generating additives are added during processing to enable the detection of gas leaks. Methane, the predominant constituent of natural gas, is one of the more stable flammable gases (Curry, 1981) but it is flammable within the limits of a 5–15% mixture with air. In comparison, hydrogen sulphide is flammable within 4–46% in air at a much lower ignition temperature. Natural gas has an energy content of 1000 Btu/scf (1 Btu/scf or 1 Btu/ft^3 = 37.3 kJ/m^3) and is very often priced in terms of its energy content rather than units of mass or of volume.

In addition to composition and thermal content (Btu/scf, J/m^3), natural gas can also be characterized on the basis of the mode of natural gas which is found in reservoirs where there is no or, at best, only minimal amounts of crude oil.

Non-associated natural gas

'Non-associated' natural gas is found in reservoirs in which there is no or, at best, only minimal amounts of crude oil. Non-associated gas is usually richer in methane but is markedly leaner in terms of the higher-molecular-weight hydrocarbons and condensate materials.

Associated natural gas

'Associated' or 'dissolved' natural gas occurs either as free gas or as gas in solution in the crude oil. Gas that occurs as a solution with the crude petroleum is 'dissolved' gas whereas the gas that exists in contact with the crude petroleum ('gas cap'; Figure 3.5) is 'associated' gas.

Associated natural gas is usually leaner in methane than the non-associated gas but will be richer in higher-molecular-weight paraffinic constituents.

The most preferential type of natural gas is the non-associated gas. This gas can be produced at high pressure whereas associated, or dissolved, gas must be separated from petroleum at lower separator pressures which usually involves increased expenditure for compression. Thus it is not surprising that such gas (under conditions that are not economically favourable) will often be flared or vented.

Gas condensate

'Gas condensate' contains relatively high amounts of the higher-molecular-weight liquid hydrocarbons, often referred to as natural gasoline (Table 3.8). These hydrocarbons may occur in the gas phase in the reservoir.

Manufactured gas

Manufactured gas is the general term that has usually been applied directly to, and used synonymously with, coal gas. As such, it is the gaseous mixture that is produced when coal is heated under a variety of conditions. The manufacturing process consists essentially of heating coal to drive off the volatile products, some of which may be gases and others liquids and tars, to leave a solid carbonaceous residue. The residue, or char, is then treated under a variety of conditions to produce other fuels which vary from a 'pure' char to different types of gaseous mixtures.

However, there are other gases that fall 'under the banner of' manufactured gas. These materials usually occur in the petroleum industry or, for that matter, in any industry where a gaseous product is manufactured with a specific end use in mind.

Therefore, it is appropriate to include such materials in this section for a better understanding of the gaseous products that might be subjected to the various gas-cleaning operations.

Coal gas

Coal gas proper ('town gas') (Table 3.9) is gas obtained by the carbonization or coking of coal at temperatures between 950°C (1740°F) and 1350°C (2460°F), either in conventional gas retorts or in coke ovens (Chapter 5).

Town gas is the gas that was so common in the early part of the century when much of domestic heating and lighting was accomplished by the use of gas from the nearby gas works. Almost every town and city of any size, and consequence, had a plant where coal was converted into gas for domestic and industrial use – as often evident from the odours in the near and sometimes far vicinity of the plant! Hence the name 'town gas'. The composition of the gas varied but usually contained the combustible gases carbon monoxide, hydrogen, some methane, and anything else that came through the limited gas cleaning process (Table 3.10).

Just as feedstock blending and product blending are standard operations in the petroleum-refining industry, so coal gas is often blended with other gases, such as natural gas, to augment supplies during peak periods. The product sold as coal gas may be a blend of gases, designed to meet the required thermal

Table 3.9 Composition of coal gas produced under different conditions

Gases	High temperature			Low tempera- ture (%)
	Horizontal retort (%)	Vertical retorts Without steam (%)	With steam (%)	
Carbon dioxide	2.0	2.2	3.4	4.5
Unsaturated hydrocarbons	3.1	2.3	1.8	3.8
Oxygen	0.5	0.4	0.7	0.2
Carbon monoxide	8.0	10.3	15.1	8.3
Hydrogen	50.6	49.5	49.3	29.1
Methane	28.1	28.5	21.2	49.1
Nitrogen	7.7	6.8	8.5	5.0

Table 3.10 General range and types of non-hydrocarbon constituents in coal gas

Type of impurity	Typical concentration (vol %)
Hydrogen sulphide	0.3–3.0
Carbon disulphide	0.016
Carbon oxysulphide	0.009
Thiopene	0.010
Mercaptans	0.003
Ammonia	1.1
Hydrogen cyanide	0.10–0.25
Pyridine bases	0.004
Nitric oxide	0.0001
Carbon dioxide	1.5–2.0

and compositional specifications. The blending of coals may also be applicable to the coal gas industry but, to refer to the 'other end' of the process, the use of coal blends for the production of specific cokes is still open to much research and development.

Until about 1940, gas produced from coal was an important part of the energy mix in the United States and Europe. This gas was rapidly replaced by natural gas in the distribution systems of gas utilities serving the residential, commercial and industrial markets. During the last two decades, there have been fluctuations in the estimations of the supplies of natural gas leading to 'glut and famine' predictions relative to the ever-increasing demand. Should the 'famine' predictions become a reality, it is more than likely that coal will, once again, be a major source of gas. Should this be the case, it might be anticipated that the older processes will be replaced by modern methods capable of very high production rates at improved efficiency and reduced cost.

Coal gasification can be accomplished by different processes, including pyrolysis or partial oxidation with air or oxygen and steam using different types of equipment (Chapter 5). But the outcome is, essentially, the same. Conversion of coal(s) to gaseous products (Table 3.9). Various processes operate as fixed beds or fluidized beds or with an entrained bed. The pressure may vary from near atmospheric to 1000 pounds per square inch (1 psi = 6895 pascals = 6.895 kPa) or more and the processes involve the use of high temperatures (650–1645°C, 1200–3000°F).

The direct products of gasification vary in heating value from 120 to 150 Btu/ft^3 (4.5 to 5.6 × 10^3 kJ/m^3) (low-Btu gas), to 300 Btu/ft^3 (11.2 × 10^3 kJ/m^3) (medium-Btu gas or synthesis gas), and to as high as 600 Btu/ft^3 (22.4 × 10^3 kJ/m^3). Low-Btu gas can be used as a fuel for industrial processes or for the production of electrical power by electrical utilities. Medium-Btu gas, which consists principally of carbon monoxide and hydrogen, can be used directly as a fuel, or it can be upgraded by catalytic methanation to essentially pure methane, which is, for all practical purposes, identical to natural gas. Such a product gas is commonly called pipeline-quality gas or substitute natural gas (SNG).

Producer gas (low-Btu gas)

Producer gas is made by the combustion of coal or coke in a limited supply of oxygen. Producer gas is the gas usually employed in the steel-making industry and as a means of supplying heat to large industrial furnaces.

In the process, a mixture of air and steam is passed upwards through a hot, deep bed of coal or coke maintained at a high temperature (over 1000°C, 1830°F) whereby the carbon in the coal reacts to produce the gaseous materials.

The gas consists mainly of carbon monoxide and nitrogen with small proportions of hydrogen (obtained by the 'water gas' reaction), methane (obtained by the partial carbonization of coal) and carbon dioxide (resulting from conditions unfavourable for the complete conversion of carbon to carbon monoxide). The gas is high in nitrogen introduced in the air and its heating value is low because of the high percentage of inert gases (Table 3.11).

Although the chemistry of producer gas formation can be conveniently represented by the simple equations

$$[2C]_{coal} + O_2 (+N_2) = 2CO (+N_2)$$
$$[C]_{coal} + H_2O = CO + H_2$$

it is, in fact, more correctly represented by a sequence of four equations (Table 3.12).

If the temperature is lower, even with intimate contact, and equilibrium is attained, a notable amount of carbon dioxide will be produced:

$$[C]_{coal} + O_2 (+N_2) = CO_2 (+N_2)$$

Table 3.11 General composition of producer gas (modified by steam)

Steam saturation, temperature of blast	60°C	70°C	80°C
Percentage composition of gas:			
Carbon dioxide	5.25	9.15	13.25
Carbon monoxide	27.30	21.70	16.05
Hydrogen	16.60	19.65	22.65
Methane	3.35	3.40	3.50
Nitrogen	47.50	46.10	44.55
Total combustibles	47.25	44.75	42.20
Cal. value of gas, Btus per } gross	185.6	177.5	169.5
cu. ft at 0° and 760 mm } net	173.0	163.3	154.3

Table 3.12 Heats of reaction for the 'common' gas-forming reactions

	$\Delta H_{1000^\circ C}$ (MJ/kg mol)
$C + air \rightarrow CO_2 + N_2$	-395.4
$CO_2 + C \rightarrow 2CO$	$+167.9$
$C + H_2O \rightarrow CO + H_2$	$+135.7$
$CO + H_2O \rightarrow CO_2 + H_2$	-32.18

If, however, the high temperature has been maintained and the carbon converted entirely to carbon monoxide, the gas will consist of 33% carbon monoxide and 67% nitrogen from the producer gas equation (above).

Producer gas, which contains carbon monoxide (Table 3.11), is used as it comes from the generator after preliminary purification. It was once the cheapest form of industrial gas and could again become important; it may be possible to produce it in modernized equipment on a large scale.

Coke oven gas (medium-Btu gas)

Coke oven gas has an intermediate heating value and is made by destructive distillation of a packed bed of coal out of contact with air. The process results in the formation of coke, which is used in the blast furnace. The gas is utilized totally within the steel-making complex.

The combination of the oil embargoes of the 1970s, the ensuing energy crises, and restrictions on air pollution resulted in the proposal of a large number of modern processes (Speight, 1983; Hessley et al., 1986; Hessley, 1990). Nearly all these processes use a mixture of steam and oxygen to burn coal. Less oxygen is used than is required for complete combustion; therefore, the products are primarily carbon monoxide and hydrogen. In some cases, depending on the end use of the gas, moderate amounts of methane and liquid products are produced.

Substitute (synthetic) natural gas (SNG; high-Btu gas)

Gas which is to be used within the natural gas transmission and distribution system must be essentially methane. The medium-Btu gas processes can be used as precursors to produce substitute natural gas (SNG). Other new processes which operate at very high pressure (1 psi = 6.895 kPa) are being developed to produce very large quantities of gas (2.5×10^8 ft^3/day of 1000 Btu/ft^3 gas, 7×10^6 m^3/day of 3.7×10^3 kJ/m^3 gas) to supplement the declining supplies of natural gas. Some of these processes use the partial combustion of oxygen as a direct source of heat, while others use air combustion in a variety of indirect modes.

Blast furnace gas

Blast furnace gas is a low-grade producer gas made by the partial combustion of the coke used in the furnace and modified by the partial reduction of iron ore. The gas contains more carbon dioxide, and less hydrogen, than normal producer gas made from coke, and has a lower calorific value.

Blast furnace gas is a by-product of the manufacture of pig iron. Like producer gas, it is derived from the partial combustion of coke. Some of the combustibles in the gases are used to reduce the iron ore; thus the final gas contains about 27% carbon monoxide and more than 70% of inert gases (carbon dioxide and nitrogen), giving it the lowest heating value (less than 100 Btu/ft^3, 3.8×10^3 kJ/m^3) of any usable fuel gas. It is used for the operation of gas engines, heating by-product coke ovens, steel plant heating, steam raising and crude heating.

Water gas

Water gas is the gaseous mixture produced by the thermal decomposition of coal or coke by steam:

$$[C]_{coal/coke} + H_2O = CO + H_2$$

The reaction of carbon monoxide with steam also occurs:

$$CO + H_2O = CO_2 + H_2$$

which results in a higher carbon dioxide content of the gas product as the temperature is lowered.

Blue water gas

Blue water gas is obtained by passing steam over red-hot coke and is called 'blue' water gas (Table 3.13) because of its blue flame, a characteristic of the combustion of carbon monoxide. It is necessary to distinguish blue water gas from 'carburetted' water gas, which contains hydrocarbon material to bring the composition up to that of town gas.

Table 3.13 Composition of 'blue' water gas

	At Trail, Canada	At Belle, WV, USA
Composition of gas (vol %):		
CO_2	11.3	16.9
CO	54.1	46.6
H_2	31.0	34.0
N_2	2.6	2.2
Calorific value	10.7	10.1
O_2 used (m^3/m^3 gas)	0.23	0.24
Steam decomposed (%)	61	53.5

Fuel gas

Fuel gas is, essentially, any fuel in the gaseous state whose potential heat energy can be readily transmitted and distributed through pipes from the point of origin directly to the place of consumption. The types of fuel gases are natural gas, liquefied petroleum gas (LPG), refinery gas, coke oven gas and blast furnace gas.

Table 3.14 Compositions of 'typical' refinery gases

Composition (wt %)	Primary fractionator gas	Powerformer tail gas	Fluid cat cracker tail gas	Hydrocracker Isomax tail gas	Steam cracker Naphtha light ends	Gas oil
H_2	–	1.5	0.6	1.4	1.2	1.3
C_1	8.5	6.0	7.9	21.8	17.4	19.6
C_2	15.4	17.5	11.5	4.4	7.0	3.5
$C_2=$	–	–	3.6	–	33.3	38.0
C_3	30.2	31.5	14.0	15.3	0.7	1.0
$C_3=$	–	–	16.4	–	27.0	19.7
C_4	45.9	43.5	21.8	57.1	13.4	16.9
$C_4=$	–	–	24.2	–		
$C_4=$	–	–	–	–		
Gas yield (wt %)	9.2	20.0	16.5	14.5	72.8	59.6

Source: Francis and Peters (1980).

Fuel gas composition varies depending upon the source of the gas and the method of production. However, most fuel gases are composed in whole or in part of the combustibles hydrogen, carbon monoxide, methane, ethane, propane, butane and oil vapours and, in some instances, of mixtures containing the inert gases nitrogen, carbon dioxide and water vapour.

Refinery gas

Refinery gas is a generic term applied to the gaseous mixture which is produced during the refining of crude oil. The mixture may contain valuable hydrocarbon products such as propanes and butanes as well as methane, ethane, hydrogen, olefins and dienes (Table 3.14) together with a spate of 'undesirable' components such as the oxides of carbon and sulphur as well as hydrogen sulphide (Table 3.15) (Speight, 1991). Efforts were made in the past to reclaim the valuable constituents of the gas and the remainder would be flared or vented to the atmosphere.

Table 3.15 General range of composition of refinery gases %V/V

H_2S	6 to 8
H_2	5 to 8
CH_4	10 to 30
C_2	10 to 20
C_3	40 to 50
C_4	5 to 30
Calorific value (MJ/m³ av.)	84.9
Specific gravity av.	1.35 (air = 1.00)

The term 'process gas' is also often used, again in a very loose generic sense, to include all of the gaseous products and by-products that emanate from a variety of refinery processes and which must be removed prior to release of the gases to the atmosphere or prior to use in another part of the refinery, i.e. as a fuel gas or as a process feedstock.

A process gas might contain hydrocarbons mixed with carbon oxides (CO_x, where $x = O$ and/or 1), sulphur oxides (SO_x, where $x = 2$ and/or 3), as well as ammonia and hydrogen sulphide. On occasion, oxides of nitrogen (NO_x, where $x = 1$ and/or 2) might also occur in the process gases.

However, as already noted above, under the usual circumstances refinery gases consist mainly of propanes (C_3 hydrocarbons) and butanes (C_4 hydrocarbons) with smaller proportions of hydrogen (H_2), methane (CH_4), ethane (C_2H_6) and hydrogen sulphide (H_2S). In fact, because of the hydrogen content the gas is often treated, by reforming (Table 3.16), to enrich this hydrogen content and then (with a hydrogen content of over 60%) sent, or recycled, to the hydrotreating/hydrocracking units.

Partial oxidation gas

The partial oxidation process is used in the petroleum industry as a means of producing hydrogen for various hydrotreating and/or hydrocracking operations. As with many processes, the chemistry can often be represented in the simplest manner:

Table 3.16 Composition of reformed refinery gas

Composition (vol %)	Refining gas, H_2S free	Reformed gas	With blow run	With blow run and cold enrichment
CO_2	0.7	2.4	6.6	5.6
O_2	0.7	0.4	0.3	0.6
CO	1.7	15.0	16.9	12.6
H_2	5.3	55.6	34.8	31.7
CH_4	26.0	20.8	13.8	10.1
C_2H_6	19.4	–	–	10.9
C_3H_8	17.8	–	–	–
C_4+	5.0	–	–	–
Unsats. (C_3H_6)	21.4	4.1	2.5	4.2
N_2	2.0	1.7	2.52	24.3
Specific gravity (air = 1.00)	1.02	0.40	0.64	0.70
Calorific value (MJ/m^3)	68.5	21.0	14.3	20.6

Source: Francis and Peters (1980).

Table 3.17 General chemistry of the gasification of refinery feedstocks

	Heat of reaction (kcal/mol of C)
$(CH_2)n + nH_2O \rightarrow 2nH_2 + nCO$	$+49.3$
$(CH_2)n + 2nH_2O \rightarrow 3nH_2 + nCO_2$	$+39.4$
$(CH_2)n + \frac{n}{2}O_2 \rightarrow nCO + nH_2$	-26.4
$(CH_2)n + nO_2 \rightarrow nCO_2 + nH_2$	-94.1
$(CH_2)n + \frac{3n}{2}O_2 \rightarrow nCO_2 + nH_2O$	-151.9
$(CH_2)n + nH_2 \rightarrow nCH_4$	-12.9

$$C_nH_{2n} + O_2 + nCO_2 + nH_2$$

but the reaction is much more complex (Table 3.17).

A gaseous mixture is produced that requires further treatment before use. The oxides of carbon need to be removed as do the various hydrocarbon gases and sulphur-containing species.

Liquefied petroleum gas

Liquefied petroleum gas (LPG) is the term applied to certain specific hydrocarbons, such as propane, butane and pentane and their mixtures, which exist in the gaseous state under atmospheric ambient conditions but can be converted to the liquid state under conditions of moderate pressure, at ambient temperature. As such, the gas has little relevance in terms of gas clean-up but there may on occasion be reference to contaminant removal before use in a refining operation.

Sewage gas

Sewage gas is the general name given to the gaseous product, mostly methane, of the 'fermentation', or biological decomposition, of sewage and other bio-organic waste products. It has received, and continues to receive, attention as a possible source of methane (Lapp, 1979; Overend, 1979; Jones and Radding, 1980; Robinson, 1980). Although it is not strictly a fossil fuel resource in the sense of the current context, there has recently been an interest in using the gas produced during the 'fermentation' of sewage as a fuel.

The decomposition products of many forms of organic matter produced by bacteria include methane as a major item. In sewage works, this gas may be recovered as a profitable by-product. In some instances where this is done, the gas is burned in gas engines to produce power. The sewage or plant matter is maintained at 26–37°C (79–99°F) in large tanks out of contact with air. The usual composition of sewage gas is 75% methane and 25% carbon dioxide, although the percentage of carbon dioxide released varies throughout the cycle.

Peat and wood gas

Analogous to coal, peat also produces a gas when subjected to thermal decomposition (Brame and King, 1955). By virtue of the composition of peat, the gas contains a higher proportion of the oxides of carbon. The extent of the use of peat as a source of heat is much less than that of coal and, therefore, is less of a concern than coal.

Wood also produces gases when thermally decomposed (Table 3.18) (Brame and King, 1955). The use of wood as an industrial source of heat is somewhat limited, other than in plants that manufacture wood charcoal and similar products. The formation of gaseous products might be formulated, chemically, in a similar way to the formation of gases from coal, but with the notable exception, as with peat, of a higher proportion of oxygenated products.

However, the use of wood for domestic heating, whether it be functional or aesthetic, has increased over the past decade. The results of this are often seen in the form of a pollutive cloud in areas where there was little or no pollution

Table 3.18 Composition of gas from wood

	Temperature of carbonization	
	$<400°C$	$>1000°C$
Yield of gas (m³/ton dry wood)	125	>550
Yield of gas (MJ/kg[1])	1.650	7.3
Calorific value of gas (MJ/m³)		
gross	13.0	12.85
Composition (vol %):		
CO_2	30	20
C_nH_m	4	2
CO	25	25
CH_4	14	12
H_2	20	35
N_2	7	6

only a few years ago. The potential to clean up the emissions for wood is limited because of the large domestic market; the construction of gas-cleaning plants as an adjunct to family homes is certainly not practical – at least at this time! However, the presence of catalytic converters on fireplaces and the option, in some fireplace units, of recycling the gases to promote a more complete combustion are becoming commonplace.

Recognition of the potential pollution problems from the use of wood for domestic heating is reflected in the passage of laws in the United States prohibiting the use of wood completely or allowing burning on certain days of the week. Other options such as the use of natural gas are being advocated.

Wood is more likely to act as a producer of pollutant gases during the production of wood-derived chemicals and in the pulp and paper industry (Austin, 1984). In fact, it should not be assumed that the only industries capable of producing polluting gases are those related to fossil fuels. It is the purpose of this text to focus on the fossil fuel industries but there are many other industries which are not the subject of this text that are capable of producing noxious gaseous products (Austin, 1984).

Reserves

In any text dealing with fossil fuels, there must be some recognition, and definition, of the terminology used to describe the amounts, or reserves, of fossil fuels available for recovery and processing. But the terminology used to describe fossil fuel resources is often difficult to define with any degree of precision.

The classification schemes often use different words which should, in theory, mean the same thing but, English being the language that it is, there is usually some difference in the way in which the terms can be interpreted. The terminology used here is that more commonly found although other systems do exist and should be treated with caution in their interpretation. Indeed, if the words themselves leave much latitude in the manner of their interpretation, how can the resource base be determined with any precision?

The reserves of fossil fuels are often classified in a variety of categories (Figure 3.6) as follows.

Proven (proved) reserves

This refers to those reserves of a fossil fuel that are actually found (proven), usually by drilling (Figure 3.7). The estimates have a high degree of accuracy and are either frequently updated as the recovery operations proceed or updated by reservoir characteristics using, for example, production data, pressure transient analysis and reservoir modelling.

However, even though the reserves of a fossil fuel may be proven, there is also the need to define resources as being recoverable (using currently available technology without assuming, often extravagantly, that new technology will miraculously appear or will be invented) and non-recoverable.

Inferred (unproved) reserves

The term 'inferred' reserves is also commonly used in addition to, or in place of, 'potential' reserves.

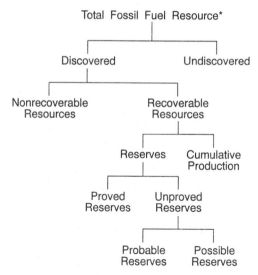

Figure 3.6 Terminology and 'classification' of various resources (*Also applicable terminology for any other naturally occurring resource)

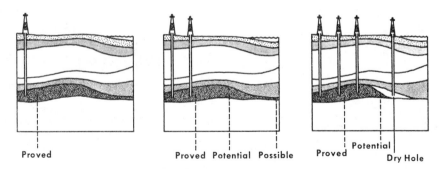

Figure 3.7 Schematic representation of various resources as proven by drilling operations

The inferred reserves are regarded as being of a higher degree of accuracy than the potential reserves and the term is applied to those reserves that are estimated using an improved understanding of reservoir frameworks (see the preceding subsection above). The term also usually includes those reserves that can be recovered by the further development of recovery technologies.

Potential (probable, possible) reserves

These reserves are the additional resources of fossil fuel that are believed to exist in the earth. The data are estimated (usually from geological evidence) but have not been substantiated by drilling operations.

Other terminology such as 'probable' and 'possible' reserves is also employed but falls into the subcategory of being unproven.

Undiscovered reserves

One major issue in the estimation of fossil fuel resources is the all-too-frequent use of the term 'undiscovered' resources.

Caution is advised when using such estimated data as they are very speculative and are regarded as many energy scientists as having little value other than being widely optimistic. The differences between the data obtained from these various estimates can be considerable but it must be remembered that any data about the reserves of natural gas (and, for that matter, about any other fuel or mineral resource) will always be open to questions about the degree of certainty (Figure 3.7).

There are three important items that outweigh the pure guesswork of 'undiscovered' natural gas resources and these are: (a) the actual discoveries of new fields; (b) the development of improved recovery technologies for already-known reserves of gas-in-place; and (c) the estimates of the resource base that are derived from known reservoir properties where the whole of the reservoir is not explored.

It should also be remembered that the total resource base of any fossil fuel (or, for that matter, of any mineral) will be dictated by economics (Figure 3.8) (Nederlof, 1988). Therefore, when resource data are quoted some attention must be given to the cost of recovering those resources. And, most important, the economics must also include a cost factor that reflects the willingness to secure total, or a specific degree of, energy independence.

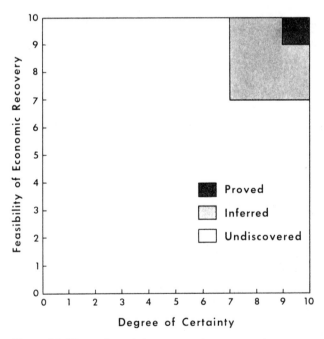

Figure 3.8 Illustration of the degree of accuracy of resources estimation

References

Austin, G.T. (1984) *Shreve's Chemical Process Industries*, McGraw-Hill, New York

Berkowitz, N. (1979) *Introduction to Coal Technology*, Academic Press, New York

Brame, J.S.S. and King, J.G. (1955) *Fuel: Solid, Liquid and Gaseous*, Edward Arnold, London

Curry, R.N. (1981) *Fundamentals of Natural Gas Conditioning*, PennWell, Tulsa, OK

Fahey, D.R. (1986) *Industrial Chemicals via C_1 Processes*, Symposium Series No. 328, American Chemical Society, Washington, DC

Francis, W. and Peters, M. (1980) *Fuels and Fuel Technology*, Pergamon Press, Oxford, England

Hessley, R.K. (1990) *Fuel Science and Technology Handbook* (ed. J.G. Speight), Marcel Dekker, New York

Hessley, R.K., Reasoner, J.W. and Rilewy, J.T. (1986) *Coal Science*, Wiley, New York

Jones, J.L. and Radding, S.B. (eds) (1980) *Thermal Conversion of Solid Wastes and Biomass*, Symposium Series No. 130, American Chemical Society, Washington, DC

Katz, D.L., Cornell, D., Kobayashi, R., Poettmenn, F.H., Vary, J.A., Elenbass, J.R. and Weinaug, C.F. (1959) *Handbook of Natural Gas Engineering*, McGraw-Hill, New York

Lapp, H.M. (1979) In *Chemistry for Energy* (ed. M. Tomlinson), Symposium Series No. 90, American Chemical Society, Washington, DC, Chapter 9

Machie, J.C. (1991) *Catalysis Reviews – Science and Engineering*, **33**, 1&2, 169

METC (1988) *An Assessment of the Natural Gas Resource Base of the United States*. Report No. DOE/W-31109-H1, Morgantown Energy Technology Center, Morgantown, WV

Nederlof, M.H. (1988) *Annual Review of Energy*, **13**, 95

Overend, R. (1979) In *Chemistry for Energy* (ed. M. Tomlinson), Symposium Series No. 90, American Chemical Society, Washington, DC, Chapter 12

Probstein, R.F. and Hicks, R.E. (1990) *Synthetic Fuels*, pH Press, Cambridge, MA

Robinson, J.S. (ed.) (1980) *Fuels from Biomass: Technology and Feasibility*, NoyesData Corp., Park Ridge, NJ

Speight, J.G. (1983) *The Chemistry and Technology of Coal*, Marcel Dekker, New York

Speight, J.G. (1990) *Fuel Science and Technology Handbook*, Marcel Dekker, New York, and references cited therein

Speight, J.G. (1991) *The Chemistry and Technology of Petroleum*, 2nd edn, Marcel Dekker, New York, and references cited therein

Wolf, E.E. (ed.) (1992) *Methane Conversion by Oxidative Processes*, Van Nostrand Reinhold, New York

4

Origin and production of gas – natural gas

Introduction

By definition (and as presented in Chapter 3) there are several types of gas, all candidates for clean-up, that exist within the industrial complex. These gaseous materials vary from natural gas produced during petroleum operations as well as from those wells that only produce gas to the variety of gases produced in industrial processes (Austin, 1984; Probstein and Hicks, 1990) and strategies are very necessary to protect the environment from unmanageable amounts of waste (Frosch and Gallopoulos, 1989).

Whilst the major constituent of natural gas is methane (Table 4.1), there are those components such as carbon dioxide (CO), hydrogen sulphide (H_2S), mercaptans (thiols; R-SH), as well as trace amounts of sundry other pollutants. The fact that methane has a foreseen, and valuable, end use makes it a desirable product, but in any one of several other situations it is considered a pollutant, having been identified as a greenhouse gas (Graedel and Crutzen, 1989).

When gases are produced as by-products of industrial processes, the story is somewhat different. With the exception of steam, it is extremely like that all of the by-products (Table 4.2) will be considered as potential pollutants. A host of such gases are produced from a variety of industries and the gas that tends to receive the most attention as a major pollutant is sulphur dioxide (Kyte, 1991).

This is very true in the current context when fossil fuels are processed. The occurrence of sulphur in coal and petroleum (the two major fossil fuels, with the exception of natural gas) leads to the production of sulphurous products such as sulphur dioxide and hydrogen sulphide. The various coal combustion processes are major producers of sulphur dioxide whereas, because of the participation of hydroprocesses, petroleum refining usually (but not always) produces other sulphur pollutants, such as hydrogen sulphide (Chapters 3 and 5).

The cumulus-like plumes of industrial gases, belching forth so prominently and unabated into the lives of many people in the industrialized urban areas of the world, were the true sign of the mighty industrial complex. Industry, if that general term can be used to illustrate all manufacturing concerns, has embraced the concept of 'clean air' and a clean environment, and a concerted effort is being made to ensure that gaseous emissions are of minimal, and

Table 4.1 General composition of natural gas associated with petroleum

Category	Component	Amount (%)
Paraffinic	Methane (CH_4)	70–98
	Ethane (C_2H_6)	1–10
	Propane (C_3H_8)	Trace – 5
	Butane (C_4H_{10})	Trace – 2
	Pentane (C_5H_{12})	Trace – 1
	Hexane (C_6H_{14})	Trace – 0.5
	Heptane and higher (C_7+)	Non–trace
Cyclic	Cyclohexane (C_6H_{12})	Traces
Aromatic	Benzene (C_6H_6), others	Traces
Non-hydrocarbon	Nitrogen (N_2)	Trace – 15
	Carbon dioxide (CO_2)	Trace – 1
	Hydrogen sulphide (H_2S)	Trace occasionally
	Helium (He)	Trace – 5
	Other sulphur and nitrogen compounds	Trace occasionally
	Water (H_2O)	Trace – 5

preferably no, detriment to the environment. Indeed, it is anticipated that the recent passage of clean air legislation, as in the United States for example, will have a buoyant effect on the gas-processing industry (Haun *et al.*, 1991) and major changes are foreseen in gas processing itself (Cannon, 1990).

But a balance must be struck between energy supply and environmental issues (Malin, 1990) or, to parrot a phrase of a decade or so ago, the population of the world may find itself freezing in the green darkness! The threat of the institution of an energy tax based on carbon content (*Oil and Gas Journal*, 1991) may, or may not, aid in the cautious development of non-renewable (fossil) energy resources but this remains to be seen. Whatever happens, improvements in gas-processing technology can be extremely valuable. And even though this will touch just one small corner of the environmental issues that are evident today, it will be a start – a small start, but a very positive start!

In order to formulate the clean-up of gaseous emissions, be they from the use of fossil fuels or for that matter from any industrial feedstock, it is beneficial to understand the formation and production of gaseous species. The first feedstock that seems to draw everyone's attention is natural gas.

The use of natural gas as a fuel as well as its potential replacement for coal and oil serve as a focus of public attention. Any product that is to be used so commonly in many parts of the world, particularly North America and Western Europe, as a major source of heat cannot escape attention. And then, of course, there are the manufactured gases.

It is the object of this chapter and the three following chapters to present an outline of the methods by which these gases are produced (Chapter 5), stored and transported (Chapter 6), and thereby to give an indication of not only the chemistry of their formation but also their composition and chemical content (Chapter 7). From this outline, it will be easier to understand which methods are more appropriate for the removal of noxious contaminants of the mixtures, and which components can be employed beneficially.

Table 4.2 Examples of contaminants produced by different processes

Industry	Air contaminants emitted	Industry	Air contaminants emitted
Aluminium reduction	Particulates, CO, SO_2, hydrocarbons and fluorides	Non-ferrous smelters Copper	SO_2, particulates
Cement plants	Particulates, CO, SO_2, NO_2	Lead	SO_2, CO, particulates
Coal preparation	Dust, smoke, particulates, sulphur oxides, H_2S	Zinc	Particulates and SO_2, CO
Coke plants	Particulates, SO_2, CO, H_2S, methane, ammonia, H_2, phenols, hydrogen cyanide, N_2, benzene, xylene	Paint and varnish manufacturing	Acrolein, other aldehydes and fatty acids, phthalic anhydride (sublimed); ketones, fatty acids, formic acids, acetic acid, glycerine, acrolein, other aldehydes, phenols and terpenes; from tall oils, hydrogen sulphide, alkyl sulphide, butyl mercaptan, and thiofene; olefins, branched-chain aromatics and ketone solvents
Fertilizer industry (chemical)	PH_3, P_2O_5, HF, SiF_4, NH_3, NH_4Cl, SiF_4, HF, NO_2		
Kraft pulp mills	Mercaptans, H_2S, H_2, organic sulphides, and disulphides	Rendering plants	SO_2, mercaptans, ammonia
Municipal and industrial incinerators	Particulates, CO, SO_2, ammonia, organic acids, aldehydes, NO_2, hydrocarbons, HCl	Steel mills	CO, particulates, SO_2, CO_2, NO_2

Occurrence

The proven reserves of natural gas are of the order of almost 4000 trillion cubic feet (1 Tcf $= 1 \times 10^{12}$ cubic feet $= 0.028 \times 10^{12}$ m^3) of which some 2000 Tcf exist in the United States, Canada, Western Europe, and the former USSR (Tables 4.3 and 4.4; Figure 4.1) (Grow, 1980; *BP Statistical Review of World Energy*, 1987; *BP Review of World Gas*, 1987; True, 1991). However, it must be recognized that these data are approximations and will vary depending upon whatever new finds or usage figures are available. And the former Soviet Union is included here because the distribution of gas reserves within the Commonwealth of Independent States is still very much open to question.

In addition, certain natural gas reserves in the United States can be inferred (see Chapter 3 for 'inferred reserves') from what is already known from various exploration programmes. If these data are taken into account, with whatever degree of trepidation, the natural gas resource base of the United States can be expanded to some 1118 Tcf and, furthermore, recovery estimates can be made on the basis of gas price (METC, 1988b).

Natural gas occurs in association with petroleum or as dry gas deposits in many locations throughout the world (Figure 4.2) that are not associated with

Table 4.3 General distribution of natural gas reserves

Country	Proved reserves (Tcf)	Share (%)
USSR	1600.0	38.02
Iran	600.4	14.27
Abu Dhabi	182.8	4.34
Saudi Arabia	180.1	4.28
USA	166.2	3.95
Qatar	163.2	3.88
Algeria	114.7	2.73
Venezuela	105.7	2.51
Canada	97.6	2.32
Iraq	95.0	2.26
Indonesia	91.5	2.17
Nigeria	87.4	2.08
Mexico	72.7	1.73
Netherlands	60.9	1.45
Norway	60.7	1.44
Malaysia	56.9	1.35
Kuwait	48.6	1.15
Libya	43.0	1.02
China	35.3	
Argentina	27.0	
India	25.1	
UK	19.8	3.97
Pakistan	19.5	
Australia	15.4	
Bangladesh	12.7	
Germany	12.4	
Subtotal	3994.6	94.92
Others	213.7	5.08
Total	4208.3	100.00

Table 4.4 General distribution of natural gas production

Country	Production (bcf)	Share (%)
USSR	28 800.0	38.1
USA	18 358.0	24.3
Canada	4 267.0	5.6
Netherlands	2 618.1	3.5
UK	1 700.7	2.2
Algeria	1 588.0	2.1
Saudi Arabia	1 401.2	1.9
Indonesia	1 345.7	1.8
Mexico	1 332.7	1.8
Romania	1 181.1	1.6
UAE	1.092.1	1.4
Norway	896.5	1.2
Venezuela	862.0	
Iran	854.4	
Argentina	816.1	
Australia	749.9	8.7
Italy	736.8	
Malaysia	569.7	
Germany	565.7	
China	517.7	
Top 20 total	70 253.4	92.9
Others	5 354.2	7.1
Total	75 607.6	100.0

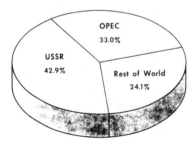

Figure 4.1 General distribution of natural gas resources

oil (Chapter 3). In fact, hydrocarbon gases are the products of the decomposition of organic materials at varying stages of maturation.

In view of the close proximity of gas and petroleum in many reservoirs, it is necessary to consider the formation of natural gas as paralleling the formation of petroleum (Tissot and Welte, 1978; Clayton, 1991). Even though there are many examples of gas occurring without the presence of petroleum, and vice versa, there may be a satisfactory reason for the separation.

It is not the purpose of this chapter to differentiate between the formation of natural gas and petroleum but to present a simple outline of the formation of natural gas compared with the production of a range of manufactured gases in

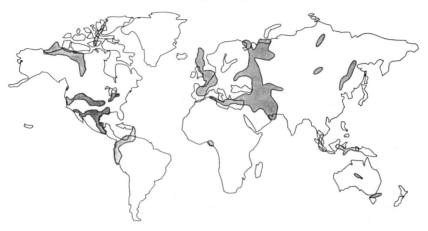

Figure 4.2 Location of major oilfields where natural gas is also known to occur

order that their chemistry might be understood. Nor is it the purpose of this chapter to present a detailed discussion of the various theories of natural gas formation. Only a brief outline will be given to introduce the reader to the concepts behind gas formation and occurrence.

However, there is a simple assumption that is necessary for the purpose of this chapter, namely: the organic detritus that serves as the precursor for liquid petroleum also yields, through the auspices of the maturation process, the components that constitute natural gas. Many theories have been proposed for the origin of petroleum and natural gas but the diversity of the precursors and the variation in the prevailing physical conditions make it difficult to explain fully the origin of the petroleum and gas in any given reservoir.

The most widely accepted theory utilizes the concept of a wide range of organic natural precursors and advocates that the hydrocarbons were generated from the organic detritus under the influence of pressure, temperature and microbial effects over geological time (see, for example, Clayton, 1991). Furthermore, variations in the character of the organic precursors and the prevailing physical conditions such as temperature and pressure play an important role in the generation of oil and/or gas (see, for example, Tissot and Welte, 1978; Speight, 1991).

There are those theories which promote the formation of natural gas and waxy crude oils from the remains of terrestrial plants whilst, on the other hand, the non-waxy crude oils were, reputedly, produced from the organic remains of aquatic organisms. Interesting though this theory may be, it is difficult to equate the formation of any one particular type of crude oil with any group of specific precursors (Speight, 1991). The transportation of organic detritus by rivers or by tidal forces is a major unknown as are the prevailing conditions that bring about the chemical changes necessary to convert the detritus into the constituents of crude oil.

There is also the theory that, because rivers do appear to have played a role in transporting terrestrial matter to the sea, formations which include ancient river deltas and ancient beaches are favourable places for gas to exist. It has been suggested that the deepest sediments are rich in organic matter that is of terrestrial origin and are overlain by marine sediments rich in matter of aquatic origin. Thus, a vertical sequence has been envisaged in which the gas-generating materials are at the bottom of the source rock and the oil-generating materials at the top.

Attractive though these two particular theories may seem, there is in fact no universally accepted theory that satisfactorily explains all aspects of the formation of petroleum and natural gas (Gold, 1984, 1985). Nevertheless, it is generally accepted that temperature and pressure conditions have played a role in the formation of the gas, and in a more general sense natural gas is considered to originate in any one, or more, of three principal ways: (a) the thermogenic process: (b) the biogenic process; and (c) the abiogenic process (Figure 4.3) (Pusey, 1973; Gold, 1985).

The thermogenic process involves the relatively slow decomposition of organic material which occurs in sedimentary basins; usually some degree of heat is required. The biogenic process involves the formation of methane by the action of living organisms (bacteria) on organic materials (Gold, 1984). The abiogenic process, unlike the other two processes, does not require the presence of organic matter as the starting material (Gold and Soter, 1982,

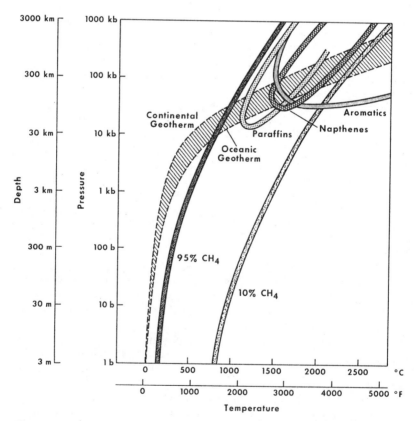

Figure 4.3 General relationship of temperature and pressure to petroleum and gas formation

1986). Such gas is believed to have originated from non-biogenic material (i.e. mineral matter) and has accumulated in the deeper traps since the time of formation of the earth. Some of this gas could also have diffused to surface formations in the intervening time.

Most of the natural gas is generally believed to be thermogenic and biogenic in origin; the origin of natural gas through abiogenic means is still considered to be highly speculative. Nevertheless, there is the belief that the abiogenic gas is that gas which could also be classified as coming from 'unconventional reservoirs'. However, such a classification of gas resources leaves much to be desired.

Once the hydrocarbons are formed, whatever the mode of origin, their direction of movement in the earth is upwards (i.e. towards the surface). It is more than likely that there are exceptions to this general rule and movement of the products in downward or sideways directions from their place of formation (source rock), as faults and formations allow, to their place of accumulation (reservoir rock) can also be envisaged. Irrespective of the direction of movement of hydrocarbons from the source rock, the movement displaces some of the sea water that originally filled the pore spaces of the sedimentary rock. This movement of the hydrocarbons is inhibited when the oil and gas reach an impervious rock that traps or seals the reservoir.

Conventional reservoirs

Natural gas occurs in various locations throughout the world. It is, therefore, worth devoting some attention to the occurrence of the gas in order that its properties might be better understood. Indeed, in parallel, it is also essential to understand the collection and transportation of natural gas so that the hazards associated with the handling of the resources might also be appreciated.

Natural gas, like petroleum, is located in the earth in reservoirs and, just as 'conventional' petroleum reservoirs can vary in character (Speight, 1991), natural gas reservoirs also vary considerably and can be classified as 'conventional' and 'unconventional'. The latter reservoirs include formations such as tight sands, tight shales, geopressured aquifers, coal beds, deep sources and gas hydrates (Meyer, 1977; Gibbons, 1985; Haas *et al.*, 1987; Sharma *et al.*, 1987; Schwochow, 1991; GRI, 1992).

There are many different types of geological structures that are capable of forming reservoirs for the accumulation of oil and gas (Figure 4.4). The depth of the reservoirs is variable and it has been noted that methane is stable at depths in excess of 40 000 ft (12 000 m) (Barker and Kemp, 1980). Clean sandstones are more favourable for methane preservation than carbonates. This concept brings up the distinct possibility that natural gas may be found at depths of 15 000–30 000 ft (4570–9140 m) that have not yet been explored.

Just as oil can vary in composition depending upon the placement of the well (and the depth of the well) in the reservoir, so each well in a natural gas reservoir may also produce gas with a different composition. In addition, the composition of the gas from each individual well is also likely to change as the reservoir is depleted. Thus, production equipment may need to be changed as the well ages to compensate for any changes in the composition of the gas.

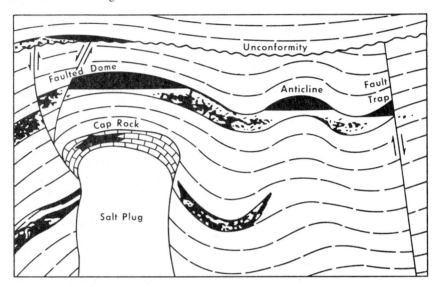

Figure 4.4 Illustration of geological structures after formation of petroleum and gas reservoirs

Properties of reservoir rocks

When natural gas is produced or stored in reservoir rock, the nature of the porous medium determines the quantity of gas content of the rock and the ability of the fluids to move through the rock.

A typical reservoir rock is sandstone, a clastic sedimentary rock comprising an aggregate of sand-sized (0.0024–0.079 inch, 0.06–2.0 mm) fragments of minerals. Many varieties of sandstone can be recognized on the basis of the relative percentage of quartz (SiO_2), feldspars (aluminosilicate minerals, $K_2O.Al_2O_3.6SiO_2$) as well as volcanic rock and other rock fragments. The sandstones, because of their interconnected pore spaces, are the most important reservoir rocks in the earth's crust. The total bulk volume and the pore volume (porosity) surpass those of any other mineral group. Briefly, by way of introduction or definition, the fraction of the total bulk volume of the rock occupied by the pore space in which the gas is stored is called the porosity of the rock. Porosity is a most important term to reservoir engineers and in forming estimates of liquid and gaseous fossil fuel resources (Chapter 3). The ability of reservoir rock to conduct the flow of fluids through it is referred to as its permeability.

Unconventional reservoirs

The emphasis is generally on those subsurface reservoirs that are equivalent to the 'typical' petroleum reservoir (Figure 4.4). In addition to the conventional sandstone and/or limestone reservoirs, there are, however, other sources of natural gas which, because of their differences to those sources noted above, are classed as 'unconventional' sources. These sources include tight sands, tight shales, geopressured aquifers, coal, deep sources and gas hydrates (Nederlof, 1988).

The unconventional reservoirs used here as examples are based on descriptions of those located in the United States, with the understanding that there are similar reservoir structures scattered throughout the world.

Tight sands

Tight sands (also variously called tight gas sands from which tight gas is produced) are those formations where natural gas occurs in rock formations of extremely low permeability. Geologically, the tight formations are very similar to the accumulations of 'conventional' formations but the main feature of tight sand resources is the rate at which the gas can be produced from the formation, and they are characterized by slow rates of production. Substantial amounts of gas occur within such formations where the porosities are in the range 5–15%, the permeability is extremely low (0.001–1.0 millidarcy) and there are irreducible water saturations.

Production wells in tight sand formations generally need some form of stimulation (such as rock fracturing around the well bore) to increase the gas flow rates. There are many examples of such reservoirs in the United States (Figure 4.5) that contain natural gas where conventional production methods cannot be used (Hass et al., 1987; METC, 1988a).

The need to develop fracturing techniques to produce gas economically from tight sands cannot be overemphasized and has been the subject of many investigations (METC, 1988a). Three major techniques have been proposed:

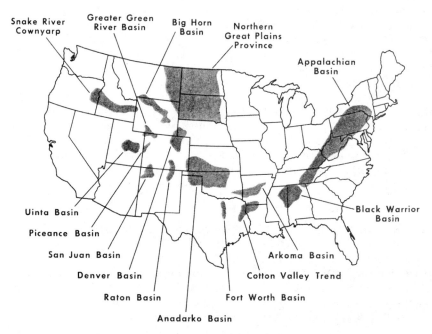

Figure 4.5 Location of 'tight sand' reservoirs within the United States

nuclear explosives, chemical explosives, and massive hydraulic fracturing (MHF).

Nuclear detonations have also been proposed to increase petroleum recovery (Atkinson and Johansen, 1964) and to recover hydrocarbon liquids from tar sand deposits (Watkins and Anderson, 1964) and from oil shale deposits (Watkins and Anderson, 1964; Lekas et al., 1967) but have not been put into practice. As above, nuclear detonations have also been proposed for the release of gas from tight reservoirs (Gevertz et al., 1965; Holzer, 1968). There is always the fear that such detonations may have unforeseen long-term consequences (such as radioactive products) that could constitute a definite danger to the user. The failure to reduce these ideas to practice is, therefore, not surprising!

Chemical explosives appear to be only effective in areas where natural fractures exist, but this is not usually the case for tight sands and massive hydraulic fracturing appears to offer the most promise (Veatch and Moschovidis, 1986). The process usually requires the injection of a fracturing fluid (water) at high pressures over a prolonged period of time to induce a fracture. This is then followed by the injection of a fluid which contains 'propping agents' (glass beads or sand) to prevent closure of the fracture when the fluids flow back into the well bore after cessation of the pumping operation.

Tight shales

Shales are finely laminated deposits that are generally rich in organic matter but, unfortunately, only have a permeability of the order of 1 millidarcy or less. The mineral content of shales is quite diverse although common mineral constituents are quartz (SiO_2), with some kaolinite [$AlSiO_2(OH_2)$], pyrite (FeS_2) and feldspars (aluminosilicate minerals, $K_2O.Al_2O_3.6SiO_2$).

Natural gas found in tight shale beds is also referred to as shale gas and is found in the pore spaces of the shale and/or adsorbed on the shale. The gas is often further defined by the formation in which it occurs, e.g. Devonian shale gas.

The Devonian shales in the Appalachian, Illinois and Michigan basins (Figure 4.6) are such gas-containing formations and can yield a gas having a heating value as high as 1250 Btu/ft^3 (46 500 kJ/m^3). These tight shales are an attractive source of gas and may contribute very significantly to gas production in the coming years. Although a commercial means of obtaining the gas from these tight shale formations has not yet been fully defined, it is believed that hydraulic fracturing may be the answer.

Geopressured aquifers

The brine in geopressured aquifers, which can form due to rapid subsidence, can contain up to 40 ft^3 (10.1 m^3) of natural gas per barrel (15.9×10^2 m^3) of water. Such geopressured aquifers in the United States occur in a 'locale' that extends onshore and offshore from Texas to Florida along the Gulf of Mexico. An estimated 1700 Tcf (1 Tcf = 1 trillion cubic feet = 1×10^{12} ft^3) of gas reserves (unproven at this time) are estimated to be in this region.

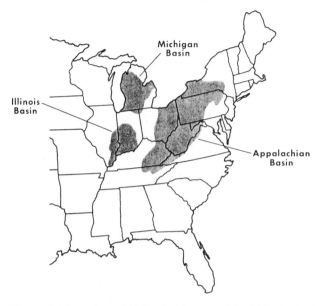

Figure 4.6 Location of 'tight shale' reservoirs within the United States

Coal seams (coal beds)

Methane is associated with many coal seams (and often referred to as 'coal bed methane') and is a common occurrence in coal mines. It is the dreaded 'firedamp' which has been the cause of many explosions with accompanying loss of life (Speight, 1983). The gas is occluded in the pores of the coal under pressure and is gradually released during mining operations. Firedamp consists of 93–99% methane, <3% ethane, <4% carbon dioxide and <6% nitrogen plus other inert gases and is removed, at considerable cost and danger, by mine ventilation.

The current thinking is to recover much of the firedamp by 'drainage' through a series of boreholes sited behind the coal face. Indeed, the technique can also be applied to the recovery of methane from thin, unworked, seams.

Such gas in seams at depths less than 3000 ft (915 m) has been estimated to be 260 Tcf in the United States but it has been estimated that practical constraints may allow production of less than 40 Tcf (Figure 4.7). Gaseous products are also obtained from coal by the various thermal processes that have been advocated for the production of liquid fuels from coal (Speight, 1983).

Deep sources

Deep source gas is natural gas that exists deep within the earth and accumulated in both 'conventional' or 'unconventional' reservoirs (Gold and Soter, 1980). There is also the possibility that deep source gas could exist in traps under 'basement' rocks. The evidence for the existence of such natural gas is mainly speculative but, if proven, it would imply that substantial amounts of natural gas may exist beneath existing shallow gas fields.

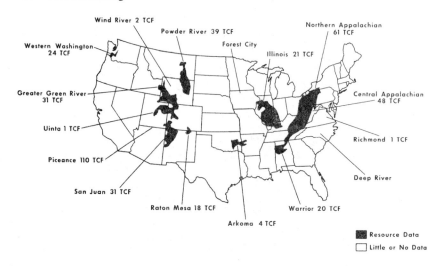

Figure 4.7 Location of 'coal seam' reservoirs within the United States

Gas hydrates

Gas hydrates (see also Chapter 7) are ice-like complexes (clathrates) of gas and water that form under prescribed conditions of temperature and pressure (Trofimuk *et al.*, 1972; Holder *et al.*, 1982). The hydrates are often found under water (at depths greater than 100 feet, 30 m) and under the permafrost and represent a potentially huge resource (Holder *et al.*, 1983, 1984). There has also been speculation that there may even be free natural gas trapped under the hydrate resource.

References

Atkinson, C.H. and Johansen, R.T. (1964) Report of Investigations No. 6494, United States Bureau of Mines, Washington, DC

Austin, G.T. (1984) *Shreve's Chemical Process Industries*, McGraw-Hill, New York

Barker, C. and Kemp, M.K. (1980) *Conference on Natural Gas Resource Development in Mid-Continent Basins: Production and Exploration Techniques, University of Tulsa, Tulsa, Oklahoma, 11–12 March*

BP Review of World Gas (1987) British Petroleum, London, September

BP Statistical Review of World Energy (1987) *British Petroleum*, London, June

Cannon, R.E. (1990) *Oil and Gas Journal*, **88**, (28), 47

Clayton, C. (1991) *Marine and Petroleum Geology*, **8**, 232, and references cited therein

Curry, R.N. (1981) *Fundamentals of Natural Gas Conditioning*, PennWell Publishing Co., Tulsa, OK

Frosch, R.A. and Gallopoulos, N.E. (1989) *Scientific American*, **261**, (3), 144

Gevertz, H., Lemon, R.F., Hollis, W.T., Lekas, M.A., Ward, D.C. and Atkinson, C.H. (1965) *Project Gasbuggy*, United States Atomic Energy Commission, Washington, DC

Gibbons, J.H. (1985) *U.S. Natural Gas Availability*, Office of Technology, Assessment, Congress of the United States, Washington, DC

Gold, T. (1984) *Scientific American*, **251** (5), 6

Gold, T. (1985) *Annual Review of Energy*, **10**, 53

Gold, T. and Soter, S. (1980) *Scientific American*, **242**, (6), 154

Gold, T. and Soter, S. (1982) *Energy Exploration and Exploitation*, **1**, (1), 89

Graedel, T.E. and Crutzen, P.J. (1989) *Scientific American*, **261**, (3), 58

GRI (1992) *Quarterly Review of Methane from Coal Seams Technology*, Vol. 9, Gas Research Institute, Chicago

Grow, G.C. (1980) *Proc. Mid-Year Meeting, American Institute of Chemical Engineers, Philadelphia, PA, June*

Haas, M.R., Brashear, J.P. and Morra, F. (1987) *Journal of Petroleum Technology*, January, 77

Haun, R.R., Ellington, E.E. and Otto, K.W. (1991) *Oil and Gas Journal*, **89**, (29), 46

Hobson, G.D. and Pohl, W. (1973) *Modern Petroleum Technology*, Applied Science Publishers, Barking, England

Holder, G.D., Angert, P.F., John, V.T. and Yen, S.L. (1982) *Journal of Petroleum Technology*, **34**, 1127

Holder, G.D., Angert, P.F. and Pereira, V. (1983) In *Natural Gas Hydrates*, Butterworths, Boston, MA

Holder, G.D., Kamath, V.A. and Godbole, S.P. (1984) *Annual Review of Energy*, **9**, 427

Holzer, F. (1968) Report No. UCRL-50386, US Atomic Energy Commission, Washington, DC

Kohl, A.L. and Riesenfeld, F.C. (1979) *Gas Purification*, Gulf Publishing Company, Houston, TX

Kyte, W.S. (1991) *Desulphurisation 2: Technologies and Strategies for Reducing Sulphur Emissions*, Institute of Chemical Engineers, Rugby, Warwickshire

Lekas, M.A., Bray, B.G., Carpenter, H.C., Aronson, H.H., Dineen, G.U. and Downen, J.D. (1967) *Project Bronco*, United States Atomic Energy Commission, Washington, DC

Maddox, R.N. (1982) *Gas Conditioning and Processing*, Gulf Publishing Company, Houston, TX

Malin, C.B. (1990) *Oil and Gas Journal*, **88**, (2), 20

Meyer, R.F. (1977) *Proc. Conf. on the Future Supply of Nature-made Petroleum and Gas, United Nations Institute for Training and Research, Laxenburg, Austria, 5–16 July*, Pergamon Press, New York

Nederlof, M.H. (1988) *Annual Review of Energy*, **13**, 95

Newman, S.A. (1985) *Acid and Sour Gas Treating Processes*, Gulf Publishing Company, Houston, TX

Oil and Gas Journal (1991) **89**, (39), 36

Probstein, R.F. and Hicks, R.E. (1990) *Synthetic Fuels*, pH Press, Cambridge, MA

Pusey, W.C. (1973) *World Oil*, **176**, (5), 137

Schwochow, S.D. (1991) *Bibliography and Index of the Quarterly Review of Methane from Coal Seams Technology*, Gas Research Institute, Chicago

Sharma, G.D., Kamath, V.A., Godbole, S.P., Patil, S.L., Paraqnjpe, S.G., Mutalik, P.N. and Nadem, N. (1987) *Development of Alaskan Gas Hydrate Resources*. Report No. DOE/FE/61114-2608, Morgantown Energy Technology Center, United States Department of Energy, Morgantown, WV, and references cited therein

Speight, J.G. (1983) *The Chemistry and Technology of Coal*, Marcel Dekker, New York

Speight, J.G. (1991) *The Chemistry and Technology of Petroleum*, 2nd edn, Marcel Dekker, New York

Tissot, B.P. and Welte, D.H. (1978) *Petroleum Formation and Occurrence*, Springer, New York

Trofimuk, A.A., Cherskiy, N.V., Makagon, Y.F. and Tsarev, V.P. (1972) *International Geology Review*, **15**, 1042

True, W.R. (1991) *Oil and Gas Journal*, **89**, (29), 41

Veatch, R.W. and Moschovidis, Z.A. (1986) *Proc. Int. Meeting on Petroleum Engineering*, Society of Petroleum Engineers of AIME, Dallas, TX, Vol. 2, p. 421

Watkins, J.W. and Anderson, C.C. (1964) Information Circular No. 8219, United States Bureau of Mines, Washington, DC

5

Origin and production of gas – process gas

Introduction

Whilst natural gas plays a major role in modern industry and must, therefore, be cleaned before use, the products of natural gas usage must also be free from pollutants. If the major product is carbon dioxide, the questions relating to the effects of this gas when it is released to the atmosphere must be addressed and suitable precautions need to be taken.

Natural gas is not the only means by which gas is produced in industrial settings. For example, coal has been a source of gas since at least the last days of the seventeenth century and continues to be a source (desirable and undesirable!) of gas for industrial use and of the accompanying emissions (desirable and undesirable).

In order to appreciate the gaseous products that are produced by the thermal decomposition of coal, a very general description of the various types of gases that can be produced from coal, and the processes by which they are produced, are given here.

Manufactured gases, or coal gases, are the gaseous mixtures that are produced when coal is thermally decomposed under a variety of conditions (Chapter 3). The processes consist essentially of heating coal to drive off the volatile products, some of which may be gases and others liquids and tars, to leave a solid carbonaceous residue:

$$[C]_{coal} + heat + (?) = [C]_{char} + C_nH_{2n+2} + CO + CO_2 + H_2$$

or

$$coal + heat + (?) = char + liquid + gas$$

The residue char is then treated under a variety of conditions to produce other fuels which vary from a 'purified' char to different types of gaseous mixtures.

The different processes by which these gaseous mixtures are produced are much more complex than the relatively simple chemical equations would indicate, since coal is an extremely complex material (see, for example, Berkowitz, 1979; Elliott, 1981; Meyers, 1981; Funk, 1983; Speight, 1983; Hessley *et al.*, 1986; Hessley, 1990). Moreover, the presence of nitrogen, sulphur and mineral matter in the coal, as well as the complex nature of the thermal degradation process, dictates the production of a gaseous mixture that is by no means pure and may even need an adjustment of the relative amounts of the different constituents before further use:

$[N]_{coal} = NO_x + HCN$, etc., where $x = 1$ or 2
$[O]_{coal} = CO_x$, etc., where $x = 1$ or 2
$[S]_{coal} = SO_x + H_2S$, etc., where $x = 1$ or 2

For convenience, the names of the gaseous mixtures are used here as originally designated, with the understanding that over the decades since their first introduction there may be differences that are evident in the means of production and in the make-up of the gaseous products.

Coal gas

In a very general sense, 'coal gas' is the term applied to the mixture of gaseous constituents that are produced during the thermal decomposition of coal at temperatures in excess of 500 °C (930 °F), often in the absence of oxygen (air) (Chapter 3). A honeycombed solid residue (coke), tars and other liquids are also produced in the process. The tars and other liquids (liquor) are removed by condensation leaving principally hydrogen, carbon monoxide and carbon dioxide in the gaseous phase. The gaseous product also contains low-boiling hydrocarbons, sulphur-containing gases, often nitrogen oxides, ammonia and hydrogen cyanide.

The proportions of gas, coke, tar and other liquids vary according to the particular method used for the carbonization (especially on the retort configuration), process temperature, as well as on the nature (rank) of the coal used. Purification of the gas is necessary (Figure 5.1).

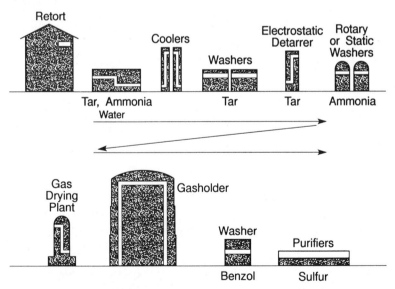

Figure 5.1 Simplified illustration of a coal gas cleaning plant

Combustion

Coal combustion is an old art and probably represents the oldest known use of this fossil fuel. However, coal combustion is more complex than would at first be thought. The complexity of coal itself and the variable process parameters all contribute to the overall complexity of the process (Field *et al.*, 1967; Levy *et al.*, 1981; Chigier, 1991).

There are two major methods of coal combustion: fixed-bed combustion and combustion in suspension. The first fixed beds (e.g. open fires, fireplaces, domestic stoves) were simple in principle. Suspension burning of coal began in the early 1900s with the development of pulverized coal-fired systems, and by the 1920s these systems were in widespread use. Spreader stokers, which were developed in the 1930s, combined both principles by providing for the smaller particles of coal to be burned in suspension and larger particles to be burned on a grate.

A major concern in the present-day combustion of coal is the performance of the process in an environmentally acceptable manner through either the use of low-sulphur coal or postcombustion clean-up of the flue gases. Thus, there is a marked trend to more efficient methods of coal combustion and, in fact, a combustion system that is able to accept coal without the necessity of postcombustion treatment or without emitting objectionable amounts of sulphur and nitrogen oxides and particulates is very desirable.

The parameters of rank and moisture content are regarded as determining factors in combustibility as it relates to both heating value and ease of reaction as well as the generation of pollutants (Levy *et al.*, 1981). Thus, whilst the lower-rank coals may appear to be more 'reactive' than higher-rank ones, though exhibiting a lower Btu/ft^3 (kJ/m^3) value and thereby implying that rank does not affect combustibility, environmental constraints arise through the occurrence of heteroatoms (i.e. non-carbon atoms such as nitrogen and sulphur) in the coal. At the same time, anthracites (with a low volatile matter content) are generally more difficult to burn than bituminous coals.

Chemistry

In direct combustion, coal is burned (i.e. the carbon and hydrogen in the coal are oxidized into carbon dioxide and water) to convert its chemical energy into thermal energy:

$$[C]_{coal} + O_2 = CO_2$$
$$[2H]_{coal} + O_2 = H_2O$$
$$[C]_{coal} + H_2O = CO + H_2$$

after which the sensible heat in the products of combustion can then be converted into steam that can be external work or directly into shaft horsepower (e.g. in a gas turbine). In fact, the combustion process actually represents a means of achieving the complete oxidation of coal.

On a more formal basis, the combustion of coal may be simply represented as the staged oxidation of coal carbon to carbon dioxide:

$$[2C]_{coal} + O_2 = 2CO$$
$$2CO + O_2 = 2CO_2$$

with any reactions of the hydrogen in the coal being considered to be of secondary importance.

The stoichiometric reaction equations are quite simple but there is a confusing variation of hypotheses about the sequential reaction mechanism which is caused to a great extent by the heterogeneous nature (solid and gaseous phases) of the reaction. But for the purposes of this text the chemistry will remain simple as shown in the above equations. Other types of combustion systems may be rate controlled due to the onset of the Boudouard reaction

$$CO_2 + C = 2CO$$

In more general terms, the combustion of carbonaceous materials (which contain hydrogen and oxygen as well as carbon) involves a wide variety of reactions among reactants, intermediates and products (Table 5.1). The reactions occur simultaneously and consecutively (in both forward and reverse directions) and may at times approach a condition of equilibrium. Furthermore, there is a change in the physical and chemical structure of the fuel particle as it burns.

The complex nature of coal as a molecular entity (Berkowitz, 1979; Meyers, 1981; Speight, 1983; Hessley et al., 1986) has resulted in the chemical explanations of coal combustion being confined to the carbon in the system and, to a much lesser extent, with only a passing acknowledgement of the hydrogen and other elements (Table 5.1). However, it must be recognized that the system is extremely complex and that the heteroatoms (nitrogen, oxygen and sulphur) can exert an influence on the combustion which can bring about the serious environmental concerns.

For example, the conversion of nitrogen and sulphur to their respective oxides during combustion is a major issue:

$$[S]_{coal} + O_2 = SO_2$$
$$2SO_2 + O_2 = 2SO_3$$
$$[2N]_{coal} + O_2 = 2NO$$
$$2NO + O_2 = 2NO_2$$
$$[N]_{coal} + O_2 = NO_2$$

The sulphur dioxide that escapes into the atmosphere is either deposited locally or converted to sulphurous and sulphuric acids by reaction with moisture in the atmosphere:

$$SO_2 + H_2O = H_2SO_3$$
$$2SO_2 + O_2 = 2SO_3$$
$$SO_3 + H_2O = H_2SO_4$$

Table 5.1 Representation of the chemical reactions involved in coal combustion

	Btu/lb,	kcal/kg
$C(s) + O_2(g) \rightarrow CO_2(g)$	−169 290,	−94.4
$2C(s) + O_2(g) \rightarrow 2CO(g)$	−95 100,	−52.8
$C(s) + CO_2(g) \rightarrow 2CO(g)$	+74 200,	+41.2
$2CO(g) + O_2(g) \rightarrow 2CO_2(g)$	−243 490,	−135.3
$2H_2(g) + O_2(g) \rightarrow 2H_2O(g)$	−208 070,	−115.6
$C(s) + H_2O(g) \rightarrow CO(g) + H_2(g)$	+56 490,	+31.4
$C(s) + 2H_2O(g) \rightarrow CO_2 + 2H_2(g)$	+38 780,	+21.5
$CO(g) + H_2O(g) \rightarrow CO_2(g) + H_2(g)$	−17 710,	−9.8

or

$$2SO_2 + O_2 + 2H_2O = 2H_2SO_4$$

Nitrogen oxides also contribute to the formation and occurrence of acid rain, in a similar manner to the production of acids from the sulphur oxides, and yield nitrous and nitric acids:

$$NO + H_2O = H_2NO_2$$
$$2NO + O_2 = 2NO_2$$
$$NO_2 + H_2O = HNO_3$$

or

$$2NO + O_2 + H_2O = 2HNO_3$$

In addition to causing objectionable stack emissions, coal ash (Table 5.2) and volatile inorganic material generated by the thermal alteration of mineral matter in coal (Figure 5.2) will adversely affect heat transfer processes by fouling heat-absorbing and radiating surfaces and will also influence the performance of the combustion system by causing corrosion. Consequently, operating procedures must therefore provide for the effective countering of all these hazards.

Table 5.2 Constituents of coal ash

	%
SiO_2	40–90
Al_2O_3	0–60
Fe_2O_3	5–25
CaO	1–15
MgO	0.5–4
$Na_2O + K_2O$	1–4

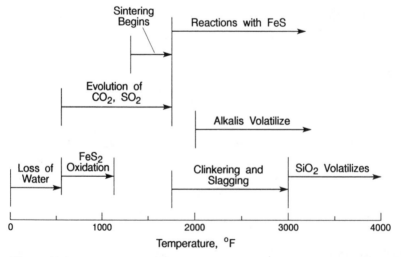

Figure 5.2 Representation of the changes occurring in mineral matter during coal combustion

Table 5.3 Air requirements for combustion systems

Fuel	Type of furnace or burner	Excess air (wt %)
Pulverized coal	Water-cooled furnace for slag-tap or dry ash removal	15–40
Crushed coal	Cyclone furnace	10–15
Coal	Spreader stoker	30–60
	Water-cooled vibrating grate stoker	30–60
	Chain grate and travelling grate stokers	15–50
Fuel oil	Oil burners, register type	5–10
	Multifuel burners and flat flame	10–20
Wood	Dutch oven (10–23% through grates) and Hofft type	20–25

Source: E.J. Hoffman (1978) *Coal Conversion*, Energon, Laramie, WY, p. 101.

Combustion systems

Combustion systems vary in nature depending upon the nature of the feedstock and the air needed for the combustion process (Table 5.3). However, the two principal types of coal-burning systems are usually referred to as layer and chambered. The former refers to fixed beds while the latter is more specifically for pulverized fuel.

Fixed (or slowly moving) beds

For fuel-bed burning on a grate, a distillation effect occurs and the result is that the liquid components which are formed will volatilize before combustion temperatures are reached; cracking may also occur. The ignition of coal in a bed is almost entirely by radiation from hot refractory arches and from the flame burning of the volatile components. In fixed beds, the radiant heat above the bed can only penetrate a short distance into the bed.

Consequently, convective heat transfer determines the intensity of warming up and ignition. In addition, convective heat transfer also plays an important part in the overall flame-to-surface transmission. The reaction of gases is greatly accelerated by contact with hot surfaces and, while the reaction away from the walls may proceed slowly, the reaction at the surface proceeds much more rapidly.

Fluidized beds

In the simplest terms, fluidized combustion occurs in expanded beds (Figure 5.3). The reaction occurs at lower temperatures (925°C, 1700°F), but high convective transfer rates exist due to the bed motion. In fact, heat loads higher than in comparably sized radiation furnaces can be effected, i.e. smaller chambers produce the same equivalent heat load. Moreover, fluidized systems can operate under substantial pressures thereby allowing more efficient gas clean-up. Fluidized-bed combustion is a means for providing high heat transfer rates, controlling sulphur, and reducing nitrogen oxide emissions as a result of the low temperatures in the combustion zone (Table 5.4).

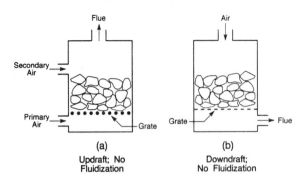

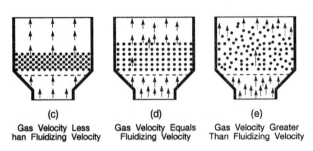

Figure 5.3 Simplified representation of (a) and (b) simple coal bed combustors and (c), (d), (e) fluidized-bed systems

Table 5.4 General summary of the fluidized-bed combustion of coal

Item	Fluidized-bed combustion
Sulphur in exit gases	10–20% of input
NO_x in exit gases	50–700 ppm
Particulates	Mechanical collectors plus precipitators required
Trace metals	Captured by calcium except for mercury and cadmium
Heat release	$1–2\,MW/m^3$
Fuel condition	Dry solids
Combustion efficiency	87–97%
Problem areas	Boiler tube corrosion and turbine blade erosion

There are, however, mechanical problems associated with fluidized combustion: difficulties are encountered with the feeding of coal and particularly the withdrawal and separation of the ash from the char or unreacted coal for recycling back to the combustion chamber. There are also problems with pollution control. While the sulphur may be removed downstream with suitable ancillary controls, it may also be captured in the bed, thereby adding to the separation and recycling problems. Capture during combustion, however, is recognized as the ideal and is a source of optimism for fluidized combustion. Moreover, ash agglomeration is not guaranteed as a means of ready separation and reducing bed carry-over; attrition occurs and particulates occur in the flue gases that require controls.

Thus, there remain problems in sulphur and particulate control along with problems with feeding, withdrawal, separation and recycling, but in spite of this fluidized combustion presents an intriguing prospect for direct firing.

A fluidized bed is an excellent medium for contacting gases with solids, and this can be exploited in a combustor since sulphur dioxide emissions can be reduced by adding limestone ($CaCO_3$) or dolomite ($CaCO_3.MgCO_3$) to the bed (see Chapters 8–10 and 12).

The sulphur oxides react to form calcium sulphate, which leaves the bed as a solid with the ash:

$$2SO_2 + O_2 = 2SO_3$$
$$SO_3 + CaCO_3 = CaSO_4 + CO_2$$

or

$$2SO_2 + O_2 + 2CaCO_3 = 2CaSO_4 + 2CO_2$$

The spent sorbent from fluidized-bed combustion may be taken directly to disposal and is much easier than the disposal of salts produced by wet limestone scrubbing. These latter species are contained in wet sludges having a high volume and a high content of salt-laden water. The mineral products of fluidized-bed combustion, however, are quite dry and in a chemically refractory state, and therefore disposal is much easier and less likely to result in pollution.

The spent limestone from fluidized-bed combustion may be regenerated thereby reducing the overall requirement for lime and decreasing the disposal problem. Regeneration is accomplished with a synthesis gas (consisting of a mixture of hydrogen and carbon monoxide) to produce a concentrated stream of sulphur dioxide:

$$CaSO_4 + H_2 = CaO + H_2O + SO_2$$
$$CaSO_4 + CO = CaO + CO_2 + SO_2$$

The calcium oxide product is supplemented with fresh limestone and returned to the fluidized bed. Two undesirable side reactions can occur in the regeneration of spent lime leading to the production of calcium sulphide:

$$CaSO_4 + 4H_2 = CaS + 4H_2O$$
$$CaSO_4 + 4CO = CaS + 4CO_2$$

which results in the recirculation of sulphur to the bed.

Entrained systems

In entrained systems, fine grinding and increased retention times intensify combustion but the temperature of the carrier and degree of dispersion are also important. In practice, the coal is introduced at high velocities which may be greater than 100 ft/s (30.5 m/s) and involve expansion from a jet to the combustion chamber.

Types of entrained systems include cyclone furnaces (which have been used for various coals) and other systems developed and utilized for the injection of coal–oil slurries into blast furnaces or for the burning of coal–water slurries.

The cyclone furnace (developed in the 1940s to burn coal having low ash-fusion temperatures) is a horizontally inclined, water-cooled, tubular furnace in which crushed coal is burned with air entering the furnace tangentially. Temperatures may be of the order of 1700°C (3100°F) and the ash in the coal is converted to a molten slag that is removed from the base of the unit. Coal fines burn in suspension while the larger pieces are captured by the molten slag and burn rapidly.

An advantage of the cyclone furnace is the low dust burden in the secondary furnace and, hence, its lesser emission of particulate matter from the stack. Most cyclone furnaces capture about 90% of the ash in the coal and convert it to molten slag. Cyclone furnaces have two major shortcomings: (a) the ash of the coal must be convertible to molten slag at furnace temperatures; and (b) the nitrogen oxide emissions are excessive (about 1000 ppm) because of the high furnace temperature. Cyclone furnaces have been widely used in areas where the coal contains ash with a low fusion temperature. For successful removal of slag, the slag viscosity cannot exceed 250 poise (m^2/s) at 1420°C (2600°F) and many cyclone furnaces may not meet this requirement. The addition of iron ore, limestone or dolomite makes it possible to flux the coal ash thereby decreasing the viscosity at furnace temperatures.

Carbonization

Next to combustion, carbonization represents one of the largest uses of coal. However, the carbonization of coal is a well-established technology (Berkowitz, 1979; Holowaty et al., 1981; Speight, 1983; Hessley et al., 1986); therefore, only a brief outline will be given here along with indications of current interests.

Carbonization is essentially a process for the production of a carbonaceous residue by thermal decomposition (with simultaneous removal of distillate) of organic substances:

$$[C]_{organic} = [C]_{coke/char/carbon} + liquids + gases$$

The process may also be referred to as destructive distillation and has been applied to a whole range of organic materials, but more particularly to natural products such as wood, sugar and vegetable matter to produce charcoal. However, in the present context, coal usually yields 'coke' which is physically dissimilar from charcoal and appears with the more familiar honeycomb-type structure.

The original process of heating coal (in rounded heaps; the hearth process) remained the principal method of coke production for over a century, although an improved oven in the form of a beehive shape was developed in the Durham–Newcastle area of England in about 1759 (Holowaty et al., 1981; Speight, 1983).

The principal use of the process was to form a carbonaceous residue by heating the coal in an atmosphere relatively deficient in oxygen:

$$[C]_{coal} = [C]_{residue} + volatile products$$

Both processes lacked the capability to collect the volatile products (liquids and gases); it was not until the mid-nineteenth century, with the introduction of indirectly heated 'slot' ovens, that it became possible to collect the liquid and gaseous products for further use.

Gases of high calorific value are obtained by the low-temperature or medium-temperature carbonization of coal. The gases obtained by the carbonization of any given coal change in a progressive manner with increasing temperature (Table 5.5). The composition of coal gas also changes during the course of carbonization at a given temperature (Table 5.6).

In terms of the production of lower-molecular-weight species (liquids and gases) from coal, carbonization is not recognized as efficient as synthesis gas production or as efficient as direct hydrogenation of coal. It is, however, a more facile process and enables a relatively inexpensive removal of the hydrocarbon-rich portion of coal. Therefore, there has been a renewal of interest in the carbonization process (coal pyrolysis) to obtain liquid and gaseous hydrocarbons from coal.

The thermal decomposition of coal is a complex sequence of chemical and physical events which can be described in terms of several important physiochemical changes, such as the tendency of the coal to soften and flow when heated. In fact, some coals become quite fluid at temperatures of the order of 400–500°C (750–930°F) and there is a considerable variation in the degree of maximum plasticity, the temperature of maximum plasticitiy, as well as the plasticity temperature range for various coals (Habermehl *et al.*, 1981).

Thus, as coal is heated to the plastic stage, it is probable that some thermal 'depolymerization' can occur with the release of liquids and gases. These

Table 5.5 Illustration of the effect of temperature on gas composition

Gas comp. and yields (vol %)	Temperature of carbonization (°C)					
	500	*600*	*700*	*800*	*900*	*1000*
CO_2	5.7	5.0	4.4	4.0	3.2	2.5
Unsats.	3.2	4.0	5.2	5.1	4.8	4.5
CO	5.8	6.4	7.5	8.5	9.5	11.0
H_2	20.0	29.0	40.0	47.0	50.0	51.0
CH_4	49.5	47.0	36.0	31.0	29.5	29.0
C_2H_6	14.0	5.3	4.5	3.0	1.0	0.5
Yield (m^3/tonne)	62.3	102	176	238	278	312
Calorific value (MJ/m^3) sat. s.t.p.	39.0	29.0	26.5	24.4	22.3	22.3
Yield (MJ/tonne)	2118	2960	4660	5810	6200	6960

Source: Francis and Peters (1980).

Table 5.6 Illustration of the effect of reaction time on gas composition

Composition and yield of gas (vol %)	Duration of carbonization at 950°C				
	0.5 h	*0.5– 1.5 h*	*1.5– 3.0 h*	*3.0– 4.5 h*	*4.5– 7.5 h*
CO_2	3.2	2.8	2.5	1.8	0.5
Unsats.	6.5	4.5	2.0	0.3	–
CO	7.5	9.0	7.0	6.5	6.0
H_2	34.0	43.5	53.8	63.5	67.5
CH_4	43.5	37.0	31.5	23.5	22.5
C_2H_6	1.5	0.5	0.2	–	–
Calorific value (MJm^3) at s.t.p. sat.	26.8	24.3	22.5	18.1	18.1
Gas yield (m^3/tonne)	37.7	92.1	63.5	34.8	83.6
MJ/tonne	1010	2240	1430	630	1510

Source: Francis and Peters (1980).

changes (including the liberation of water, carbon dioxide and traces of hydrocarbons) can occur at temperatures below 300°C (570°F) and are accelerated markedly when a temperature of 350°C (660°F) is reached. The maximum evolution of tarry material, hydrocarbons and combustible gases usually occurs over the temperature range 450–500°C (840–930°F) whereas products such as hydrogen sulphide and ammonia are released at temperatures of 250–500°C (480–930°F) or at temperatures in excess of 500°C (930°F), respectively. At temperatures of the order of 550°C (1020°F) the visible changes to the coal are virtually complete and a semicoke residue is formed. At temperatures in excess of 550°C (1020°F), the semicoke hardens and shrinks to form coke with the evolution of methane, hydrogen and traces of carbon monoxide and carbon dioxide.

In more general terms, coal carbonization processes (Table 5.7) are generally regarded as 'low temperature' when the temperature of the process does not exceed 700°C (1290°F) or 'high temperature' if the temperature of the process is at, or in excess of, 900°C (1650°F); the temperature of a 'medium-temperature' carbonization process needs no further definition.

Table 5.7 Simplified classification of coal carbonization procedures

Carbonization process	Final temp.		Products	Processes
	°C	°F		
Low temperature	500–700	930–1290	Reactive coke and high tar yield	Rexco (700 C) made in cylindrical vertical retorts; Coalite (650 C) made in vertical tubes
Medium temperature	700–900	1290–1650	Reactive coke with high gas yield, or domestic briquettes	Town gas and gas coke (obsolete); Phurnacite, low-volatile steam coal, pitch-bound briquettes carbonized at 800 C
High temperature	900–1050	1650–1920	Hard, unreactive coal for metallurgical use	Foundry coke (900 C), blast furnace coke (950–1050 C)

Source: Pitt and Millward (1979, p. 52).

Low-temperature carbonization

Low-temperature carbonization was mainly developed as a process to supply 'town' gas for lighting purposes as well as to provide a 'smokeless' (devolatilized) solid fuel for domestic consumption (Seglin and Bresler, 1981). However, the process by-products (tars) were also found to be valuable in so far as they served as feedstocks for an emerging chemical industry and were also converted to gasolines, heating oils and lubricants (Aristoff et al., 1981).

The coals that were preferred for low-temperature carbonization were usually lignites or subbituminous (as well as highly-volatile bituminous) coals which yield porous solid products over the temperature range 600–700°C

(1110–1290°F). The reactivities of the semicoke products were usually equivalent to those of the parent coals. Certain of the higher-rank (caking) coals were less suitable for the process (unless steps were taken to destroy the caking properties) because of their tendency to adhere to the walls of the carbonization chamber.

On a commercial scale, the low-temperature carbonization of coal was employed extensively in the industrialized nations of Europe but suffered a major decline after 1945 as oil and natural gas became more widely available. However, the subsequent rapid escalation in oil prices as well as newer and more restrictive environmental regulations have stimulated (and reactivated) interest in the recovery of hydrocarbon liquids from coal by low-temperature thermal processing.

The options for efficient low-temperature carbonization of coal include vertical and horizontal retorts which have been used for batch and continuous processes. In addition, stationary and revolving horizontal retorts have also been operated successfully and there are also several process options employing fluidized or gas-entrained coal. During the last half century, coke production from batch-type carbonization of coal has been supported by a variety of continuous retorting processes which allow much greater through-put rates than were previously possible. These processes employ rectangular or cylindrical vessels of sufficient height to carbonize the coal while it travels from the top of the vessel to the bottom and usually employ the principle of heating the coal by means of a countercurrent flow of hot combustion gas. Most notable of these types of carbonizers are the Lurgi–Spulgas retort and the Koppers continuous 'steaming' oven (Speight, 1983).

High-temperature carbonization

When heated at temperatures in excess of 700°C (1290°F), low-temperature chars lose their reactivity through devolatilization and also suffer a decrease in porosity. High-temperature carbonization is, therefore, employed for the production of coke (Eisenhut, 1981). As with the low-temperature processes, the tars produced in high-temperature ovens are also sources of chemicals and chemical intermediates (McNeil, 1981).

Documented efforts at coke making date from the late sixteenth century (1584, to be precise) (Fess, 1957) and have seen various adaptations of conventional wood-charring methods to the production of coke with the eventual evolution of the (self-descriptive) beehive oven, which by the mid-nineteenth century had become the most common vessel for the coking of coal (Speight, 1983). The heat for the process was supplied by burning the volatile matter released from the coal and, consequently, the carbonization would progress from the top of the bed to the base and the coke was retrieved from the side of the oven at completion of the process.

Some beehive ovens (with various improvements and additions of waste heat boilers, thereby allowing heat recovery from the combustion products) may still be in operation but generally they have been replaced by wall-heated, horizontal chamber ('slot') ovens in which higher temperatures can be achieved as well as better control over the quality of the coke. Modern slot-type coke ovens are approximately 50 ft (15 m) long and 20 ft (6 m) high, with the width chosen to suit the carbonization behaviour of the coal to be processed. For example, the most common widths are 18 and 20 in (about

0.5 m), but some ovens may be as narrow as 12 in (0.3 m) and others 22 in (0.6 m) wide.

Several of these chambers (usually 20 or more, alternating with similar cells that contain heating ducts) are constructed in the form of a battery over a common firing system through which the hot combustion gas is conveyed to the ducts. The flat roof of the battery acts as the surface for a mobile charging car from which the coal (25–40 tons, 25–40 × 10³ kg) is released to each oven through three openings along the top. The coke product is pushed from the rear of the oven through the opened front section on to a quenching platform or into rail cars that move the coke through water sprays. The gas and tar by-products of the process are collected for further processing or for on-site use as fuel.

Most modern coke ovens operate on a regenerative heating cycle in order to obtain as much surplus gas as possible for use on the works, or for sale. If coke-oven gas is used for heating the ovens, the majority of the gas is surplus to requirements. If producer gas is used for heating, much of the coke-oven gas is surplus.

The main difference between gas works and coke-oven practice is that, in a gas works, maximum gas yield is a primary consideration whereas in the coke works the quality of the coke is the first consideration. These effects are obtained by choosing a feedstock (coal) that is suitable to the task. For example, use of lower-volatile coals in coke ovens, compared with coals used in gas works, will produce lower yields of gas when operating at the same temperatures.

In addition, the choice of heating (carbonizing) conditions and the type of retort also play a major role (Tables 5.8–5.11).

Table 5.8 Product yields per ton of coal

	High temperature (kg)	Low temperature (kg)
Furnace coke	715	–
Coke breeze	46.5	–
Semicoke (12% volatiles)	–	720
Tar	39	75
Ammonium sulphate[a]	10	9
Light oil (removed from gas by oil scrubbing)	10	8
Gas	1750	125

[a] Note that 4 kg of technical ammonium sulphate is equivalent to 1 kg of ammonia.
Source: *Industrial Engineering Chemistry* (1956) **48**, 352.

Table 5.9 Variation of gas composition with temperature of carbonization

Gas	Coking temperature, 500°C (%)	Coking temperature 1000°C (%)
CO_2	9.0	2.5
C_nH_m	8.0	3.5
CO	5.5	8.0
H_2	10.0	50.0
CH_4 and homologues	65.0	34.0
N_2	2.5	2.0

Source: Francis and Peters (1980).

Table 5.10 Composition and properties of coal gas from high-temperature retorts

Analysis of gas (vol %)	Horizontal retorts	Process Vertical retorts without steaming	Process Vertical retorts 5% steaming	Narrow coke ovens
O_2	0.4	0.4	0.4	0.4
CO_2	2	3	4	2
Unsaturated	3.5	3	2	2.5
CO	8	9	18	7.5
H_2	52	53.5	49.5	54
CH_4	30	25	24.5	28
Calorific value (CV) (MJ/m^3) at s.t.p.	21.7	20.9	19.4	20.5
Net CV (MJ/m^3)	19.4	18.6	17.4	18.2
Density, air = 1	0.4	0.43	0.43	0.38

Source: Francis and Peters (1980).

Table 5.11 Variation of gas yield and thermal properties with retort type

	Horizontal retort	Intermittent vertical ovens	Continuous vertical retorts 5% steam
Coke (wt %)	70–75	65–70	60–65
Tar and oils (dm^3/t)	55–65	50–60	65–75
Gas (m^3/tonne)	320	400	435
Gas calorific value (MJ/m^3)	21.7	20.5	19.4
Gas (MJ/tonne)	6940	8200	8420

Source: Francis and Peters (1980).

Gasification

The gasification of coal is essentially the conversion of coal (by any one of a variety of processes) to produce combustible gases which may be of low-, medium- or high-Btu content depending upon the defined use (Table 5.12) (Bodle and Huebler, 1981: Rath and Longanbach, 1991). High-Btu gas consists predominantly of methane with a heating value of approximately 1000 Btu/ft^3 (37.3×10^3 kJ/m^3) and is compatible with natural gas in so far as it may be mixed with, or substituted for, natural gas.

On the other hand, medium-Btu gas consists of a mixture of methane, carbon monoxide, hydrogen and various other gases. The heating value of medium-Btu gas usually falls in the range 300–700 Btu/ft^3 ($11–26 \times 10^3$ kJ/m^3) and is suitable as a fuel for industrial consumers. Finally, low-Btu gas consists of a mixture of carbon monoxide and hydrogen and has a heating value of less than 300 Btu/ft^3 (11.2×10^3 kJ/m^3). This gas is of interest to industry as a fuel gas or even, on occasion, as a raw material from which ammonia, methanol and other compounds may be synthesized.

The importance of coal gasification as a means of producing fuel gas(es) for industrial use cannot be understated. But coal gasification systems also have undesirable features in so far as a range of undesirable products are also

Table 5.12 Differences in required gas composition/properties for synthetic natural gas plants and for power plants

Characteristic	Required for synthetic natural gas	Required for power
Methane content	High – less synthesis required	Low – probably means no tars
H_2/CO ratio	High – less shifting required	Low – CO more efficient fuel
Moisture content	High – steam required for shift	Low – lower condensate treatment costs
Outlet temperature	Low – maximizes methane – minimizes sensible heat loss – leads to high cold gas efficiency	High – precludes tar formation – provides for steam generation – reduces cold gas efficiency
Gasifier oxidant	O_2 only – cost of N_2 removal excessive	Air or O_2 – low-Btu gas acceptable fuel

Table 5.13 Desirable and undesirable products in coal gasifier product streams

Solids
  Overheat fines
	Nitrogen compounds
	  Ammonia
	  Hydrogen cyanide

Organic liquids
  Tars
  Heavy Oils
  BTX or naphtha
  Phenols and cresylic acids
    (water soluble)
	Thiocyanates
	  Heavier ring-substituted compounds
	
	Halogen compounds
	  HCl
	  HF

Light hydrocarbons
  Paraffins (ethane, propane)
  Olefins
    Ethylene
    Acetylene
    Propylene
	Trace elements (of primary importance)[a]
	  Antimony
	  Arsenic
	  Beryllium
	  Boron
	  Cadmium

Aqueous liquids
  Foul water
	Chromium
	  Lead
	  Lithium

Sulphur compounds
  H_2S
  COS
  CS_2
  Mercaptans (RSH)
  Organic sulphides (RS or R_2S)
  Thiophenes (C_4H_4S)
  Heavier ring-substituted
    sulphur compounds
	Manganese
	  Mercury
	  Molybdenum
	  Selenium
	  Tellurium
	  Thallium
	  Tin
	  Vanadium
	  Zinc

[a] May be present as various compounds such as chlorides, hydrides, sulphides, etc.

produced (Table 5.13) which must be removed before the products are used to provide fuel and/or to generate electric power (Figure 5.4) (Alpert and Gluckman, 1986).

Chemistry

Coal gasification involves the thermal decomposition of coal and the reaction of the carbon in the coal and other pyrolysis products with oxygen, water and hydrogen to produce fuel gases such as methane either directly

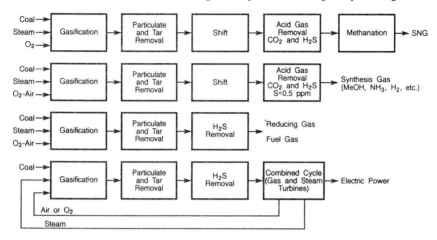

Figure 5.4 Illustration of variations in gasification procedures for different end products

$$[C]_{coal} + [H]_{coal} = CH_4$$

or through the agency of added hydrogen

$$[C]_{coal} + 2H_2 = CH_4$$

although the reactions are more numerous and more complex (Table 5.14).

The presence of oxygen, hydrogen, water vapour, carbon oxides and other compounds in the reaction atmosphere during pyrolysis may either support or inhibit numerous reactions with coal and with the products evolved. The distribution of weight and the chemical composition of the products are also influenced by the prevailing conditions (i.e. temperature, heating rate, pressure, residence time, etc.) and, last but not least, by the coal feedstock.

If air is used as a combustant, the product gas will have a heating of approximately 150–300 Btu/ft^3 (5.6–11.2 × 10^3 kJ/m^3), depending on design characteristics, and will contain undesirable constituents such as carbon dioxide, hydrogen sulphide and nitrogen. The use of pure oxygen, although

Table 5.14 Reactions involved in coal gasification

$2C + O_2 \rightarrow 2CO$
$C + O_2 \rightarrow CO_2$
$C + CO_2 \rightarrow 2CO$
$CO + H_2O \rightarrow CO_2 + H_2$ (shift reaction)
$C + H_2O \rightarrow CO + H_2$ (water gas reaction)
$C + 2H_2 \rightarrow CH_4$
$2H_2 + O_2 \rightarrow 2H_2O$
$CO + 2H_2 \rightarrow CH_3OH$
$CO + 3H_2 \rightarrow CH_4 + H_2O$ (methanation reaction)
$CO_2 + 4H_2 \rightarrow CH_4 + 2H_2O$
$C + 2H_2O \rightarrow 2H_2 + CO_2$
$2C + H_2 \rightarrow C_2H_2$
$CH_4 + 2H_2O \rightarrow CO_2 + 4H_2$

expensive, results in a product gas having a heating value of 300–400 Btu/ft^3 ($11.2 - 14.9 \times 10^3$ kJ/m^3) with carbon dioxide and hydrogen sulphide as by-products.

If a high-Btu gas (900–1000 Btu/ft^3, 33.6–37.3 kJ/m^3) is required, efforts must be made to increase the methane content of the gas:

$$CO + H_2O = CO_2 + H_2$$
$$CO + 3H_2 = CH_4 + H_2O$$
$$2CO + 2H_2 = CH_4 + CO_2$$
$$CO + 4H_2 = CH_4 + 2H_2O$$

The reaction rates for these reactions are relatively slow (with negative heats of formation, Table 5.15) and catalysts may be necessary for complete reaction (Berkowitz, 1979; Meyers, 1981; Speight, 1983; Hessley et al., 1986; Hessley, 1990).

Table 5.15 Thermodynamic data for methanation reactions

	Temperature		Reaction[a]				
	K	°C	1	2	3	4	5
A	Heat of reaction, ΔH_f (kcal)						
	300	27	−9.838	−49.298	−39.460	−59.136	−41.227
	400	127	−9.710	−50.360	−40.650	−60.070	−41.434
	500	227	−9.518	−51.297	−41.779	−60.815	−41.499
	600	327	−9.292	−52.084	−42.792	−61.376	−41.460
	700	427	−9.050	−52.730	−43.680	−61.780	−41.350
	800	527	−8.799	−53.248	−44.449	−62.047	−41.190
	900	627	−8.549	−53.654	−45.105	−62.203	−40.996
	1000	727	−8.304	−53.957	−45.653	−62.261	−40.729
B	Free energy of reaction, $\Delta F°$ (kcal)						
	300	27	−6.827	−33.904	−27.077	−40.731	−28.621
	400	127	−5.841	−28.610	−22.769	−34.451	−24.385
	500	227	−4.894	−23.062	−18.168	−27.954	−20.111
	600	327	−3.991	−17.338	−13.347	−21.329	−15.836
	700	427	−3.127	−11.493	−8.366	−14.620	−11.574
	800	527	−2.298	−5.567	−3.269	−7.865	−7.332
	900	627	−1.500	+0.594	+1.921	−1.079	−3.108
	1000	727	−0.729	−6.444	+7.173	+5.715	+1.090
C	Equilibrium constant, log K_p						
	300	27	4.973	24.698	19.724	29.670	20.849
	400	127	3.191	15.630	12.44	18.822	13.322
	500	227	2.139	10.080	7.940	12.219	8.790
	600	327	1.453	6.314	4.861	7.768	5.768
	700	427	0.976	3.588	2.611	4.564	3.613
	800	527	0.628	1.512	0.893	2.148	2.003
	900	627	0.364	−0.144	−0.466	0.261	0.755
	1000	727	0.159	−1.408	−1.568	−1.248	−0.238

[a]1 $CO + H_2O \rightarrow CO_2 + H_2$
2   $CO + 3H_2 \rightarrow CH_4 + H_2O$
3   $2H_2 + 2CO \rightarrow CH_4 + CO_2$
4   $CO_2 + 4H_2 \rightarrow CH_4 + 2H_2O$
5   $2CO \rightarrow CO_2 + C$
Source: G.A. Mills and F.W. Steffgen (1975) Catalysis Reviews, **15**, 165.

Process parameters

Primary gasification involves thermal decomposition of the raw coal and many schemes to produce mixtures containing various proportions of carbon monoxide, carbon dioxide, hydrogen, water, methane, hydrogen sulphide and nitrogen, and typical products of thermal decomposition such as tar, oils and phenols. A solid char product may also be produced, and often represents the bulk of the weight of the original coal. The type of coal being processed determines (to a large extent) the amount of char produced and the composition of the gas product.

Secondary gasification usually involves gasification of the char from the primary gasifier. This is usually done by reacting the hot char with water vapour to produce carbon monoxide and hydrogen:

$$[C]_{char} + H_2O = CO + H_2$$

The gaseous product from a gasifier generally contains large amounts of carbon monoxide and hydrogen, plus lesser amounts of other gases (Table 5.16). Carbon monoxide and hydrogen (if they are present in the mole ratio of 1:3) can be reacted in the presence of a catalyst to produce methane:

$$CO + 3H_2 = CH_4 + H_2O$$

Table 5.16 Variation of product gas composition with gasifier type

	Lurgi	Koppers–Totzek	Winkler
H_2	38.0	36.7	41.8
CO	20.2	55.8	33.3
CO_2	28.6	6.2	20.5
CH_4	11.4	0.0	3.0
C_2H_6	1.0	0.0	0.0
H_2S or COS	0.5	0.3	0.4
N_2	0.3	1.0	1.0

Usually, however, some adjustment to the ideal (1:3) is required and, to accomplish this, all or part of the stream is treated according to the waste gas shift (shift conversion) reaction. This involves reacting carbon monoxide with steam to produce carbon dioxide and hydrogen whereby the desired 1:3 mole ratio of carbon monoxide to hydrogen may be obtained:

$$CO + H_2O = CO_2 + H_2$$

The most notable effects in gasifiers are those of pressure (Figure 5.5) and coal character. With regard to the latter, some initial processing of the coal feedstock may be required, with the type and degree of pretreatment being a function of the process and/or the type of coal.

For example, some coals display caking, or agglomerating, characteristics when heated and these coals are usually not amenable to treatment by gasification processes employing fluidized-bed or moving-bed reactors; in fact, caked coal is difficult to handle in fixed-bed reactors. The pretreatment involves a mild oxidation treatment which destroys the caking characteristics of the coal and usually consists of low-temperature heating of the coal in the presence of air or oxygen.

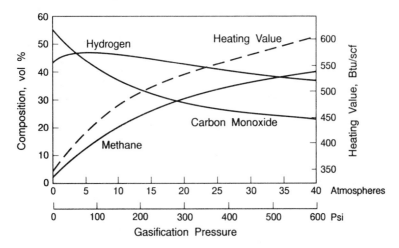

Figure 5.5 Variation of gas composition and heating value with gasifier pressure

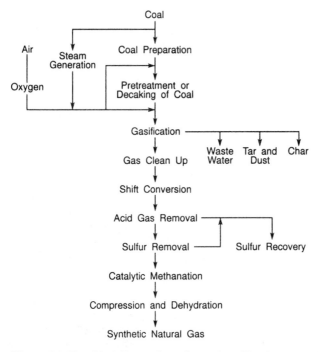

Figure 5.6 Simplified illustration of a coal gasification operation

Depending on the type of coal being processed and the composition of the gas product desired, some or all of the following processing steps (Figure 5.6) will be required: (a) pretreatment of the coal (if caking is a problem); (b) primary gasification of the coal; (c) secondary gasification of the carbonaceous residue from the primary gasifier; (d) removal of carbon dioxide, hydrogen sulphide and other acid gases; (e) shift conversion for adjustment of the carbon monoxide/hydrogen mole ratio to the desired figure; and (f) catalytic methanation of the carbon monoxide/hydrogen mixture to form methane. If high-Btu gas is desired, all of these processing steps are necessary since coal gasifiers do not yield methane in the concentrations required.

Not all high-Btu gasification technologies depend entirely on catalytic methanation, i.e. the direct addition of hydrogen to coal under presure to form methane:

$$[C]_{coal} + 2H_2 = CH_4$$

The hydrogen-rich gas for hydrogasification can be manufactured from steam by using the char that leaves the hydrogasifier. Appreciable quantities of methane are formed directly in the primary gasifier and the heat released by the methane formation is at a sufficiently high temperature to be used in the steam–carbon reaction to produce hydrogen so that less oxygen is used to produce heat for this reaction. Hence, less heat is lost in the low-temperature methanation step, thereby leading to a higher overall process efficiency.

Some mention must also be made here of the reactor types that have been used (or proposed) for coal gasification processes. There are three fundamental reactor types: (a) a gasifier reactor, (b) a devolatilizer and (c) a hydrogasifier (Figure 5.7), with the choice of a particular design (e.g. whether or not two stages should be involved; Figure 5.8) depending on the ultimate product gas desired (Table 5.16). Reactors may also be designed to operate at either atmospheric pressure or at high pressure.

There has been a general tendency to classify gasification processes by virtue of the Btu content of the gas which is produced. It is also possible to classify gasification processes according to the type of reactor vessel and whether or not the system reacts under pressure. However, for the purposes of the present

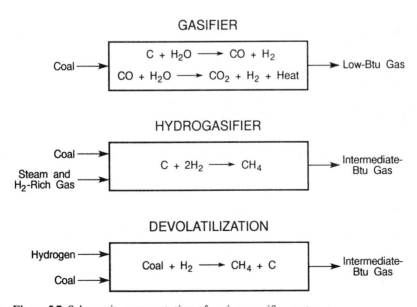

Figure 5.7 Schematic representation of various gasifier systems

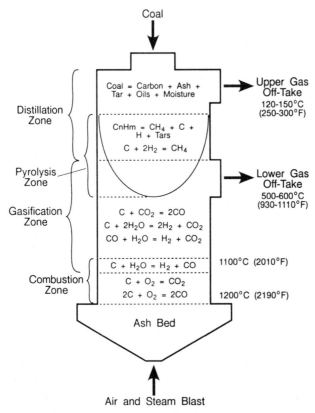

Figure 5.8 Illustration of the chemical reactions occurring in a two-stage gasifier

text the gasification processes have been segregated according to the bed types, which differ in their ability to accept (and use) caking coals (Table 5.17). Thus, gasification processes can generally be divided into four categories based on reactor (bed) configuration: (a) fixed bed, (b) moving bed (c) fluidized bed, and (d) entrained bed (Table 5.17).

In a fixed-bed process the coal is supported by a grate and combustion gases (steam, air, oxygen, etc.) pass through the supported coal whereupon the hot

Table 5.17 Operational parameters for different gasifier beds

| | Moving bed | | | Entrained |
Function	Dry ash	Slagging	Fluidized bed	flow
Capacity	Low	High	Intermediate	High
Ability to handle caking coals without pretreatment	Moderate	Shown at 300 t/d scale	Shown on small scale	Excellent
Temperature of operation	800–2000°F	800–2800°F	1600–1900°F	1700–3000°F

gases produced exit from the top of the reactor. Heat is supplied internally or from an outside source, but caking coals cannot be used in an unmodified fixed-bed reactor.

In the moving-bed system (Figure 5.9) coal is fed to the top of the bed and ash leaves the bottom with the product gases being produced in the hot zone just prior to being released from the bed.

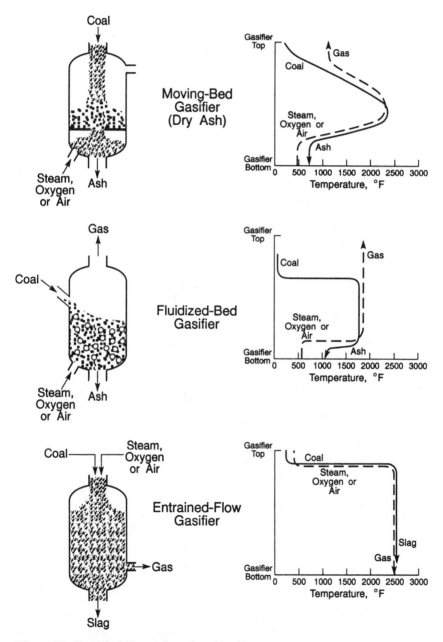

Figure 5.9 Simplified illustration of coal gasifier operations

The fluidized-bed system (Figure 5.9) uses finely sized coal particles and the bed exhibits liquid-like characteristics when a gas flows upwards through it. Gas flowing through the coal produces turbulent lifting and separation of particles and the result is an expanded bed having a greater coal surface area to promote the chemical reaction, but such systems have only a limited ability to handle caking coals.

An entrainment system (Figure 5.9) uses finely sized coal particles blown into the gas stream prior to entry into the reactor and combustion occurs with the coal particles suspended in the gas phase. The entrained system is suitable for both caking and non-caking coals.

Petroleum processing

Process gas

The terms 'refinery gas' and 'process gas' are also often used to include all of the gaseous products and by-products that emanate from a variety of refinery processes (Hobson and Pohl, 1973; Gary and Handwerk; 1975; Speight, 1991). Further to this definition, there are components of the gaseous products that must be removed prior to release of the gas to the atmosphere or prior to use of the gas in another part of the refinery, i.e. as a fuel gas or as a process feedstock.

With the exception of some of the more viscous crude oils, processing crude petroleum almost invariably involves a primary distillation of the hydrocarbon mixture, which results in its separation into fractions differing in carbon number, volatility, specific gravity and other characteristics.

The most volatile fraction, which contains most of the gases which are generally dissolved in the crude, is referred to as pipestill gas or pipestill light ends and consists essentially of hydrocarbon gases ranging from methane to the butane(s), or sometimes the pentane(s). It varies in composition and volume, depending on crude origin, and also quite frequently on any additions to the crude made at the loading point; it is not uncommon to reinject light hydrocarbons such as propane and butane into the crude before dispatch by tanker or pipeline. This results in a higher vapour pressure of the crude, but it allows one to increase the quantity of light products obtained at the refinery. Since light ends in most petroleum markets command a premium, while in the oil field itself propane and butane may have to be reinjected or flared, the practice of 'spiking' crude with LPG is becoming fairly common.

In addition to the gases obtained by distillation of crude petroleum, further highly volatile products result from the subsequent processing of naphtha and middle distillate to produce gasoline, from desulphurization processes involving hydrogen treatment of naphthas, distillates and residual fuels, and from the coking or similar thermal treatments of vacuum gas oils and residual fuels.

The most common processing step in the production of gasoline is the catalytic reforming of hydrocarbon fractions in the heptane (C_7) to decane (C_{10}) range.

In a series of processes commercialized under the names of the various reforming processes (e.g. Platforming, Powerforming, Catforming, Ultraforming, etc.), paraffinic and naphthenic (cyclic, non-aromatic) hydrocarbons are converted in the presence of hydrogen and a catalyst into aromatics, or isomerized to more highly branched hydrocarbons.

Catalytic reforming processes thus not only result in the formation of a liquid product of higher octane number, but also produce substantial quantities of gases. The latter are rich in hydrogen, but also contain hydrocarbons from methane to butanes, with a preponderance of propane, n-butane $(CH_3.CH_2.CH_2.CH_3)$ and isobutane $[(CH_3)_3CH]$. Their composition will vary in accordance with reforming severity and reformer feedstock; since all catalytic reforming processes require a substantial recycle of hydrogen it is normal to separate reformer gas into a propane $(CH_3.CH_2.CH_3)$ and/or butanes $[CH_3.CH_2.CH_2.CH_3/(CH_3)_3CH]$ stream, which becomes part of the refinery LPG production, and a lighter gas fraction, part of which is recycled. In view of the excess of hydrogen in the gas all the catalytic reforming products are saturated and there are, usually, no olefinic gases present in either gas stream.

A second group of refining operations which contribute to gas production are the catalytic cracking processes, such as fluid-bed catalytic cracking (FCC), Thermofor catalytic cracking (TCC) and other variants in which heavy gas oils are converted into cracked gas, LPG, catalytic naphthas, fuel oils and coke, by contacting the heavy hydrocarbon with the hot catalyst.

Both catalytic and thermal cracking processes, the latter being now largely used for the production of chemical raw materials, result in the formation of unsaturated hydrocarbons, particularly ethylene $(CH_2{=}CH_2)$, but also propylene (propene, $CH_3CH{=}CH_2$), isobutylene [isobutene, $(CH_3)_2C{=}CH_2$] and the n-butenes $(CH_3.CH_2.CH{=}CH_2$, $CH_3.CH{=}CH.CH_3)$ in addition to hydrogen (H_2), methane (CH_4) and smaller quantities of ethane $(CH_3.CH_3)$, propane $(CH_3.CH_2.CH_3)$ and butanes $[(CH_3)_3CH, CH_3.CH_2.CH_2.CH_3]$. Diolefins, such as butadiene $(CH_2{=}CH.CH{=}CH_2)$ are also present.

Additional gases are produced in refineries with coking or visbreaking facilities for the processing of their heaviest crude fractions.

In a visbreaker, fuel oil is passed through externally fired tubes and undergoes liquid-phase cracking reactions, which result in the formation of lighter fuel oil components; oil viscosity is thereby reduced and, in addition, some gases, mainly hydrogen, methane and ethane, are formed. Substantial quantities of both gas and solid carbon are also formed in coking processes – both fluid-bed and delayed coking – in addition to middle distillate and naphtha, which are the main objectives. When coking a residential fuel oil or heavy gas oil the feed is preheated and contacted with hot carbon; this results in extensive cracking of the long-chain hydrocarbon molecules to form lighter products ranging from methane, via LPG(s) and naphtha, to gas oil and heating oil. Coker products tend to be unsaturated and in coker tail gases olefinic components predominate.

A further source of refinery gas is hydrocracking, a catalytic high-pressure pyrolysis process in the presence of fresh and recycled hydrogen.

The feedstock is again heavy gas oil or residual fuel oil, and the process is mainly directed at the production of additional middle distillates and gasoline. Since hydrogen is to be recycled the gases produced in this process again have to be separated into lighter and heavier streams; any surplus recycled gas and the LPG product gas from hydrocracking are both saturated.

Both hydrocracker and catalytic reformer tail gases are commonly used in catalytic desulphurization processes; in the latter, feedstocks ranging from light to vacuum gas oils are passed at pressures of 500–1000 psi ($3.5–7.0 \times 10^3$ kPa) with hydrogen over a hydrofining catalyst; this results mainly in the conversion of organic sulphur compounds into hydrogen sulphide

$$[S]_{feedstock} + H_2 = H_2S + \text{hydrocarbons}$$

but also produces some light hydrocarbons by hydrocracking.

Thus, refinery and natural gas streams, whilst ostensibly being hydrocarbon in nature, may contain large amounts of acid gases such as hydrogen sulphide and carbon dioxide. Most commercial plants employ hydrogenation to convert organic sulphur compounds into hydrogen sulphide. Hydrogenation is effected by means of recycled hydrogen-containing gases or external hydrogen over a nickel molybdate or cobalt molybdate catalyst (Ranney, 1975).

In summary, refinery process gas, in addition to hydrocarbons, may contain other contaminants such as carbon oxides (CO_x, where $x=0$ and/or 1), sulphur oxides (SO_x, where $x=2$ and/or 3) as well as ammonia (NH_3), mercaptans (R-SH) and carbonyl sulphide (COS).

The presence of these impurities may eliminate some of the sweetening processes since some processes remove large amounts of acid gas but not to a sufficiently low concentration. On the other hand, there are those processes that are not designed to remove (or are incapable of removing) large amounts of acid gases, but they are capable of reducing the acid gas impurities to very low levels when the acid gases are present only in low-to-medium concentrations.

Acid gases corrode refining equipment, harm catalysts, pollute the atmosphere, and prevent the use of hydrocarbon components in petrochemical manufacture. When the amount of hydrogen sulphide is large, it may be removed from a gas stream and converted to sulphur or sulphuric acid. Some natural gases contain sufficient carbon dioxide to warrant its recovery as dry ice (Bartoo, 1985).

Gases prepared from petroleum feedstocks are also used widely in the chemical industry for the production of ammonia, methanol, oxo-alcohols and synthetic hydrocarbons. The steps involved in the manufacture of synthesis gas are similar to those used for the production of town gas from hydrocarbons, i.e. the feed is purified, reacted with steam over a catalyst, generally under pressure, and the gaseous product containing hydrogen, carbon monoxide and carbon dioxide is adjusted in composition to the correct hydrogen/carbon monoxide ratio for the particular synthesis envisaged. This composition change is effected by passing the gas with additional steam through a shift conversion step, in which carbon monoxide is catalytically converted into carbon dioxide, which can be removed by absorption, and additional hydrogen is formed (see page 93). A wide range of synthesis gases from almost pure hydrogen to gases with substantial concentrations or carbon monoxide can thus be made.

Partial oxidation gas

It is often desirable to augment the supply of naturally occurring or by-product gaseous fuels or to produce gaseous fuels of well-defined composition and combustion characteristics. This is particularly true in areas devoid of natural gas where the manufacture of fuel gases, originally from coal and more recently from liquid petroleum fractions, has become well established.

Almost all petroleum fractions can be converted into gaseous fuels, although conversion processes for the heavier fractions require more elaborate technology and a higher investment to achieve the necessary purity and uniformity of the manufactured gas stream. In addition, the thermal yield from the gasification of heavier feedstocks is invariably lower than that of gasifying light naphtha or LPG(s) since, in addition to the production of hydrogen, carbon monoxide and gaseous hydrocarbons, heavy feedstocks also yield some tar and coke.

Thus, another refinery process, other than the 'usual' process gas, is the partial oxidation process which is used as a means of producing hydrogen for the various hydrotreating or hydrocracking operations (Table 5.18).

As with many processes where the chemistry can often be represented in the simplest manner

$$C_nH_{2n} + O_2 = nCO_2 + nH_2$$

the reaction is much more complex. A gaseous mixture is produced that does require treatment before further use. Oxides of carbon need to be removed as do the various hydrocarbon gases and sulphur-containing species.

Non-catalytic partial oxidation, generally using oxygen or enriched air as the gasifying agent, is less selective as regards feedstock, and heavier fractions, including fuel oil, can be gasified with oxygen or steam/oxygen mixtures.

Both steam reforming and partial oxidation of hydrocarbons are, advantageously, carried out under pressure, since pressurization of the liquid feeds is cheaper than compression of the product gases and also since one can thereby substantially reduce the size of the reaction vessels. The only type of gasification plant which invariably operates at atmospheric pressure is the cyclic gasifier, a piece of equipment widely used in the gas industry for the reforming of light hydrocarbons. Here the heat of reaction is supplied by burning part of the feed on a refractory regenerator, which sometimes also provides the catalyst support; steam/reformer feed and fuel/combustion air are charged in alternating cycles, and a set of interconnected valves ensure that only fuel gases are collected, whereas combustion gases are exhausted through a stack.

Table 5.18 Composition of gaseous products obtained by partial combustion (partial oxidation)

	Using air and gas oil	Using oxygen and heavy fuel oil
Composition of gas (vol %):		
CO_2	6	4
CO	5.5	47
O_2	0.5	–
Unsats.	16	–
CH_4	8.5	0.3
H_2	1.5	48.5
N_2	62	0.2
Calorific value (MJ/m^3) s.t.p. sat.	19.7	13.0
Specific gravity	1.03	0.55
Gas yield (MJ/MJ oil)	0.79	0.84

Liquefied petroleum gas

Liquefied petroleum gas (LPG) is the term applied to certain specific hydrocarbons, such as propane, butane and pentane, and their mixtures, which exist in the gaseous state under atmospheric ambient conditions but can be converted to the liquid state under conditions of moderate pressure, at ambient temperature. As such, the gas has little relevance in terms of gas clean-up but there may on occasion be reference to contaminant removal before use in a refining operation.

The processes that have been developed to accomplish gas purification vary from a simple once-through wash operation to complex multistep recycling systems. In many cases, the process complexities arise because of the need for recovery of the materials used to remove the contaminants or even recovery of the contaminants in the original, or altered, form (Kohl and Riesenfeld, 1979: Newman, 1985).

Gas treating

The overall gas and LPG yield in a refinery depends on a number of variables. Refineries with extensive conversion facilities will produce large quantities of gases for each barrel of crude oil. Similarly, petrochemical refineries, in which most of the naphtha and middle distillates are used as chemicals feedstock, tend to produce more gas than fuel refineries.

The gases obtained from the various process options have to be purified before further treatment; this involves the removal of hydrogen sulphide (H_2S) and mercaptans (R-SH) by means of alkaline absorbents, particularly alkanolamines ($H_2N.R'.OH$) (see Chapters 9, 10 and 12).

While large numbers of processes for the removal of sulphur from gases have been developed over the years (Chapters 9, 10 and 12), several have found commercial acceptance, but the majority of refineries in which sulphur-containing crude oils and gases are handled employ variations of the alkanolamine process (Chapter 12) for the absorption and recovery of hydrogen sulphide, followed, where this is economically justified or required for reasons of air pollution, by sulphur production in a Claus unit (see page 282).

Solvents used for hydrogen sulphide absorption include aqueous solutions of ethanolamine (also called monoethanolamine, MEA), diethanolamine (DEA) and diisopropanolamine (DIPA) amongst others. These solvents differ in volatility and selectivity for the removal of hydrogen sulphide (H_2S), mercaptans (R-SH) and carbon dioxide (CO_2) from gases of different composition. Other alkaline solvents used for the absorption of acidic components in gases include potassium carbonate (K_2CO_3) solutions (combined with a variety of activators and solubilizers to improve gas/liquid contacting).

In most of these processes, sulphur is separated as concentrated hydrogen sulphide gas; in the Stretford process (Chapters 9 and 12), on the other hand, a hydrogen-sulphide-saturated anthraquinone disulphonate solution is oxidized with air to produce elemental sulphur, which is precipitated and removed by filtration.

While most alkaline solvent absorption processes result in gases of acceptable purity for most purposes, it is now often essential to remove the last traces of residual sulphur compounds from gas streams. This is in addition to ensuring product purity by the removal of water, higher hydrocarbons and dissolved elemental sulphur from LPG. This can be done by passing the gas over a bed of molecular sieves, which are synthetic zeolites commercially available in several proprietary forms. Impurities are retained by the packed bed, and when the latter is saturated it can be regenerated by passing hot clean gas or hot nitrogen, generally in a reverse direction.

There are many variables in treating refinery gas or natural gas and the precise area of application of a given process is difficult to define, although several factors must be considered: (a) the types and concentrations of contaminants in the gas; (b) the degree of contaminant removal desired; (c) the selectivity of acid gas removal required; (d) the temperature, pressure, volume and composition of the gas to be processed; (e) the carbon dioxide/hydrogen sulphide ratio in the gas; and (f) the desirability of sulphur recovery due to process economics or environmental issues.

Process selectivity indicates the preference with which the process removes one acid gas component relative to another. For example, some processes remove both hydrogen sulphide and carbon dioxide; other processes are designed to remove hydrogen sulphide only. It is important to consider the process selectivity for, say, hydrogen sulphide removal compared with carbon dioxide removal that ensures minimal concentrations of these components in the product – hence the need to consider the carbon dioxide/hydrogen sulphide ratio in the gas stream.

Processing refinery gas therefore involves the use of several different types of processes but there is always overlap between the various processing concepts. In addition, the terminology used for gas processing can often be confusing and/or misleading because of this overlap (Nonhebel, 1964; Curry, 1981; Maddox, 1982).

References

Alpert, S. and Gluckman, M.J. (1986) *Annual Review of Energy*, **11**, 315

Aristoff, E., Rieve, R.W. and Shalit, H. (1981) In *Chemistry of Coal Utilization. Second Supplementary Volume* (ed. M.A. Elliot), Wiley, New York, Chapter 16

Bartoo, R.K. (1985) In *Acid and Sour Gas Treating Processes* (ed. S.A. Newman), Gulf Publishing Company, Houston, TX

Berkowitz, N. (1979) *Introduction to Coal Technology*, Academic Press, New York

Bodle, W.W. and Huebler, J. (1981) In *Coal Handbook* (ed. R.A. Meyers), Marcel Dekker, New York, Chapter 10

Curry, R.N. (1981) *Fundamentals of Natural Gas Conditioning*, PennWell Publishing Company, Tulsa, OK

Eisenhut, W. (1981) In *Chemistry of Coal Utilization. Second Supplementary Volume* (ed. M.A. Elliott), Wiley, New York, Chapter 14

Elliott, M.A. (ed.), (1981) *Chemistry of Coal Utilization. Second Supplementary Volume*, Wiley, New York

Fess, F.M. (1957) *History of Coke Oven Technology*, Gluckauf, Essen

Field, M.A., Gill, D.W., Morgan, B.B. and Hawksley, P.G.W. (1967) *Combustion of Pulverized Coal*, British Coal Utilization Research Association, Leatherhead, Surrey

Francis, W. and Peters, M. (1980) *Fuels and Fuel Technology*, Pergamon Press, Oxford, England

Funk, J.E. (1983) In *Riegel's Handbook of Industrial Chemistry* (ed. J.A. Kent), Van Nostrand Reinhold, New York, Chapter 3

Gary, J.H. and Handwerk, G.E. (1975) *Petroleum Refining: Technology and Economics*, Marcel Dekker, New York

Habermehl, D., Orywal, F. and Beyer, H.-D. (1981) In *Chemistry of Coal Utilization. Second Supplementary Volume* (ed. M.A. Elliott), Wiley, New York, Chapter 6

Hessley, R.K. (1990) In *Fuel Science and Technology Handbook* (ed. J.G. Speight), Marcel Dekker, New York

Hessley, R.K., Reasoner, J.W. and Riley, J.T. (1986) *Coal Science*, Wiley, New York

Holowaty, M.O., Phelps, R.G., DuBroff, W. and Landsly, G.L. (1981) In *Coal Handbook* (ed. R.A. Meyers), Marcel Dekker, New York, Chapter 9

Kohl, A.L. and Riesenfeld, F.C. (1979) *Gas Purification*, Gulf Publishing Company, Houston, TX

Levy, A., Barrett, R.E., Giammar, R.D. and Hazard, H.R. (1981) In *Coal Handbook* (ed. Robert A. Meyers), Marcel Dekker, New York, Chapter 8

McNeil, D. (1981) In *Chemistry of Coal Utilization. Second Supplementary Volume* (ed. M.A. Elliott), Wiley, New York, Chapter 17

Meyers, R.A. (ed.) (1981) *Coal Handbook*, Marcel Dekker, New York, Chapter 1

Maddox, R.N. (1982) *Gas Conditioning and Processing*, Gulf Publishing Company, Houston, TX

Newman, S.A. (1985) *Acid and Sour Gas Treating Processes*, Gulf Publishing Company, Houston, TX

Nonhebel, G. (1964) *Gas Purification Processes*, George Newnes, London

Pitt, G.J. and Millward, G.R. (eds) (1979) *Coal and Modern Coal Processing: An Introduction*, Academic Press, New York

Ranney, M.W. (1975) *Desulfurization of Petroleum*, Noyes Data Corp., Park Ridge, NJ

Rath, L.K. and Longanbach, J.R. (1991) *Energy Sources*, **13**, 443

Seglin, L. and Bresler, S.A. (1981) In *Chemistry of Coal Utilization. Second Supplementary Volume* (ed. M.A. Elliott), Wiley, New York, Chapter 13

Speight, J.G. (1983) *The Chemistry and Technology of Coal*, Marcel Dekker, New York

Speight, J.G. (1991) *The Chemistry and Technology of Petroleum*, 2nd edn, Marcel Dekker, New York

6

Recovery, storage and transportation

Introduction

Further to the discussion of the origin and occurrence of gases (Chapters 4 and 5), it is beneficial to understand the various exploration techniques for natural gas as well as the methods of storage and transportation for the various types of natural and industrial gases. This will offer an additional dimension in understanding the behaviour, properties and handling of the different gases as well as in understanding where, in the process of handling, 'Murphy's law' applies and emergency measures may need to be applied.

As might be expected, the type of exploration techniques that are employed for the location of natural gas occurrences depends upon the nature of the site. In other words, as for many current environmental operations, the recovery techniques applied to a specific site are dictated by the nature of the site and must therefore be regarded as being site specific.

For example, in areas where little is known about the subsurface, preliminary reconnaissance techniques will be necessary to identify potential reservoir systems that require further investigation. Techniques for reconnaissance include satellite and high-altitude imagery, magnetic and gravity surveys, all of which can be employed to make inferences about the subsurface structure.

Once an area has been selected for further investigation, more detailed methods (such as the seismic reflection method) are brought into play. Drilling is the final stage of the exploratory programme and is, in fact, the only method by which a natural gas reservoir can be conclusively identified. However, in keeping with the concept of site specificity, in some areas drilling may be the only means to commence the project. The risk involved in drilling depends upon what is known about the subsurface at the site.

A classification scheme (Figure 6.1) has been devised so that exploratory wells may be categorized according to the relationship of the site to known petroleum reservoirs.

Exploration

The oldest scientific method of exploration, surface geology, attempts to find out what is below from surface observations. Aerial photographs, the study of rock outcrops, MAD (Magnetic Anomaly Detection) from the air, gravity

Objective of Drilling			Initial Classification When Drilling is Started	Final Classification After Completion or Abandonment		
				Successful	Unsuccessful	
Drilling for a New Field on a Structure or in an Environment Never Before Productive			1. New-Field Wildcat	New-Field Discovery Wildcat	Dry New-Field Wildcat	
Drilling for a New Pool on a Structure or in a Geological Environment Already Productive	New Pool Tests	Drilling Outside Limits of a Proved Area of Pool	2. New-Pool (Pay) Wildcat	New-Pool Discovery Wells (Sometimes Extension Wells) — New-Pool Discovery Wildcat	Dry New-Pool Tests — Dry New-Pool Wildcat	
		Drilling Inside Limits of Proved Area of Pool — For a New Pool Above Deepest Proven Pool	3. Deeper Pool (Pay) Test	Deeper Pool Discovery Well	Dry Deeper Pool Test	
		For a New Pool Below Deepest Proven Pool	4. Shallower Pool (Pay) Test	Shallower Pool Discovery Well	Dry Shallower Pool Test	
Drilling for Long Extension of a Partly Developed Pool			5. Outpost or Extension Test	Extension Well (Sometimes a New-Pool Discovery Well)	Dry Outpost or Dry Extension Test	
Drilling to Exploit or Develop a Hydrocarbon Accumulation Discovered by Previous Drilling			6. Development Well	Development Well	Dry Development Well	

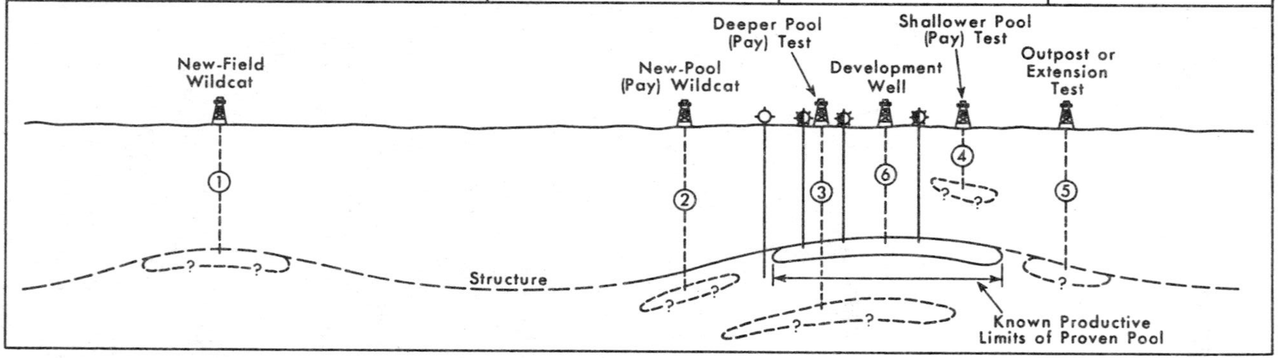

Figure 6.1 A system of well classification. (Reprinted by permission of the American Association of Petroleum Geologists, AAPG)

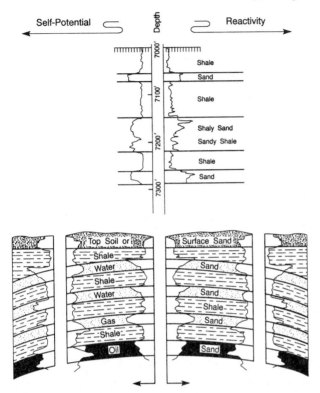

Figure 6.2 Illustration of correlation of electrical logs with subterranean formations

surveys and remote sensing are typical tools used in conjunction with the geological aspects of exploration (Haun and LeRoy, 1958). Because surface geology can only provide partial answers at best, subsurface geology comes into the exploration picture. One basic tool for acquiring subsurface data in a given area is core drilling. The diamond-bit core drill permits the capture of continuous rock specimens in sequence drilled from the surface.

Exploratory holes drilled in the ground are usually surveyed by electrical equipment (producing electrical logs) that will aid in the correlation of the various strata. For example, a common log, the self-potential–resistivity log (Figure 6.2), allows a differentiation to be made between sands and shales occurring at various depths. Electrical logs obtained from wells at different locations may be used to correlate several strata of water sand, oil sand, gas sand, and shale across a wide area (Figure 6.2).

In geophysical prospecting, one of the more recent and successful methods for determining subsurface structural features is seismic exploration. The data on reflected waves are analysed to separate the random noise and leave a coherent signal which is indicative of subsurface structure, revealing such anomalies as anticlines, faults and traps.

Drilling

The first test hole (wildcat well) drilled in a given area may find gas (or oil); the risk is then reduced and any subsequent wells are called development wells. Once a drill site is selected, the rig and equipment must be assembled. A modern drilling rig stands 100 ft (30 m) tall and can be assembled in two or three pieces. The power is provided by diesel, gasoline or electric engines.

Conventional drilling rigs are classified as cable tool and rotary.

Cable tool drilling

The cable tool, first used in China (spring-pole drilling, *ca.* 259 BC), is based upon the principle of impacting the earth repeatedly with a heavy guillotine-like tool hanging from the end of a steel cable. As the hole progresses due to the crushing action of the tool (called a trepan, the same term as used in medicine for the instrument employed to drill through the skull!), the hole is filled with mud. The mud is bailed out, the bailer being a cylindrical bucket designed to remove the cuttings and well-bore fluids.

The cable tool drill exposes sediments to the atmosphere in an open-hole fashion. This method is usually not used in holes over 5000 ft (1500 m) deep.

Rotary drilling

The rotary drill (Figure 6.3) is the standard drilling rig. It is fast, efficient and economical and can reach depths up to about 30 000 ft (9150 m). The rig structure has a crown block at the top (crow's nest) and a pulley that connects to the travelling block and to a block-and-tackle arrangement designed to lift the mud hose and swivel block.

The various drill pipes are mechanically joined across drill collars and the last pipe has the drill bit attached to it. The drill bit consists of three rotating cones studded with sharp teeth made of tough alloy steel. The drill pipe assembly (drill string) is made of steel of high tensile strength and is usually 4 to 5 inches (10–125 cm) in diameter.

The total weight of all successive drill pipes extending 20 000 ft (6100 m) from the kelly (drill item) would be of the order of 400 000 lbs (179 000 kg).

Core drilling

Core drilling involves drilling through an area in the shape of concentric annuli and recovering a cylindrical core of the material drilled at the middle. Core drilling uses diamond-studded core drill bits and is much more expensive than other methods. Besides conventional core bits, retrievable wire-line core bits and side-wall cutting tools are occasionally used.

Air–gas drilling

In order to achieve a high drilling rate, low pressure is required at the bit. This is sometimes realized by aerating the column or circulating drilling mud. Lifting the mud out of the hole by high-pressure air or gas has been tried with limited success in areas where no water intrudes into the hole.

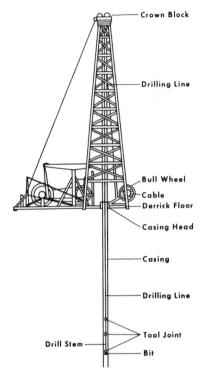

Figure 6.3 General illustration of a drilling rig

Directional drilling

Directional drilling is a special technique which permits the drilled hole to be slanted through a desired number of degrees. It is very specialized, expensive and sometimes hazardous. Some holes are known to be whipstocked almost 90°.

This technique permits the reaching of desired pay horizon locations which have inaccessible surface features, such as reaching the sides of salt domes likely to have oil reservoirs without having to drill through the massive salt body.

Well completion

Well completion involves the final finishing of drilled, cased and cemented holes.

Wells are completed by either having the casing set on top of the pay horizon (open hole) or being perforated by shaped charges after the producing formation has been cemented across.

Well stimulation

Well stimulation involves increasing the productivity of the well by fracturing and acidizing techniques. The initial cleaning of the wells involves mud–acid cleaning. The mud–acid is usually 15% hydrochloric and hydrofluoric acid containing corrosion inhibitors.

Drill-stem testing

Drill-stem testing is the term used to designate a special well-flow test run during drilling in order to evaluate productive potential or a designated producing interval. It involves the conduction and interpretation of both unsteady-state drawdown and buildup tests.

Recovery

Once the drilling operations have conclusively established that a reservoir exists and the resource is sufficient for economical recovery, a production and processing system is constructed.

A 'typical' gas production and processing system (Figure 6.4) may be considered but site specificity (i.e. specific geological and other constraints that might be applicable to any given site) may dictate the need for an 'atypical' gas production and processing system. Nevertheless, the system, like a petroleum refinery (Speight, 1991), is an integrated collection of different units that will ultimately produce a purified gas product.

The manner in which the gas is produced from the reservoir depends upon the properties of the reservoir rock and whether or not the gas is associated with petroleum in the reservoir. In general terms, the gas in the reservoir rock will migrate to the producing well because of the pressure differential between the reservoir and the well (Figure 6.5). The rate at which the gas migrates depends upon the permeability of the reservoir rock; low-permeability rocks may be fractured either by explosives or by hydraulic methods to yield better access to the well.

Production can continue as long as there is adequate pressure within the reservoir to produce the gas. The reservoir pressure will usually decrease as the natural gas is extracted and an additional method of recovery may have to be employed. For example, water may be injected into the reservoir to displace the gas from the pores of the reservoir rock (Figure 6.6). Such an operation helps to maintain reservoir pressure during the recovery operation and will improve the recoverability of the natural gas.

Once the gas is produced from a well, it is passed through a multiphase separator (Lockhart *et al.*, 1987), mixed with the gas from other nearby wells and processed to obtain the residue gas that will be sold to the consumer.

There are other options for the gas; for example, when gas occurs in association with oil, it may be reinjected into the reservoir to maintain pressure for the optimal recovery of the petroleum. On the other hand, the gas may be reinjected into the reservoir for storage if market demand is low, or pipeline facilities for transport are not available.

Generally, the gas is gathered at low pressure and must be compressed to

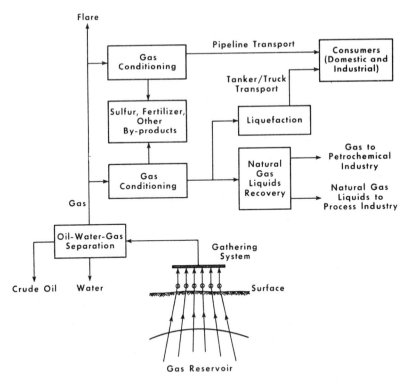

Figure 6.4 Simple illustration of a gas production, collection and processing system

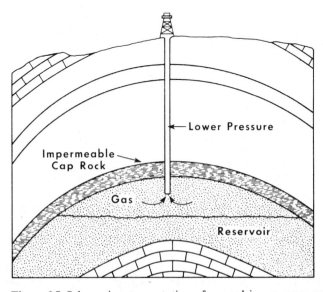

Figure 6.5 Schematic representation of a gas-drive recovery system

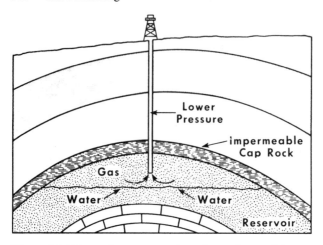

Figure 6.6 Schematic representation of a gas/water-drive recovery system

the processing pressure which is usually the residue gas sales pressure (Figure 6.7). If the distance between the wells and the gas-collecting system is short, it is often an advantage to locate compressors at the plant where they can be attended for more efficient operation. For greater distances, field compressors, designed for unattended operation, should be used.

For a true gas field rather than where the gas is produced from an oil field, the gathering lines are usually of a size that is suitable to deliver the gas at utilization pressure. However, as the well-head pressure declines, compression is necessary to maintain the flow of gas. In the later stages of the operation of a field when the production volume of the natural gas is on the decline and gathering lines may be oversized, it may be possible to bypass the field compressors.

Hydrate formation in gas-gathering lines (see also Chapters 7, 9 and 10) can be a serious problem, particularly if high pressures, low temperatures and long gathering line distances are involved. The gas hydrates, which have a crystalline (clathrate) structure and can form at temperatures above the freezing point of water, will tend to deposit and plug the lines and valves.

To combat hydrate formation in the gathering lines, it may be necessary to implement any one of the following procedures: (a) use of temperature and pressure regime at which hydrates cannot form by, for example, the use of well-head gas heaters to heat the gas entering the gathering system; (b) removal of water from the gas at the source before it enters the gas-gathering line; (c) injection of hydrate inhibitors (such as methanol or ethylene glycol) into the gathering lines; (d) reinjection of gas back into the formation from which it is produced or injection of additional gas from other formations into the formation (Kennedy, 1972).

Such techniques will ensure maximum recovery by maintaining the field pressure and/or preserving the gas that cannot be sold because an outlet is not available since venting may not be permitted owing to local environmental regulations (Moran *et al.*, 1986). Injection pressures of 3000–4000 psi (20 700–27 600 kPa) are often employed for either of the above situations.

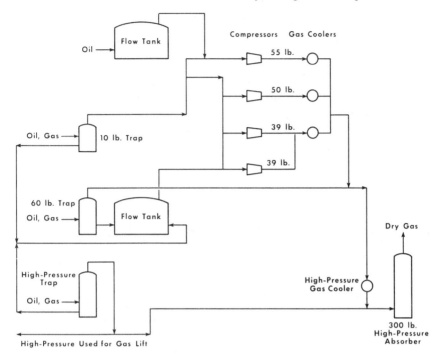

Figure 6.7 Schematic representation of a gas-gathering system

Storage

Gas storage allows a bridge to be made between constant production and the variable demand of markets, which are subject to the weather, for engineering and economic advantage (Figure 6.8).

The storage of natural and industrial gases is a major aspect of gas marketing and can involve above- or below-ground systems. The former is more suited to industrial gases whilst the latter is more suited to natural gas. Having a supply on hand and the option to store the product when market demand is low are major contributions to the economics of gas production.

Surface storage

There are three types of gas-holder used for surface storage, namely: (a) wet, (b) dry, and (c) liquefied.

Wet gas-holders

The wet gas-holder is essentially an inverted metal dome which floats in a water tank, with inlet and outlet gas connectors (Figure 6.9). Such simple gas-holders, usually for industrial (coal) gas, are made in small sizes (up to a capacity of about $18\,000\,\text{ft}^3$, $500\,\text{m}^3$). The pressure on the gas is controlled by adjusting the balancing weights.

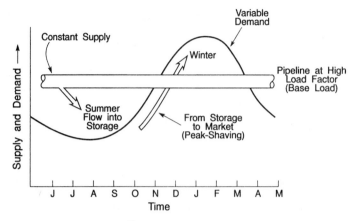

Figure 6.8 Illustration of how underground storage can help meet market demand

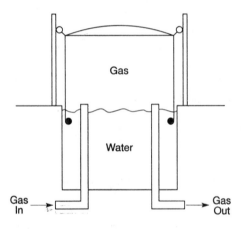

Figure 6.9 Cross-section of a simple water-sealed gas-holder

On the other hand, larger gas storage holders (able to hold up to $10\,000\,000\,ft^3$, $300\,000\,m^3$, of gas) are telescopic. The connections between each section (the 'lift', often four to six in number) are water sealed. Steel columns, framework and guide wheels are provided to ensure vertical movement and stability. In some holders, spiral guides are provided to each lift; the guides rotate during movement and are self-aligning.

Traces of hydrogen sulphide and other sulphur compounds present in small quantities in the gas can cause corrosion of holders but can be prevented by adding a suspension of zinc oxide which prompts the formation of a precipitate of zinc sulphides.

Natural gas is usually distributed at a low dewpoint and wet gas-holders find little use in the storage and distribution system. Wet gas-holders are more often than not seen in use for the various industrial gases, such as coal gas.

Dry gas-holders

Gases can be stored dry in a number of ways. At atmospheric pressure (or near to atmospheric pressure) the gases are contained in piston-type vertical gas-holders. These holders consist of a vertical steel-plate cylindrical tank with a moving piston floating on the stored gas.

It is, however, much cheaper to store gas at high pressures – particularly if it is being piped to the site at high pressure. The large pipelines used for gas transportation serve as reservoirs of high-pressure gas. The use of cylindrical or spherical pressure vessels provides a further addition if necesary.

Liquefied storage

Another method for the storage of gas is to liquefy it (Figure 6.10) and to place the liquid in storage containers.

Gases delivered as liquids at high pressure and atmospheric temperature are usually kept under pressure and stored in large, spherical, above-ground vessels called Hortonspheres. These are a common sight at petroleum refineries where products such as butane and pentane are usually stored this way.

If the gas is delivered as a low-temperature atmospheric-pressure liquid, it is usually kept at low temperatures for storage. However, this precludes long-distance transport by pipeline to the store owing to the heat gain in the pipes. With this type of storage system, the gas is usually kept by allowing more of it to evaporate and to extract latent heat.

Transportation of gas in liquid form by tanker is generally economical and it is easier to store the gas subsequently if it is maintained as a liquid.

In all instances, special care must be taken to ensure that the containers are safe and free from leaks (Considine, 1977).

Underground storage

Underground storage facilities are usually classified as market or field storage; the former is located near to major consuming (i.e. usually urban) areas where the variable demand resulting from a variety of seasonal (weather) fluctuations may need to be serviced by a combination of pipeline gas and storage gas. In field storage situations, it is the variable supply to the major market pipelines that is supplemented by the availability of storage gas.

Reservoirs (i.e. anticlinal systems, faults, stratigraphic lenses and salt domes; Figure 6.11) are essentially systems for gas (and crude oil) storage until they are 'tapped' and production begins (Speight, 1990, 1991). Thus, it is not surprising that they can also be considered to be systems that offer storage capacities and capabilities to the gas industry. In considering depleted gas reservoirs for prospective underground storage (Tek, 1987; Soni, 1988), the usual order of preference is dry gas fields, depleted condensate fields and depleted oil fields. The depleted dry gas field is the best prospect because it is closest to the natural environment for dry pipeline-quality storage gas. The parameters to be considered for prospective underground storage locations are size, location and access to major pipeline supplies.

Other underground storage areas employ salt domes and aquifers. Salt domes are found throughout the world and generally give the appearance of a

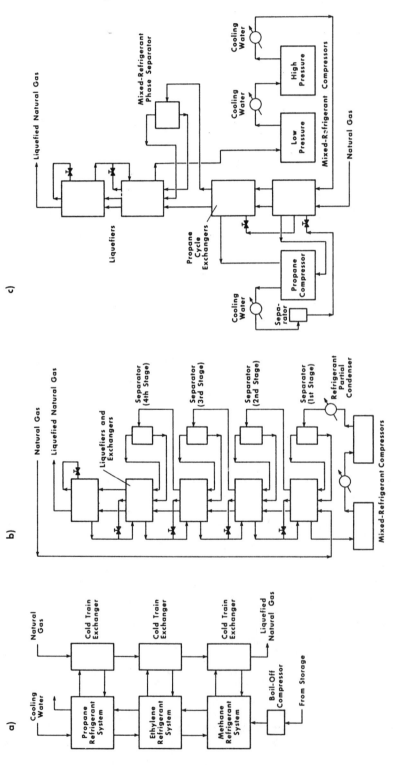

Figure 6.10 Schematic representation of gas liquefaction systems: (a) cascade system; (b) single-pressure system; (c) propane system

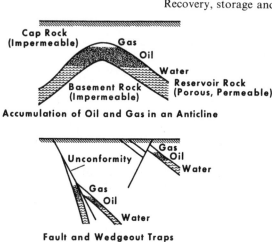

Accumulation of Oil and Gas in an Anticline

Fault and Wedgeout Traps

Anticline, Wedgeout, and Lens Traps

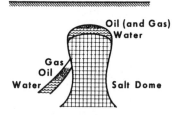

Traps Associated with a Salt (Mainly Sodium Chloride) Dome

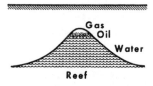

Oil (and Gas) Entrapment in a Limestone Reef

Figure 6.11 Schematic representation of different reservoir types

placement plug in which the adjacent strata are intruded by the salt and may be ruptured and sheared (Pettijohn, 1957; Fairbridge, 1972; Hamblin, 1975). In many instances, salt recovery (usually by water injection, collection of the brine, and evaporation to yield the salt) leaves cavities in the earth which are suitable for the storage of natural gas through a series of injection wells.

An aquifer is a lithologic unit (or even a combination of such units) which is porous and has a greater water transmissability than neighbouring lithologic units (Fairbridge, 1972). The aquifer is, therefore, capable of storing and transmitting water.

For storage purposes (Figure 6.12), the gas is introduced (by means of wells) under pressure into the aquifer and, as the pressure increases, the gas displaces the water and commences to fill the pore spaces of the aquifer. When high-pressure aquifers are used for gas storage, there is the possibility of an opportunity to recover some of the indigenous dissolved gas.

As already noted (Chapter 4), natural gas is also found in unconventional reservoirs such as tight sands and shales, as well as in geopressured reservoirs and coal seams. These unconventional reservoirs are unlikely to provide an opportunity for effective underground storage in the foreseeable future.

To commence the storage operation, the gas is pumped into the old wells by means of compressors similar to those employed to move gas throughout the pipeline system (Kumar, 1987). Natural gas is usually stored under the same

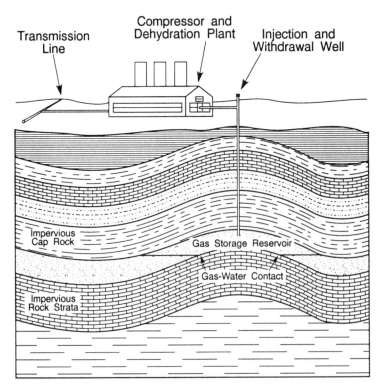

Figure 6.12 Illustration of gas storage in an aquifer

pressure conditions that originally existed in the field; such a protocol preserves the integrity of the formations that make up the field and reduces the potential for gas loss due to formation disturbance that may be induced by increased pressures.

Transportation

The means by which gas will be transported depends upon several factors: (a) the physical state of the gas to be transported, i.e. whether it is in the gaseous or liquid phase; (b) the distance over which the gas will be moved; (c) features such as the geological and geographical characteristics of the terrain – including land and sea operations; (d) the complexity of the distribution systems; and (e) any environmental regulations that are directly associated with the mode of transportation.

In the last case, the possibility of pipeline rupture as well as the effect of the pipeline itself on the ecosystems are two such factors that need to be addressed. In general and apart from any economic factors, it is possible to construct, and put in place, a system for gas transportation in the gaseous or liquid phase that allows system flexibility.

Pipelines

Pipeline transportation (Kumar and Chilingarian, 1987) is a common means of gas transportation and there are many pipeline systems throughout the world and the United States (Considine, 1977; Francis and Peters, 1980) where there are environmental and legal standards to be met (Moran et al., 1986). Construction of a pipeline system involves not only environmental and legal considerations but also compliance with the permitting regulations of the local, state and/or federal authorities.

In general, many of the available pipeline systems use pipe material up to 36 inches (0.9 m) in diameter (although lately larger-diameter pipe, up to 54 inches (1.4 m) in diameter, has become more favourable) and sections of pipe may be up to 40 feet (12.2 m) long. Protective coatings are usually applied to the pipe to prevent corrosion from outside influences.

The gas pressure in long-distance pipelines is usually up to 1500 psi (10 300 kPa) but can vary up to 5000 psi (34 500 kPa). To complete the pipeline, it is necessary to install a variety of valves/regulators that can be opened or closed to adjust the flow of gas and also to shut down a section of the system where an unexpected rupture may be caused by natural events (such as the weather) or even by unnatural events (such as sabotage). Most of the valves/regulators in the pipeline system can be operated nowadays by remote control so that, in the event of a rupture, the system can be closed down. This is especially valuable where it may take a repair crew considerable time to reach the site of the breakdown.

The trend in recent years has been to expand the pipeline system into marine environments where the pipeline is actually under a body of water. This has arisen mainly because of the tendency for petroleum and natural gas companies to expand their exploration programmes to the sea. Lines are now laid in marine locations where depths exceed 500 feet (152 m) and cover distances of several hundred miles (1 mile = 1.61 kilometres) to the shore.

Tankers

Just as petroleum can be transported by sea-going tankers, so can natural gas be transported by sea-going vessels. The gas is transported either under pressure at ambient temperatures (e.g. propane and butanes) or at atmospheric pressure but with the cargo under refrigeration (e.g. LPG). For safety reasons, petroleum tankers are constructed with several independent tanks so that the rupture of any one tank will not necessarily drain the whole ship – unless it is a severe bow-to-stern (or stern-to-bow) rupture. Similarly, natural gas tankers also contain several separate tanks (Figure 6.13).

On occasion, usually for material such as LPG(s), road and/or rail transportation may be employed to move the product to market (Considine, 1977). The road and/or rail transportation of coal gas is less likely; such products will probably be used locally through the agency of gas transmission lines from the producer to the consumer.

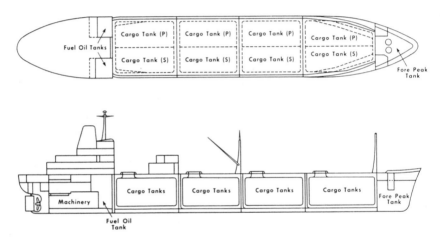

Figure 6.13 Illustration of a sea-going tanker equipped for gas transportation

References

Considine, D.M. (1977) *Energy Technology Handbook*, McGraw-Hill, New York

Fairbridge, R.W. (1972) *The Encyclopedia of Geochemistry and Environmental Sciences*. Encyclopedia of Earth Sciences Series, Vol. IVA, Dowden, Hutchinson and Ross, Stroudsberg, PA

Francis, W. and Peters, M.C. (1980) *Fuels and Fuel Technology*, Pergamon, New York

Hamblin, W.K. (1975) *The Earth's Dynamic Systems*, Burgess, Minneapolis, MN

Haun, J.D. and LeRoy, L.W. (eds) (1958) *Subsurface Geology in Petroleum Exploration*, Colorado School of Mines, Golden, CO

Kennedy, J.L. (1972) *Oil and Gas Journal*, **70**, (9), 69

Kumar, S. (1987) *Gas Production Engineering*, Gulf Publishing Company, Houston, TX

Kumar, S. and Chilingarian, G.V. (1987) In *Surface Operations in Petroleum Production*, Elsevier, New York, Chapter 8 and references cited therein

Lockhart, F.J., Chiligarian, G.V. and Kumar, S. (1987) In *Surface Operations in Petroleum Production*, Elsevier, New York, Chapter 3.

Moran, J.M., Morgan, M.D. and Wiersma, J.H. (1986) *Introduction to Environmental Science*, Freeman, New York

Pettijohn, F.J. (1957) *Sedimentary Rocks*, Harper, New York

Soni, A.K. (1988) *Indian Mining Engineering Journal*, **27**, (9), 9

Speight, J.G. (1990) *Fuel Science and Technology Handbook*, Marcel Dekker, New York

Speight, J.G. (1991) *The Chemistry and Technology of Petroleum*, 2nd edn, Marcel Dekker, New York

Tek, M.R. (1987) *Underground Storage of Natural Gas*, Vol. 3, Gulf Publishing Company, Houston, TX

Composition and properties

Natural gas

Natural gas is a complex mixture of essentially straight-chain paraffinic hydrocarbons and certain impurities. The constituents of natural gas are usually methane, ethane, propane, butanes and pentanes, plus smaller amounts of hexanes, heptanes, octanes and even higher-molecular-weight hydrocarbons.

The properties of these materials (Reid *et al.*, 1977) are very worthy of study and, on this note, there are standard tests that can be employed to determine the properties and character of natural gas and, to a lesser extent because of their composition, manufactured gases. However, other methods are available that can be applied to manufactured gases (Francis and Peters, 1980).

The tests acknowledged in this text are those tests that are accepted by the American Society for Testing and Materials (ASTM), the British Standards Institution (BSI), the Institute of Petroleum (IP) and the International Organization for Standardization (ISO). The tests given here are referenced by their number. Some of them may have been withdrawn but they are, nevertheless, worthy of note not only because of historical interest but because they may still find use in some laboratories.

In addition to the main components mentioned above, there may, on occasion, be carbon dioxide, hydrogen sulphide, nitrogen and water vapour. Small quantities of aromatic compounds (e.g. benzene derivatives) and helium or argon have also been known to occur in natural gas (Tables 7.1 and 7.2).

Natural gas contains constituents other than methane (Hertweck and Fox, 1984; Moore and Hamack, 1985) and a knowledge of the occurrence and concentration of these constituents is a necessary prerequisite to any step in the processing sequence (Nonhebel, 1964; Lowenheim and Moran, 1975; Kohl and Riesenfeld, 1979; Trusell, 1985; Willis, 1986). A knowledge of the terminology that is recognized for application to gaseous fuels (ASTM D 4150) is also of value.

For example, many wells will produce gas that contains hydrogen sulphide and/or low-molecular-weight thiols (mercaptans; R-SH). The hydrogen sulphide concentration in natural gas will usually vary from barely detectable quantities to more than 0.30% (3000 ppm). Other sulphur derivatives are not usually present in significant quantities and may occur in trace amounts (ASTM D 1266; IP 272; ISO 2192). Thus, a sulphur removal process must be very precise since natural gas contains only a small quantity of sulphur-

Table 7.1 General composition of associated natural gas

Category	Component	Amount (%)
Paraffinic	Methane (CH_4)	70–98
	Ethane (C_2H_6)	1–10
	Propane (C_3H_8)	Trace – 5
	Butane (C_4H_{10})	Trace – 2
	Pentane (C_5H_{12})	Trace – 1
	Hexane (C_6H_{14})	Trace – 0.5
	Heptane and higher (C_7+)	None–trace
Cyclic	Cyclohexane (C_6H_{12})	Traces
Aromatic	Benzene (C_6H_6), others	Traces
Non-hydrocarbon	Nitrogen (N_2)	Trace – 15
	Carbon dioxide (CO_2)	Trace – 1
	Hydrogen sulphide (H_2S)	Trace occasionally
	Helium (He)	Trace – 5
	Other sulphur and nitrogen compounds	Trace occasionally
	Water (H_2O)	Trace – 5

containing compounds that must be reduced by several orders of magnitude – most consumers of natural gas require less than 4 ppm in the gas.

A characteristic feature of natural gas that contains hydrogen sulphide is the presence of carbon dioxide (generally in the range of 1 to 4% v/v). In cases where the natural gas goes not contain hydrogen sulphide, there may also be a relative lack of carbon dioxide.

Specifications for natural gas for the consumer ('sales gas') include one or more of the following: water content, hydrocarbon content, heating value, specific gravity (this term has recently been replaced by the term 'relative density'), acid gas content, temperature and pressure. As with any property measurement, the value of any specification depends on the availability of reliable test methods to determine the specific property as well as on the instrumentation (Sharples and Panhill, 1985; Kumar, 1987; ASTM; Yon, 1988).

Process gases

The composition of natural gas may be generally (with a degree of tongue-in-cheek speculation and at the risk of immediately being contradicted!) considered to be consistent in so far as the major constituents are hydrocarbons, particularly methane (ASTM D 3956). And perhaps in a very general manner, this is not an unreasonable assumption for the general purposes of this chapter. Any variations in composition may be assumed to be due to the inclusion of non-hydrocarbon materials in the gaseous mix. However, the same general (tongue-in-cheek) reasoning cannot be applied to process gases.

The composition of process gases is very dependent upon the process, be it coal gas manufacture, gases from coal carbonization or gases from coal combustion, and numerous process gases are produced by the petroleum and petrochemical industries as well as a host of other manufacturing industries (Tables 7.3–7.5).

Table 7.2 Composition and properties of selected natural gases

	Examples of untreated gases			Examples of some of the major natural gases as marketed				
	Non-associated gas		'Associated gas'	UK	Holland	USA		
	of relatively older period	of geologically recent period						
	1 France, Lacq	*2* Salt Lake, Utah	*3* Persian Gulf area, Aga Jari	*4* North Sea, Leman Bank	*5* Groningen	*6* Monroe, Louisiana	*7* Amarillo, Texas	*8* Ashland, Kentucky
Composition (% v/v):								
Methane	69.1	96.0	66.0	94.7	81.2	94.7	72.9	75.0
Ethane	2.8	–	14.0	3.0	2.9	2.8	19.0	24.0
Propanes	0.8	–	10.5	0.5	0.4	–	–	–
Butanes	1.5	–	5.0	0.2	0.1	–	–	–
C_5^+	0.6	–	2.0	0.2	0.1	–	–	–
Hydrogen sulphide	15.4	–	–	–	–	–	–	–
Carbon dioxide	9.7	3.7	1.5	0.1	0.9	0.2	0.4	–
Nitrogen	–	0.3	1.0	1.3	14.4	2.3	7.7	1.0
Helium	–	–	–	<0.1	<0.1	–	–	–
Argon	–	–	–	–	–	–	–	–
Spec. grav. (dry gas/dry air)	0.75	0.59	0.87	0.59	0.64	0.58	0.68	0.67
Calorific value, gross								
Btu/ft³ (s.t.p. dry)	910	952	1403	1037	843	988	1054	1161
MJ/m³ (s.t.)	33.9	35.4	52.3	38.6	31.4	36.8	39.2	43.2

Source: Francis and Peters (1980).

Table 7.3 Composition and heating value of various fuel gases

Fuel gas	Natural gas (mid-continent)	Natural gas (Pennsylvania)	Coke oven gas	Blue water gas	Carburetted water gas	Bituminous producer gas	Fuel oil gas
Carbon monoxide	–	–	6.3	42.8	33.4	27.0	46.1
Carbon dioxide	0.8	–	1.8	3.0	3.9	4.5	4.3
Hydrogen	–	–	53.0	49.9	34.6	14.0	48.0
Nitrogen	3.2	1.1	3.4	3.3	7.9	50.9	0.3
Oxygen	–	–	0.2	0.5	0.9	0.6	
Methane	96.0	67.6	31.6	0.5	10.4	3.0	0.4
Ethane	–	31.3	–				
Illuminants	–	–	3.7	–	8.9		
Gross MJ/m^3	36	46	21	11	20	5.5	

Table 7.4 Properties of various coal-derived gases

	Coke oven gas	Producer gas	Water gas	Carburetted water gas	Synthetic coal gas
Reactant system	Pyrolysis	Air plus steam	Steam (cyclic air)	Steam (cyclic air)	Oxygen plus steam at pressure
Analysis (vol %)[a]					
Carbon monoxide, CO	6.8	27.0	42.8	33.4	15.8
Hydrogen, H$_2$	47.3	14.0	49.9	34.6	40.6
Methane, CH$_4$	33.9	3.0	0.5	10.4	10.9
Carbon dioxide, CO$_2$	2.2	4.5	3.0	3.9	31.3
Nitrogen, N$_2$	6.0	50.9	3.3	7.9	
Other[b]	3.8	0.5	0.5	9.8	2.4
Fuel value (MJ/m^3, Btu/ft^3)	22.0	5.6	11.5	20.0	10.5

[a] Analyses and fuel values vary with the type of coal and operating conditions.
[b] Other contents include hydrocarbon gases other than methane, hydrogen sulphide and small amounts of other impurities.
Source: Francis and Peters (1980).

Table 7.5 Composition and properties of various refinery gases

	Refinery gases			Commercial propane		Example of commercial butane
	Low CV	Inter-mediate CV	High CV	For enriching	For gas making	
Analysis (mol %):						
CH_4	12.0	21.4	17.2	nil	nil	0.1
C_2H_6	13.0	18.6	17.8	1.5	nil	7.2
C_2H_4	nil	2.5	4.0	nil	nil	nil
C_3H_8	13.0	16.0	20.7	45.0	92.5	0.5
C_3H_6	2.0	5.4	14.9	52.0	5.0	4.2
C_4H_{10}	2.0	12.4	21.9	1.5	2.5	87.0
C_4H_8	1.0	nil	nil			1.0
CO_2	nil	0.1	0.3	nil	nil	hil
H_2	56.0	21.8	0.2	nil	nil	nil
N_2	1.0	1.8	3.0	nil	nil	nil
Total	100.0	100.0	100.0	100.0	100.0	100.0
Gross CV						
(Btu/ft³)	1000	1620	2181	2386	2476	3052
MJ/m³	37.3	60.4	81.3	88.9	92.3	113.7
Spec. grav.						
(air = 1)	0.54	0.96	1.35	1.47	1.53	1.96
Characteristics as liquid:						
Spec. grav.						
(H_2O = 1)	–	–	–	0.52	0.51	0.58
Gross CV						
(Btu/lb)	–	–	–	21 440	21 250	21 350
MJ/kg	–	–	–	49.9	49.4	49.7

CV = calorific value.

But no matter what the source of the gas, there is always the need to determine its constituents before processing so that the most efficient and economical methods may be employed for gas-cleaning operations.

In summary, and at the risk of some repetition, the gas streams produced by various processes from coal (see Chapter 5) are the gaseous mixtures produced when coal is heated under a variety of conditions (Brame and King, 1955). The processes consist essentially of heating coal to drive off the volatile products, some of which may be gases and others liquids and tars, to leave a solid carbonaceous residue. The residue/char is then treated under a variety of conditions to produce other fuels which vary from a 'purified' char to different types of gaseous mixtures:

$$coal + heat + (?) = char + liquid + gas$$

or

$$[C]_{coal} + heat + (?) = [C]_{char} + C_nH_{2n+2} + CO + CO_2 + H_2$$

The different processes by which these gaseous mixtures are produced are much more complex than the relatively simple chemical equations would indicate. Moreover, the presence of nitrogen, sulphur and mineral matter in the coal, as well as the complex nature of the thermal degradation process, dictates the production of a gaseous mixture that is by no means pure and may even need an adjustment of the relative amounts of the different constituents before further use:

$$[N]_{coal} = NO_x + HCN, \text{ etc., where } x = 1 \text{ or } 2$$
$$[O]_{coal} = CO_x, \text{ etc., where } x = 1 \text{ or } 2$$
$$[S]_{coal} = SO_x + H_2S, \text{ etc., where } x = 1 \text{ or } 2$$

For example, in direct combustion coal is burned (i.e. the carbon and hydrogen in the coal are oxidized into carbon dioxide and water) to convert the chemical energy in the coal into thermal energy:

$$[C]_{coal} + O_2 = CO_2$$
$$[2H]_{coal} + O_2 = H_2O$$

On a more formal basis, the combustion of coal may be simply represented as the staged oxidation of coal carbon to carbon dioxide

$$[C]_{coal} + O_2 = 2CO$$
$$2CO + O_2 = 2CO_2$$

with any reactions of the hydrogen in the coal being considered to be of secondary importance. But the conversion of nitrogen and sulphur to their respective oxides during combustion is a major issue that is often ignored in simple chemical explanations of coal combustion:

$$[S]_{coal} + O_2 = SO_2$$
$$2SO_2 + O_2 = 2SO_3$$
$$[2N]_{coal} + O_2 = 2NO$$
$$2NO + O_2 = 2NO_2$$
$$[N]_{coal} + O_2 = NO_2$$

Coal carbonization is essentially a process for the production of a high-carbon-containing residue by thermal decomposition (with simultaneous removal of distillate) of organic substances:

$$[C]_{organic} = [C]_{coke/char/carbon} + \text{liquids} + \text{gases}$$

The chemistry might be represented simply (but with the same degree of trepidation) as for coal combustion, recognizing that a large proportion of the heteroatoms (oxygen, nitrogen and sulphur) will also occur in the gaseous product mix.

Coal gasification involves the thermal decomposition of coal and the reaction of the carbon in the coal and other pyrolysis products with oxygen, water and hydrogen to produce fuel gases such as methane

$$[C]_{coal} + [H]_{coal} = CH_4$$

or with added hydrogen

$$[C]_{coal} + 2H_2 = CH_4$$

If a high-Btu gas (900–1000 Btu/ft^3, 33.5–37.3 × 10^3 kJ/m^3) is required, efforts must be made to increase the methane content of the gas:

$$CO + H_2O = CO_2 + H_2$$
$$CO + 3H_2 = CH_4 + H_2O$$
$$2CO + 2H_2 = CH_4 + CO_2$$
$$CO + 4H_2 = CH_4 + 2H_2O$$

On the other hand, refining petroleum is a somewhat different scenario to coal conversion and refinery gas streams often contain substantial amounts of acid gases such as hydrogen sulphide and carbon dioxide (Chapter 5), but more particularly hydrogen sulphide which arises from the hydrodesulphurization of feedstocks which contain organic sulphur (Bland and Davidson, 1967; Speight, 1981, 1991):

$$[S]_{feedstock} + H_2 = H_2S + hydrocarbons$$

Acid gases corrode refining equipment, harm catalysts, pollute the atmosphere and prevent the use of hydrocarbon components in petrochemical manufacture. When the amount of hydrogen sulphide is large, it may be removed from a gas stream and converted to sulphur or sulphuric acid. Some natural gases contain sufficient carbon dioxide to warrant its recovery as dry ice (Bartoo, 1985).

Other industrial gas streams will vary depending upon the nature of the industry and the starting materials. But, be that as it may, there will always be the need to clean up the effluent gas before the remotest chance of being allowed to discharge the gas to the atmosphere becomes a reality.

The properties of the constituents of gases (Tables 7.6–7.8) give some indication of what might be expected when the individual gases occur in a mixture, assuming that the properties are additive and there is no 'interference' between the constituents as they occur in the mixture.

However, knowing the properties of the individual constituents is not the only answer to understanding the nature of gaseous products. A series of standard (ASTM) and non-standard tests have been developed that can be applied to gaseous products. It is the purpose of this chapter to summarize these tests briefly as an aid to understanding the nature of gaseous products.

Gas analysis and testing

The majority of the standard methods for the analysis and testing of gases were predominantly formulated and developed for testing natural gas or liquefied petroleum gases and may be somewhat less applicable for industrial gases because of the nature and intent of the tests (ASTM, 1963; Rawlinson and Ward, 1973; Trusell, 1985; Willis, 1986). Whereas it may be of some consequence to test for and determine the amount of acid gases in natural gas, the test may have less significance when applied to an industrial gas that consists mainly of acid gases with only minor amounts of hydrocarbon material. Fortunately, there are a variety of other test methods that have been developed over the decades which, while being only infrequently applied to natural gas, are often applied to industrial gases (Johnson, 1983).

Table 7.6 Selected properties of hydrocarbon gases

	Methane	Ethane	Propane	Isobutane	Butane	Pentane
Molecular volume of gas (cu. ft)	378.7	375.8	372.7	366.7	365.4	
Molecular weight of gas	16.04	30.07	44.09	58.12	58.12	72.15
Gal/lb mole at 60°F	6.4	9.64	10.41	12.38	11.94	13.71
Weight:						
% carbon	74.88	79.88	81.72	82.66		
% hydrogen	25.12	20.12	18.28	17.34	17.34	
Specific gravity:						
Of liquid (water=1)	0.248	0.377	0.508	0.563	0.584	0.631
Of liquid. API	340	247	147	120	111	93
Of gas (air=1)	0.555	1.048	1.550	2.077	2.084	2.490
Weights and volumes:						
Lb/gal liquid	2.5	3.145	4.235	4.694	4.873	5.250
Cu. ft gas/gal liquid	59.0	39.69	36.28	30.65	31.46	27.67
Cu. ft gas/lb liquid	24.8	12.50	8.55	6.50		
Ratio, gas volume to liquid volume	443	293.4	272.7	229.3	237.8	207.0
Initial boiling point (atmospheric pressure)	−259	−128.2	−43.7	10.9	31.1	97
Heat value (gross):						
Btu/cu. ft gas	1012	1786	2522	3163	3261	4023
Btu/lb liquid	23885	22323	21560	20732	21180	21110
Btu/gal liquid		70210	91500	103750	102600	110800
Vapour pressure (lb/sq. in abs):						
At −44°F		88	0	−9	−12	−14
At 0°F		206	38	12	−7	−13
At 33°F		343	54	17	0	−11
At 70°F		563	124	45	31	−6
At 90°F		710	165	62	44	
At 100°F			189	72	52	4
At 130°F			275	110	81	11
At 150°F			346	138	87	21
Latent heat of vaporization at boiling point:						
Btu/lb	221	211	185	158	167	153
Btu/gal	553	664	785	742	808	802
Specific heat:						
Of liquid. at C_p and 60°F, Btu/(lb)(°F)		0.780	0.588	0.560	0.549	
Of gas, at C_p and 60°F, Btu/(lb)(°F)	0.526	0.413	0.390	0.406	0.396	0.402
Of gas, at C_r and 60°F, Btu/(lb)(°F)	0.402	0.347	0.346	0.373	0.363	0.376

Table 7.7 Properties of the constituents of natural gas and manufactured gases

	Mol. wt	Calc. spec. grav. air = 1	Weight + volume relationships		Calorific value (MJ/m³ at s.t.p. sat.) Gross	Combustion requirements (v/v)		Compustion products with theoretical air		
			m^3/kg^1 at s.t.p. dry	kg/m^3 at s.t.p. dry		O_2	Air	Total dry	Total wet	$\%CO_2$ in dry flue gases
Carbon monoxide. CO	28	0.966	0.797	1.250	12.5	0.5	2.38	2.88	2.88	34.7
Hydrogen. H_2	2	0.069	11.068	0.0896	12.6	0.5	2.38	1.88	2.88	nil
Ethylene. C_2H_4	28	0.967	0.797	1.252	61.4	3.0	14.28	13.28	15.28	15.05
Benzene. C_6H_6	78	2.694	0.286	3.486	147.2	7.5	35.70	34.20	37.20	17.55
Hydrogen sulphide	34	1.176	0.658	1.52	25.2	1.5	7.14	6.64	7.64	34.7 (SO_2)

Table 7.8 Heats of combustion of natural gas constituents

Substance	MJ/m^3	Btu/ft^3
Methane	37.3	1000
Ethane	65.7	1763
Propane	93.5	2510
Butane	121.1	3248
Pentane	140.0	3752

Qualitative identifications and quantitative determinations of gaseous substances are essential for the evaluation of air quality in the ambient air and in the industrial workplace. Any pollutant that becomes obvious either through human sensations or through instrument monitoring must be removed, usually after some method of measurement. But removal is often effected at the source, i.e. the process where the pollutant is created.

The conventional methods of gas analysis used in research, process control and medical technology were not adequate to meet the demands for techniques to analyse air contaminants, some of which are recognized as having adverse effects on health, welfare and property. The requirements for gas analyses of gaseous pollutants can be quite demanding, ranging from the qualitative identification of a few microlitres of an unknown gas to the quantitation measurement of a trace contaminant at the parts-per-billion (10^9) level, with new emphasis on the measurement of contaminants at the parts-per-trillion (10^{12}) level!

Qualitative identification

The qualitative identification of the components of gaseous mixtures, especially the constituents classified as pollutants, can be extremely complex. It is the purpose of this subsection to indicate how gases may be analysed and tested for a variety of constituents. Several methods have evolved over the decades and, in fact, further development of some of the methods has led to their use as on-line analytical techniques (Cowper and Wallis, 1984).

Procedures for the qualitative identification of the components of gas streams may require the use of several instruments to provide complementary information about composition and structure, since the entire sample (ASTM D 1145 for natural gas, ASTM D 1247 for manufactured gas, and ASTM D 1265 for liquefied petroleum gas; IP 181) (see also Caffey, 1985; Hefley et al., 1985) is often limited to milligram or microgram quantities. In fact, the classical identification methods, such as boiling-point and refractive index determinations, functional group tests, combustion analyses and derivative preparations, have been largely replaced by instrumental methods. Information for identification purposes is now often obtained from various instrumental techniques of which gas chromatography and mass spectroscopy are only two examples but will serve for the purposes of illustration.

Mass spectroscopy

Mass spectroscopy is an extremely useful technique for the qualitative identification of volatile compounds (ASTM D 2650) and has also found use in the identification of environmental contaminants.

Mass spectrometry yields information about molecular weight and the presence of other atoms, such as nitrogen, oxygen and halogens, within the molecule. In addition, the fragmentation pattern often provides a unique so-called fingerprint of a molecule, allowing positive identification. If the gas is a mixture, the interpretation of the mass spectral data is difficult since the fragmentation patterns may be superimposed.

However, interfacing a mass spectrometer to a gas chromatograph provides a solution to this problem. A mass spectrometer located at the end of a gas chromatographic column can be used to analyse each component separately as it leaves the column.

Gas chromatography

A gas chromatograph is essentially a highly efficient apparatus for separating a complex mixture into individual components.

When a mixture of components is injected into a chromatograph equipped with a column suitable for the task at hand, each of the components passes through the column at different rates and reaches the end of the column at different times. In addition to the column, a detector which is calibrated for the specific components can be used for the qualitative identification and quantitative analysis of the mixture (IP 337 and IP 345).

In summary, the gas chromatograph allows a complex mixture to appear as a series of pure components (Cox, 1985).

Quantitative analysis

Once the identification of the components of a gaseous mixture has been established, further work usually focuses on measuring the amounts of pollutant in the mixture. The methods that are used for quantification can be conveniently classified into direct and indirect procedures (ASTM D 1071).

Direct-reading instruments analyse their results in real time, but often take seconds or minutes to display the data. Operation in a continuous or semicontinuous mode is also possible.

Indirect methods involve the collection and storage of a sample for subsequent analysis.

Both direct and indirect methods have inherent advantages and disadvantages. For example, by using indirect methods, samples with several pollutants can be simultaneously collected from a number of different sites with relatively inexpensive collection devices and analysed later at a central laboratory. On the other hand, direct methods may require one instrument for each pollutant at each sampling site thereby increasing the cost of the analysis and identification. Alternatively, there may be no option but to install different instruments on process equipment for process-monitoring requirements.

Direct methods

These consist of methods utilizing colorimetric indicating devices and instrumental methods.

Gas chromatography

The gas chromatograph is widely used not only for the qualitative analysis of gaseous mixtures, but also for quantitative measurements both for the analysis of collected samples and as a semicontinuous direct-reading instrument. The availability of different detectors capable of measuring the effluent from the chromatographic column and the development of valves for automatic injection of samples and for directing sample flow have further extended the versatility of this instrument.

The technique finds use in determining the components of various gaseous mixtures (ASTM D 1717, ASTM D 4424, ASTM D 1945, ASTM D 1946 and ASTM D 2163; IP 194, IP 264, IP 337 and IP 345; Clement et al., 1986).

Colorimetric indicators

Three types of direct-reading colorimetric indicators have been utilized: liquid reagents, chemically treated papers, and glass tubes containing solid chemicals (detector tubes). The simplest of these methods is the detector tube which is constructed by filling a glass tube with silica gel coated with colour-forming chemicals. For use, the ends of the sealed tube are broken and a specific volume of air, typically $6\,in^3$ ($100\,cm^3$), is drawn through the tube at a controlled rate.

These tubes rely on the fact that the capacity of the packing to absorb a pollutant is limited and that, after saturation, successive molecules penetrate farther down the tube; hence the length of the stain is concentration dependent. Detector tubes often utilize the same colour-forming chemical to detect several different gases, and therefore may be non-specific for mixtures of these gases, so the degree of accuracy may leave much to be desired! Detector tubes for analysing approximately 200 different gases are commercially available.

The standard test (ASTM D 1826) for the calorific value of gases employs a continuous-recording calorimeter. The calorific value of gases can also be calculated (ASTM D 3588).

Direct-reading instruments

Direct-reading instruments capable of measuring gases directly in the parts-per-billion (10^9) range usually contain a sampling system, electronics for processing signals, a portable power supply, a display system and a detector.

The most sensitive, and commonly used, detectors are based on electrical or thermal conductivity, ultraviolet or infrared absorption, mass spectrometry, electron capture, flame ionization, flame photometry, heat of combustion and chemiluminescence. Many of these detectors respond to the presence of $10^{-9}\,g$ quantities, and even to $10^{-12}\,g$ levels. In addition to improved accuracy, precision and analysis time, another advantage is that most instruments produce an electrical signal which can be fed into a computer for process control, averaging and record keeping.

Chemiluminescence

The phenomenon of chemiluminescence (an emissive process which occurs when all or part of the energy of a chemical reaction is released as light rather than heat) is employed for the determination of levels of ozone, oxides of nitrogen and sulphur compounds. Ozone levels in the range from 1 to 100 parts per billion can be determined by measuring the emission at 585 nm which occurs when ozone is mixed with excess ethylene. Similarly, nitric oxide (NO) levels from 10 ppb to 5000 ppm can also be measured by a chemiluminescence method.

For example, the analysis for nitrogen dioxide can be performed by reducing the nitrogen dioxide (NO_2) to nitric oxide (NO) in a catalytic converter and measuring the nitric oxide emission. Similarly the amount of nitrogen dioxide in mixtures of nitrogen dioxide and nitric oxide can also be measured by chemiluminescence. The nitric oxide level of the mixture is first measured by chemiluminescence, the nitrogen dioxide is then reduced to nitric oxide, and the measurement of the nitric oxide level is repeated.

Indirect methods

For indirect methods, the main collection devices are freeze traps, bubblers, evacuated bulbs, plastic bags and solid sorbents. Solid sorbents, because of their convenience, generally dominate many of the collection procedures. This approach has been highly successful for several classes of chemical compounds.

Properties

Practical engineering problems involving the production or storage of gas often include calculations related to density, pressure–volume relationships, specific transport properties, and hydrates (see also Chapters 6, 9 and 10).

Density

Natural gas is a mixture of mostly methane (ASTM D 3956), some ethane (ASTM D 3984) and propane (ASTM D 4362), with smaller amounts of butanes (ASTM D 4650, ASTM D 4651), pentanes, hexanes and heavier (higher-molecular-weight) hydrocarbons.

Density (ASTM D 1070, ASTM D 3588; IP 59, IP 235) is an important property since, in underground storage and pipeline flow applications (Starling, 1985), a gas may often be identified by specific gravity.

The density of an ideal gas can be derived as

$$d = (MW) \, p/RT$$

where

d = density of the ideal gas, lb_m/ft^3
p = pressure of the gas, lb_f/ft^2
MW = molecular weight, $lb_m/lb\text{-mole}/ft \times lb_f$
R = gas constant, 1544 (ft × lb)/lb-mole °R abs.
T = absolute temperature, °R abs. = 460 + °F

Since the specific gravity of a gas is the ratio of the density of that particular gas to the density of the reference gas, air, at a given temperature and pressure, it' also follows that the specific gravity (ASTM D 3588) is the ratio of the molecular·weight of the gas to the molecular weight of air:

$$SG = (MW)/29$$

Therefore

$$d = 29(SG)(p/RT)$$

For a real gas

$$d = 29(SG)p/zRT$$

where z is the non-ideality factor, also called the compressibility factor.

Compressibility factor

For a pure component, z is a function of reduced temperature and pressure (Figure 7.1):

$$z = z(T_r P_r)$$
$$T_r = T/T_c$$
$$p_r = p/p_c$$

where T_c and P_c represent the critical temperature and pressure for natural gas (see also Starling and Kumar, 1985).

Gas hydrates

When free water is present, the natural gas–water mixtures will form solids, called hydrates, at high pressures and temperatures above the freezing point of water (see also Chapters 6, 9 and 10). Since water is almost invariably present

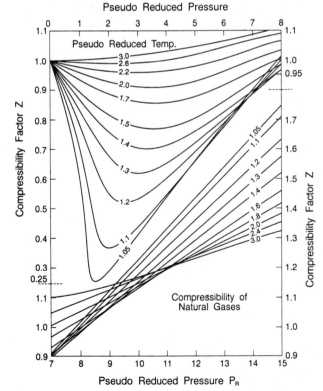

Figure 7.1 Compressibility factor for natural gas

in the production of hydrocarbons, water–hydrocarbon systems are of interest all the way from production, through gathering, processing and pipelining, to distribution of natural gas (Kohl and Riesenfeld, 1979). The formation of gas hydrates, which physically resemble ice or wet snow, is undesirable because they can cause plugging in transportation lines.

More scientifically, gas hydrates are solutions of gas in crystalline solids and it is important to realize that most hydrocarbons in natural gas form hydrates when sufficient free-liquid water is present. The general chemical formula assigned to hydrates is

$$C_nH_{2n+2}.mH_2O$$

For example

$$C_3H_8 + 17H_2O = C_3H_8.17H_2O$$

There exist several practical correlations directly giving the hydrate threshold loci on the temperature–pressure gradient (Figure 7.2) as well as for finding the quantity of methanol needed to provide a given lowering of the hydrate-forming temperature for a given gas (Figure 7.3). Since methanol is an economical agent for this task, the procedure for finding the methanol requirement uses the lowering of the freezing point of the hydrate vs. methanol concentration in water (Figure 7.3).

Water forms in the liquid phase when a gas saturated within the reservoir is cooled, as when coming up the well bore or flowing in the gathering lines. The additional of alcohol must be sufficient to provide an alcohol concentration in the water that will prevent hydrates from forming. To do this, there must be enough alcohol injected to provide for that alcohol which vaporizes into the

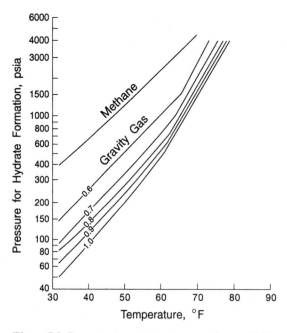

Figure 7.2 Pressure–temperature curves for predicting gas hydrate formation

gas phase, as well as that required to maintain a desired concentration in the liquid phase.

Since the partial pressure of methanol depends on the temperature of the aqueous solution and the methanol concentration, these variables are

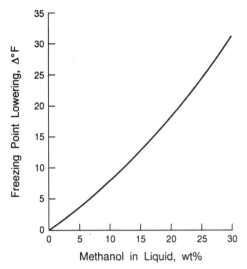

Figure 7.3 Gas hydrate freezing point depression in relation to the amount of methanol (wt/wt) percentage in the liquid

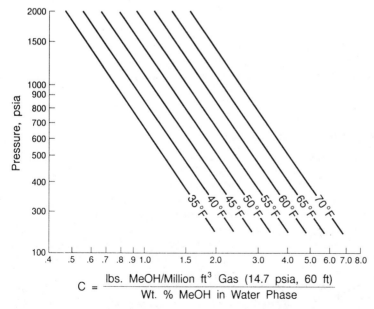

Figure 7.4 Ratio of methanol vapour to liquid composition versus pressure

Table 7.9 Physical data required to calculate the amount of methanol necessary to prevent hydrate formation

Step no.	Quantity	Units	Instructions	Example data and calculations
1	Line pressure	psia	Field data	700
2	Gas temperature at saturation	°F	Field data	60
3	Temperature at which line must be protected from hydrates (minimum gas temperature)	°F	Field data	40
4	Water content of gas at saturation	ib/MMCF*		23
5	Water content of gas at minimum temperature	ib/MMCF*		11
6	Change in water content	lb/MMCF*	4 minus 5	12
7	Hydrate temperature of gas	°F		58.0
8	Freezing point lowering required	°F	7 minus 3	18.0
9	Weight % methanol in liquid	%		20
10	Vapour to liquid composition ratio	lb/MMCF*/ wt %	Using 1 and 3	1.17
11	Methanol concentration in gas	lb/MMCF*	Multipy 9 by 10	23.4
12	Methanol to saturate liquid	lb/MMCF*	Line 6 × line 9/ 100 − line 9	3.0
13	Methanol injection rate	lb/MMCF*	11 plus 12	26.4
14	Methanol injection rate (pure methanol at 68°F)	gal/MMCF*	Line 13/6.59 lb/gal	4.0

* Million million cubic feet.

required from the chart, the quantity of the methanol depends upon the total pressure of the gas (Figure 7.4) and the methanol requirements for a given natural gas under specified conditions can also be calculated (Table 7.9).

Tests

Hydrocarbon content

The hydrocarbon content of natural gas is usually obtained indirectly by measurement of either the heating value (ASTM D 1826) or the specific gravity (ASTM D 1070, ASTM D 3588). However, it must be remembered that the composition of natural gas can vary widely, but because natural gas is a multicomponent system, neither property may be changed significantly.

In some instances, the hydrocarbon dewpoints may be specified or limits may be placed on gas enrichment with reference to specific components. Of particular importance, in this respect, are the hexanes and higher-molecular-weight hydrocarbons which may condense in the gas-gathering and/or distribution systems. If significant amounts of carbon dioxide or nitrogen are present in the natural gas, neither gravity nor heating value alone will indicate hydrocarbon content. If both of these properties are measured, the presence of either carbon dioxide or nitrogen will be reflected in a higher specific gravity and lower heating value.

On the issue of hydrocarbon content, it is also necessary to remember that the flammability of a gas can also be equated to the flammability of its individual constituents, particularly the hydrocarbon constituents (for well-documented early work see Jones, 1929; Coward and Jones, 1952).

Water content

The water content of natural gas is usually expressed as pounds of water per million cubic feet of gas or by use of dewpoint temperature and pressure (ASTM D 1142) (Jamieson and Sikkenga, 1986; Tramel, 1985; Gates and Scelzo, 1987; Kahmann, 1987; Mayeaux, 1987). The two methods have a definite relationship as shown by curves of water content as a function of saturation temperature and pressure (NGPSA, 1966). Common specifications are 1, 4 or 7 lb gas (i.e. lb water/1000 ft^3 gas) depending on the conditions to which the gas will be exposed.

The amount of water held in equilibrium in natural gas can be correlated as a function of temperature and pressure (Figures 7.5 and 7.6). If the temperature and pressure of the gas in the reservoir are known, it is possible, by calculating the temperature and pressure at the well-head, to calculate the amount of water which will condense out of the gas per million cubic feet.

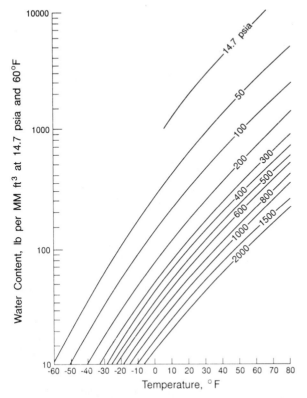

Figure 7.5 Water vapour content of natural gas at saturation conditions

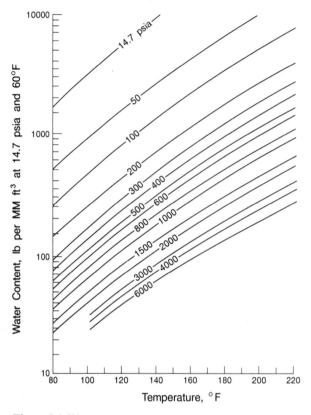

Figure 7.6 Water vapour content of natural gas at saturation conditions

Acid gas content

Hydrogen sulphide and carbon dioxide are the acid gases that are associated with natural gas. They are termed acid gases because solutions of them in water are acidic in nature. For manufactured gases, where measurements are applicable, sulphur dioxide would also be included as an acid gas.

Besides emitting a foul odour at low concentrations, hydrogen sulphur is deadly poisonous and at concentrations above 600 ppm it can be fatal in a matter of minutes and has a toxicity comparable with hydrogen cyanide. Thus, it cannot be tolerated in gas that would be used as domestic fuel. In addition, hydrogen sulphide is corrosive to all metals normally associated with gas transportation, processing and handling systems (although it is less corrosive to stainless steel), and may lead to premature failure of most of these systems.

Hydrogen sulphide forms sulphur dioxide during combustion

$$2H_2S + 3O_2 = 2H_2O + 2SO_2$$

which is usually highly toxic and corrosive. Hydrogen sulphide and other sulphur compounds can also cause catalyst poisoning in refinery processes.

Carbon dioxide has no heating value and its removal may be required in some instances (where acidic properties are of a lesser issue) to increase the energy content (Btu/ft^3, kJ/m^3) of the gas. For gas being sent to cryogenic plants, the removal of carbon dioxide is necessary to prevent its solidification.

Both hydrogen sulphide and carbon dioxide promote hydrate formation, and the presence of carbon dioxide may be less desirable for this reason. However, if none of these situations is encountered, there may be no need to remove the carbon dioxide.

Acid gas content is specified according to the particular impurity. The usual acid gas specification is for hydrogen sulphide (ASTM D 2385, ASTM D 2420, ASTM D 2725, ASTM D 4084 and ASTM D 4810; IP 103 and IP 272). Sulphur compounds may also be present in natural gas (ASTM D 1072, ASTM D 2385 and ASTM D 3031; BS 4386 (2000–104); IP 104) but may not often present a major problem if they occur only in small amounts since mercaptans are added as a warning odorant for natural gas. On the other hand, the total sulphur content of natural gas can also be determined (ASTM D 3031, ASTM D 4468).

Carbon dioxide content may also be specified (ASTM D 1945, ASTM D 1946); an upper limit is commonly 5% v/v. There are instances of carbonyl sulphide (carboxy sulphide, COS) in natural gas.

Gas liquids

Natural gas is composed predominantly of methane (CH_4), and hence natural gas liquids are the higher-molecular-weight hydrocarbons. Several of these constituents are not true liquids in so far as they are not usually in liquid form at ambient temperature and pressure. Thus, natural gas liquids are defined as: (a) ethane; (b) liquefied petroleum gas; or (c) natural gasoline (Table 7.10).

The liquefied petroleum gas is usually composed of propane (C_3H_8), butanes (C_4H_{10}) and/or mixtures thereof; small amounts of ethane and pentane may also be present as impurities (ASTM D 2600; IP 262). On the other hand, natural gasoline (like refinery gasoline) consists mostly of pentane

Table 7.10 Composition of natural gasoline(s) from natural gas

| Reid vapour pressure | Ventura gasoline plant | | | Ten-section gasoline plant |
	38 psia	60 psia	100 psia	22 psia
Ethane	Trace	0.5	0.7	0
Propane	1.1	16.0	43.8	0
Isobutane	19.0	16.0	10.7	0.2
n-Butane	41.0	34.7	23.0	22.7
Isopentane	13.2	11.2	7.4	24.1
n-Pentane	11.3	9.5	6.3	21.0
Hexane	6.8	5.7	3.8	12.6
Heptane	5.3	4.4	2.9	13.7
Octane	1.2	1.0	0.7	4.1
Nonane	1.1	1.0	0.7	1.2
Decane	Trace	Trace	Trace	0.4
	100.0	100.0	100.0	100.0

(C_5H_{12}) and higher-molecular-weight hydrocarbons. The term 'natural gasoline' has also on occasion in the gas industry been applied to mixtures of liquefied petroleum gas (ASTM D 1835, ASTM D 1837), pentanes and higher-molecular-weight hydrocarbons. Caution should be taken not to confuse the term 'natural gasoline' with the term 'straight-run gasoline' (often also incorrectly referred to as natural gasoline) which is the gasoline that is distilled unchanged from petroleum.

There are also standards for the liquid content of natural gas that are usually set by mutual agreement between the buyer and the seller, but such specifications do vary widely and can only be given approximate limits. For example, ethane may have a maximum methane content of 1.5% v/v and a maximum carbon dioxide content of 0.28% v/v. On the other hand, propane will be specified to have a minimum of 95% v/v propane, a maximum of 1–2% butane, and a maximum vapour pressure (ASTM D 1267; IP 161; ISO 4256) which limits the ethane content. For butane, the percentage of one of the butane isomers is usually specified along with the maximum amounts of propane and pentane.

Other properties

Other properties that may be specified are heat of combustion (IP 12) (Francis and Peters, 1980; Hall *et al.*, 1984), vapour pressure (IP 161), corrosivity (ASTM D 130, ASTM D 1838; BS 4351 (2000–154); IP 154; ISO 2160) and gas residues (ASTM 2148; IP 317). The specifications for the propane–butane mixtures will have limits on the amount of non-hydrocarbons and, in addition, the maximum isopentane content is usually stated.

Natural gasoline may be sold on the basis of vapour pressure or on the basis of actual composition, which is determined from the Reid vapour pressure–composition curves prepared for each product source (ASTM D 323).

More recent investigations have focused on the occurrence and detection of mercury in natural gas (Bingham, 1990).

References

ASTM. *Annual Book of ASTM Standards. Standards Relating to Gaseous Fuels, Coal, and Coke*, American Society for Testing and Materials, Philadelphia, PA

ASTM (1963) *Manual on Hydrocarbon Analysis*. Special Technical Publication No. 332, American Society for Testing and Materials, Philadelphia, PA

Bartoo, R.K. (1985) In *Acid and Sour Gas Treating Processes* (ed. S.A. Newman), Gulf Publishing Company, Houston, TX

Bingham, M.D. (1990) *SPE Production Engineering*, May, p. 120

Bland, W.F. and Davidson, R.L. (1967) *Petroleum Processing Handbook*, McGraw-Hill, New York

Brame, J.S.S. and King, J.G. (1955) *Fuel: Solid, Liquid, and Gaseous*, Edward Arnold, London

BSI. British Standards Institution, London, England

Caffey, B.R. (1985) *Proceedings. International School of Hydrocarbon Measurement*, University of Oklahoma Press, Norman, OK, p. 316

Clement, R.E., Onuska, F.I., Yang, F.J., Eiceman, G.A. and Hill, H.H. Jr (1986) *Analytical Chemistry*, **58**, (5), 321R

Coward, H.F. and Jones, G.W. (1952) *Limits of Flammability of Gases and Vapors*.

Bulletin No. 503, Bureau of Mines, United States Department of the Interior, Washington, DC

Cowper, C.J. and Wallis, P.A. (1984) *Chromatographia*, **19**, 85

Cox, L. (1985) *Proceedings. International School of Hydrocarbon Measurement*, University of Oklahoma Press, Norman, OK, p. 338

Francis, W. and Peters, M.C. (1980) *Fuels and Fuel Technology*, Pergamon, New York, Section C, Gaseous Fuels

Gates, L.M. and Scelzo, M.J. (1987) *Proceedings. International School of Hydrocarbon Measurement*, University of Oklahoma Press, Norman, OK, p. 166

Hall, K.R., March, K.N., Eubank, P.T. and Holste, J.C. (1984) *Proceedings. Annual Meeting*, American Gas Association, p. 212

Hefley, C.G., Hines, W.J. and Sattler, F.W. (1985) *Proceedings. Annual Meeting*, Gas Processors Association, p. 388

Hertweck, F.R. Jr and Fox, D.D. (1984) Information Circular No. IC 8993, Bureau of Mines, United States Department of the Interior, Washington, DC

IP. *Methods for Analysis and Testing*, Institute of Petroleum, London

ISO. International Organization for Standardization, Geneva

Jamieson, A.W. and Sikkenga, H.J. (1986) *Gas Quality* (ed. G.J. van Rossum), Elsevier, Amsterdam, p. 289

Johnson, D.P. (1983) *Proceedings. Annual Meeting*, American Gas Association, T80

Jones, G.W. (1929) *Inflammability of Mixed Gases*. Technical Paper No. 450, Bureau of Mines, United States Department of Commerce, Washington, DC

Kahmann, A.R. (1987) *Proceedings. International School of Hydrocarbon Measurement*, University of Oklahoma Press, Norman, OK, p. 170

Kohl, A.L. and Riesenfeld, F.C. (1979) *Gas Purification*, Gulf Publishing Company, Houston, TX

Kumar, S. (1987) *Gas Production Engineering*, Gulf Publishing Company, Houston, TX

Lowenheim, F.A. and Moran, M.K. (1975) *Industrial Chemicals*, Wiley, New York

Mayeaux, D.P. (1987) *Proceedings. International School of Hydrocarbon Measurement*, University of Oklahoma Press, Norman, OK, p. 153

Moore, B.J. and Hamack, J.E. (1985) Information Circular No. IC 9046, Bureau of Mines, United States Department of the Interior, Washington, DC

NGPSA (1966) *Engineering Data Book*, Natural Gas Processors Suppliers Association, Tulsa, OK

Nonhebel, G. (1964) *Gas Purification Processes*, George Newnes, London

Rawlinson, D. and Ward, E.R. (1973) In *Criteria for Quality of Petroleum Products* (ed. J.P. Allinson), Wiley, New York, Chapter 3

Reid, R.C., Prausnitz, J.M. and Sherwood, T.K. (1977) *The Properties of Gases and Liquids*, McGraw-Hill, New York

Sharples, R.J. and Panhill, W. (1985) *Oil and Gas Journal*, **83**, (26), 47

Speight, J.G. (1981) *The Desulfurization of Heavy Oils and Residua*, Marcel Dekker, New York

Speight, J.G. (1991) *The Chemistry and Technology of Petroleum*, 2nd edn, Marcel Dekker, New York

Starling, K.E. (1985) *Proceedings. International School of Hydrocarbon Measurement*, University of Oklahoma Press, Norman,. OK, p. 351

Starling, K.E. and Kumar, K.H. (1985) *Proceedings. Annual Meeting*, Gas Processors, Association, p. 147

Tramel, T.Y. (1985) *Proceedings. International School of Hydrocarbon Measurement*, University of Oklahoma Press, Norman, OK, p. 277

Trusell, F.C. (1985) *Analytical Chemistry*, **57**, (5), 191R

Willis, P.A. (1986) *Trends in Analytical Chemistry*, **5**, (3), 63

Yon, M.C. (1988) *Proceedings. International School of Hydrocarbon Measurement*, University of Oklahoma Press, Norman, OK, p. 388

8

Emissions control prior to gas-cleaning operations

Introduction

Before engaging in a general description of the commercial methods by which process gases might be cleaned, it is of interest to give some consideration to the available methods by which process gases can be cleaned at the source, i.e. integrated methods.

Briefly, gas-producing processes have previously (Chapter 5) been defined as 'external' processes, i.e. those processes in which contaminant gases are removed from product streams and from flue gas streams by application of a cleaning technology after their formation (Figure 8.1). There are also those processes in which the gases are removed *in situ*, i.e. the technology is integrated into the system in so far as the gases are removed at the time they are formed or before they emerge from the reactor in which they are formed. Of course, the definition, being very general, is open to question in terms of the timing, but it conveys the general principle. And it is the latter type of process, i.e. the *in situ*, or integrated, process, that will be addressed in this chapter. In addition, and because of the nature of the feedstock, it is the coal industry to which the definition is mainly applied.

The issues arising from the occurrence of acid rain, owing to the emission of sulphur oxides and nitrogen oxides from coal combustion, are forcing many regional, national and international re-examinations of the means by which these emissions can be controlled (Mohnen, 1988; United States Department of Energy, 1990, 1991; United States General Accounting Office, 1990) (see also Chapter 1). Whilst there are a variety of options that may offer some alleviation of the problems at the 'gas-emissions end' of the plant, there are also several options that can be employed at the 'front end' of the plant.

The formation and occurrence of acid rain is a complex problem, still speculative in many senses and full of uncertainties. As a necessary start to understanding this phenomenon, it is necessary to understand the chemistry of power station emissions (Hart, 1981).

For example, there are still differences of opinion as to how much acid deposition actually contributes to the environmental difficulties for which it has been blamed. There is also the question of how much benefit would be realized by reducing the emissions by half. Proponents consider that enough is known to warrant the institution of extreme control measures whilst opponents prefer to review the situation and determine carefully what the next

"END-OF-PIPE" SYSTEM

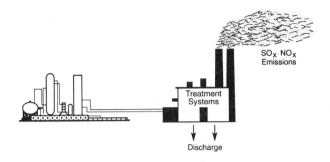

SO$_x$ NO$_x$
Emissions

Treatment
Systems

Discharge

INTEGRATED SYSTEM

Emission Control

Figure 8.1 Representation of the two types of gas-cleaning operations: (a) 'end-of-the-pipe' or external system; (b) 'integrated' or internal system

steps should be. Perhaps both are correct. But in the meantime the environment appears to be under attack! The answer appears to lie in a reduction of emissions that contribute to the acid rain issue.

The generation of electricity at plants fuelled by oil, coal or gas accounts for a considerable portion of the carbon dioxide emissions as well as for the pollutants that cause acid rain. Therefore, emissions control is a major aspect of this utility industry.

Clean combustion consists of removing the pollutants from coal as it is burned. This can be accomplished by controlling the combustion parameters (fuel, air/oxygen and temperature) to minimize the formation of pollutants and/or by injecting pollutant-absorbing substances into the combustion chamber to capture the pollutants as they are formed.

In general for coal combustion, coal, pulverized into particles small enough to form a combustible cloud, is injected with hot air into burners along the lower portion of the furnace box or boiler. The coal burns in a long, luminous flame at temperatures of 1480 °C (2700 °F) or greater. The heat is transferred to water-filled tubes attached to the sides of the boiler. The boiling water in the tubes creates steam to spin a turbine generator, which produces electricity.

In conventional coal combustion plants, the principal design goal is high efficiency to extract the most energy from a unit of coal. With advanced combustion technologies, the goal is to reduce emissions while retaining high efficiencies by altering the combustion process.

Advanced combustion systems control or remove sulphur dioxide, nitrogen oxide(s), and/or particulate matter from coal combustion gases before they enter a steam generator or heater (Pierce *et al.*, 1991). Polluants are controlled by the combustion parameters and/or sorbents used during the combustion process. Nitrogen oxide(s) is normally controlled through staged combustion, natural gas reburning, coal reburning, or a method of controlling combustion flame temperature. Sulphur dioxide is controlled through sorbent injection in the combustion chamber. Ash can be controlled (since failure to do so can result in the replacement of expensive equipment) by operating at high temperatures and converting it into molten slag, although such high temperatures do not appear to be conducive to the simultaneous capture of nitrogen oxide(s) and sulphur dioxide.

Some advanced combustion systems are designed to reduce only the emissions of nitrogen oxide(s) while others are designed to reduce or capture several pollutants (sulphur dioxide, nitrogen oxide and ash). Depending on the specific technology, these systems are capable of reducing the emissions of nitrogen oxide(s) by 50–70%, sulphur dioxide by 50–95% and ash by 50–90% when compared with conventional technology.

Emissions reduction at the front end is open to several options which, for reductions in the sulphur dioxide emissions, include coal switching, coal cleaning, unit retirement, and making changes in the process such as installing limestone injection or replacing existing equipment with fluidized-bed boilers (Pierce *et al.*, 1991). Each option may be site specific and may not be a panacea in so far as successful implementation of an option at one plant may not offer the same, if any, degree of success at another plant, either from the emissions reduction potential or the economic potential.

It is the purpose of this chapter to examine these 'front-end' options that offer the attractive capability of reducing the emissions of nitrogen oxides and sulphur oxides by, literally, preventing their formation or capturing them as they are formed in the combustor.

Emissions control

One of the questions posed for existing plants is whether all the gas from all the plants should be treated to get the specified partial reduction, or whether part of the gas should be treated with a highly efficient process to achieve the same overall reduction. In other words, should all the gas be treated by a method that removes 50% of the sulphur dioxide or should a little over half the gas be treated with a process that removes nearly 100%?

Thus, it is not surprising that the United States Department of Energy Clean Coal Program consists of several sections, notably (a) postcombustion cleaning, (b) coal conversion, (c) precombustion cleaning, and (d) clean combustion (Table 8.1), as only one of many national attempts to enable coal combustion to achieve the various emissions standards for sulphur dioxide (Table 8.2). Many process options are under serious consideration and a brief summary is presented here as a focal point.

Traditionally, research to improve precombustion cleaning has concentrated on two major categories of cleaning technology: physical cleaning and chemical cleaning. Recently a new category, biological cleaning, has attracted interest as advances have been made in microbial and enzymatic techniques for liberating sulphur and ash from coal (Bielaga and Kilbane, 1990; Frost, 1991; Miller, 1991).

Techniques that convert coal into another form of fuel bypass the conventional coal 'fuel path' altogether. In the most commonly envisioned systems, coal is converted into a gaseous fuel; in other techniques, a liquid product is made; while in still others, a combination of gases, liquids and solids is produced.

In more general terms, coal-cleaning methods notwithstanding, there is no single solution that will satisfy the widely varying fuel and design conditions of existing coal-fired plants. There are basic emission control approaches, such as: (a) use of different or cleaned coals; or (b) changes in firing practices, burners and combustion conditions; or (c) introduction of reagents which will inhibit or capture pollutants in combustion gases (Pierce *et al.*, 1991).

Coal cleaning

Coal cleaning, or beneficiation, is the means by which coal is treated prior to use, usually at a power plant, to remove unwanted mineral matter and pyrite (Eliot, 1978; Deurbrouck and Hucko, 1981; Speight, 1983; Hessley, 1990). Indeed, it is recognized that coal characteristics can affect the operation of equipment.

In the historical sense, coal cleaning has been limited in its application to coal used in industrial processes, such as steel making. Here, the major motivation for the application of coal-cleaning methods was to effect a reduction in the amount of ash produced, i.e. it was a demineralization process, and not necessarily applied with the goal of sulphur reduction.

Virtually all current methods for coal cleaning involve the use of one or more physical techniques, some of which have been used for more than a century. Physical cleaning can remove only matter that is physically distinct from the coal, such as small dirt particles, rocks and pyritic sulphur (iron pyrites, FeS_2). It cannot remove sulphur that is chemically combined with the coal (organic sulphur), nor can it remove nitrogen, another source of pollution. Currently, physical cleaning can remove 30–50% of the pyritic sulphur (or 10–30% of the total sulphur) and about 60% of the ash-forming minerals (Table 8.3).

Advanced physical cleaning techniques are expected to be significantly more effective than older techniques. Increased effectiveness is achieved by first grinding the coal into much smaller sizes at which it releases more of the ash and pyrite. Then special separation technologies can be used to assure good coal recovery, and thermal treatment can be used to reduce moisture and modify the surface characteristics to prevent reabsorption. New coal-cleaning processes can remove more than 90% of the pyritic sulphur and undesirable minerals from the coal.

Coal cleaning by physical methods removes only pyritic sulphur (FeS_2), by virtue of its density or other physical properties, and leaves much of the organic sulphur in place. Chemical cleaning processes are required to remove the chemically bound, organic, sulphur from coal.

Whilst physical cleaning methods might, and usually do, reduce the inorganic sulphur content of the coal and its potential to form sulphur oxides during combustion

$$FeS_2 + 3O_2 = FeSO_4 + SO_2$$

Table 8.1 United States Clean Coal Technology (CCT) demonstration projects (listed by category of application)

Application category	Sponsor	Project	Solicitation
Advanced electric power generation systems	ABB Combustion Engineering, Inc.	Combustion Engineering IGCC Repowering Project	CCT-II
	Alaska Industrial Development and Export Authority	Healy Clean Coal Project	CCT-III
	The City of Tallahassee	Arvah B. Hopkins Circulating Fluidized-Bed Repowering Project	CCT-I
	Clean Power Cogeneration Limited Partnership	Air-Blown/Integrated Gasification Combined-Cycle Project	CCT-III
	Colorado-Ute Electric Association, Inc.	Nucla CFB Demonstration Project	CCT-I
	DMEC-1 Limited Partnership	PCFB Demonstration Project	CCT-III
	The Ohio Power Company	Tidd PFBC Demonstration Project	CCT-I
	The Ohio Power Company and The Appalachian Power Company	PFBC Utility Demonstration Project	CCT-II
	Sierra Pacific Power Company	Piñon Pine IGCC Power Project	CCT-IV
	TAMCO Power Partners	Toms Creek IGCC Demonstration Project	CCT-IV
	Wabash River Coal Gasification Repowering Project Joint Venture	Wabash River Coal Gasification Repowering Project	CCT-IV
High-performance pollution control devices	ABB Combustion Engineering, Inc.	SNOX Flue Gas Cleaning Demonstration Project	CCT-II
	AirPol, Inc.	10-MW Demonstration of Gas Suspension Absorption	CCT-III
	The Babcock & Wilcox Company	Demonstration of Coal Reburning for Cyclone Boiler NO_x Control	CCT-II
	The Babcock & Wilcox Company	Full-Scale Demonstration of Low-NO_x Cell Burner™ Retrofit	CCT-III
	The Babcock & Wilcox Company	LIMB Demonstration Project Extension and Coolside Demonstration	CCT-I
	The Babcock & Wilcox Company	SOX-NOX-ROX Box Flue Gas Cleanup Demonstration Project	CCT-II
	Bechtel Corporation	Confined Zone Dispersion Flue Gas Desulfurization Demonstration	CCT-III
	Coal Tech Corporation	Advanced Cyclone Combustor with Internal Sulfur, Nitrogen, and Ash Control	CCT-I
	Energy and Environmental Research Corporation	Enhancing the Use of Coals by Gas Reburning and Sorbent Injection	CCT-I
	Energy and Environmental Research Corporation	Evaluation of Gas Reburning and Low-NO_x Burners on a Wall-Fired Boiler	CCT-III
	LIFAC–North America	LIFAC Sorbent Injection Desulfurization Demonstration Project	CCT-III

Category	Company	Project	
	MK-Ferguson Company	Commercial Demonstration of the NOXSO SO_2/NO_x Removal Flue Gas Cleanup System	CCT-III
	New York State Electric & Gas Corporation	Milliken Clean Coal Technology Demonstration Project	CCT-IV
	Public Service Company of Colorado	Integrated Dry NO_x/SO_2 Emission Control System	CCT-III
	Pure Air on the Lake, L.P.	Advanced Flue Gas Desulfurization Demonstration Project	CCT-II
	Southern Company Services, Inc.	Demonstration of Advanced Combustion Techniques for a Wall-Fired Boiler	CCT-II
	Southern Company Services, Inc.	Demonstration of Innovative Applications of Technology for the CT-121 FGD Process	CCT-II
	Southern Company Services, Inc.	Demonstration of Selective Catalytic Reduction Technology for the Control of NO_x Emissions from High-Sulfur-Coal-Fired Boilers	CCT-II
	Southern Company Services, Inc.	180-MWe Demonstration of Advanced Tangentially Fired Combustion Techniques for the Reduction of NO_x Emissions for Coal-Fired Boilers	CCT-II
	Tennessee Valley Authority	Micronized Coal Reburning Demonstration for NO_x Control on a 175-MWe Wall-Fired Unit	CCT-IV
	Union Carbide Chemicals and Plastics Company Inc.	Demonstration of the Union Carbide CANSOLV™ System at the ALCOA Generating Corporation Warrick Power Plant	CCT-IV
Coal processing for clean fuels	ABB Combustion Engineering, Inc., and CQ, Inc.	Development of the Coal Quality Expert	CCT-I
	Air Products and Chemicals, Inc.	Commercial-Scale Demonstration of the Liquid-Phase Methanol (LPMEOH") Process	CCT-III
	Cordero Mining Company	Cordero Coal Upgrading Demonstration Project	CCT-IV
	Custom Coals International	Self-Scrubbing Coal: An Integrated Approach to Clean Air	CCT-IV
	ENCOAL Corporation	ENCOAL Mild Coal Gasification Project	CCT-III
	Rosebud SynCoal Partnership	Advanced Coal Conversion Process Demonstration	CCT-I
Industrial applications	Bethlehem Steel Corporation	Blast Furnace Granulated Coal Injection System Demonstration Project	CCT-III
	Bethlehem Steel Corporation	Innovative Coke Oven Gas Cleaning System for Retrofit Applications	CCT-II
	Passamaquoddy Tribe	Cement Kiln Flue Gas Recovery Scrubber	CCT-II
	ThermoChem, Inc.	Demonstration of Pulse Combustion in an Application for Steam Gasification of Coal	CCT-IV

Clean Coal Demonstration Program, Program Update 1991, Department of Technology. Energy Publication DOE/FE-0247P, February 1992.

Table 8.2 Current national emission standards for sulphur (mg SO$_2$/m^3) (IEA, 1991)

Country	New plants	Existing plants
Austria	200–400	200–400
Belgium[c]	400–2000(250)[d]	
Canada[a]	740	
Denmark[c]	400–2000	
Finland	400–660	660
France[c]	400–2000	
FRG[c]	400–2000	400–2500
Greece[c]	400–2000	
Ireland[c]	400–2000	
Italy[c]	400–2000	400–2000
Japan	[b]	[b]
Luxembourg[c]	400–2000	
Netherlands[c]	200–700	400–700
Poland	540	2700–4000 (1800–3000)
Portugal[c]	400–2000	
Spain[c]	2400–9000	2400–9000
Sweden	290	290–570
Switzerland	400–2000	400–2000
Taiwan	2145–4000	2145–4000
Turkey	400–2000	
UK[c]	400–2000	
USA	740–1480[e]	[e]

[a] Guidelines.
[b] Set on a plant-by-plant basis according to nationally defined formulae.
[c] EC countries.
[d] From 1995.
[e] Clean Air Act Amendments, 1990.
Source: Kyte (1991).

the removal of organic sulphur is usually not affected and remains unresolved; the potential for sulphur dioxide production from such sulphur is very real

$$[S]_{organic} + O_2 = SO_2$$

unless high-sulphur coal can be removed as part of the process.

Application of advanced coal-cleaning technology offers some alleviation in terms of the reduction of sulphur dioxide emissions from existing coal-fired power plants.

There are various methods by which sulphur dioxide emissions, and yields of ash, can be reduced (Argonne National Laboratory, 1984) and there is the real potential for significant benefits to accrue by application of improved coal-cleaning systems.

These methods are worthy of note and, in order to place them in perspective, a brief mention is made of each relevant method here.

Heavy-liquids cyclone

The use of 'heavy liquids' such as freon, a fluorocarbon refrigerant, as a coal-cleaning liquid effects the separation of the particles by gravity and by use of a cyclone (Figure 8.2).

Table 8.3 Minerals commonly associated with coal which will produce a combustion residue (ash)

Group	Species	Formula
Shale	Muscovite	$(K, Na, H_3O, Ca)_2 (Al, Mg, Fe, Ti)_4$
	Hydromuscovite	$(Al, Si)_8 O_{20}(OH, F)_4$ (general formula)
	Illite	$(HO)_4 K_2(Si_6.Al_2) Al_4 O_{20}$
	Montmorillonite	$Na_2(Al\ Mg)Si_4 O_{10}(OH)_2$
Kaolin	Kaolinite	$Al_2(Si_2O_5)(OH)_4$
	Livesite	$Al_2(Si_2O_5)(OH)_4$
	Metahalloysite	$Al_2(Si_2O_5)(OH)_4$
Sulphide	Pyritic	FeS_2
	Marcasite	FeS_2
Carbonate	Ankerite	$CaCO_3.(Mg, Fe, Mn) CO_3$
	Calcite	$CaCO_3$
	Dolomite	$CaCO_3.MgCO_3$
	Siderite	$FeCO_3$
Chlorite	Sylvite	KCl
	Halite	$NaCl$
Accessory minerals	Quartz	SiO_2
	Feldspar	$(K, Na)_2O.Al_2O_3.6SiO_2$
	Garnet	$3CaO.Al_2O_3.SiO_2$
	Hornblende	$CaO.3FeO.4SiO_2$
	Gypsum	$CaSO_4.2H_2O$
	Apatite	$9CaO.3P_2O_5.CaF_2$
	Zircon	$ZrSiO_4$
	Epidote	$4CaO.3Al_2O_3.6SiO_2.H_2O$
	Biotite	$K_2O.MgO.Al_2O_3.3SiO_2.H_2O$
	Augite	$CaO.MgO.2SiO_2$
	Prochlorite	$2FeO.2MgO.Al_2O_3.2SiO_2.2H_2O$
	Diaspore	$Al_2O_3.H_2O$
	Lepidocrocite	$Fe_2O_3.H_2O$
	Magnetite	Fe_3O_4
	Kyanite	$Al_2O_3.SiO_2$
	Staurolite	$2FeO.5Al_2O_3.4SiO_2.H_2O$
	Topaz	$2AlFO.SiO_2$
	Tourmaline	$3Al_2O_3.4BO(OH).8SiO_2.9H_2O$
	Hematite	Fe_2O_3
	Penninite	$5MgO.Al_2O_3.3SiO_2.2H_2O$
	Sphalerite	ZnS
	Chlorite	$10(Mg, Fe)O.2Al_2O_3.6SiO_2.8H_2O$
	Barite	$BaSO_4$
	Pyrophillite	$Al_2O_3.4SiO_2.H_2O$

Froth flotation

Froth flotation treats finely ground coal with an oil-based substance which adheres to the coal and relies upon differences in surface characteristics to separate the coal from impurities.

In this process coal is fed into a 'froth flotation cell' where mechanical agitation creates air bubbles which become attached to the oil-coated coal and cause it to rise to the surface; any impurities remain in the tank.

Selective coalescence

Small coal particles, in the presence of an appropriate medium and additives, can be caused to undergo selective coalescence (i.e. agglomerate) into larger

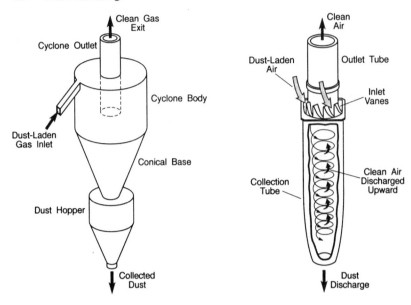

Figure 8.2 Schematic representation of two cyclone separators

particles which can then be separated from the undesirable impurities that do not coalesce.

Various non-aqueous media, such as liquid carbon dioxide, have the potential to remove better than 90% of the mineral matter (i.e ash precursors) and pyritic sulphur.

Chemical cleaning

The pretreatment of coal with chemicals, such as aqueous sodium hydroxide to remove inorganic sulphur and/or acid washing to reduce the mineral matter content, has been claimed to be successful (Meyers *et al.*, 1972; Powell and Ulmer, 1974; Hall *et al.*, 1975; Worthy, 1975).

Alkali displacement chemistry, applied by means of either a molten bath or microwave heating, has the potential to remove in excess of 90% of the total sulphur and 95% of the mineral matter from selected coals.

Chemical/biological cleaning

Removing organic sulphur that is chemically bound to the coal is a more difficult proposition than removing pyritic sulphur by physical means.

Biological cleaning represents one of the most exotic methods for coal cleaning and uses naturally occurring bacteria that can 'eat' the organic sulphur in coal. Other approaches involve using fungi, rather than bacteria, and injecting sulphur-digesting enzymes directly into the coal. The biological approach appears to be capable of removing as much as 90% of the total sulphur (pyritic and organic) in coal.

Coal switching

One approach to emissions control involves altering the fuel that is burned through switching (Rupinskas and Hiller, 1991), blending (Evans *et al.*, 1991) or cleaning (Meyers, 1981).

Changing the source of incoming coal, referred to as switching, has the objective of changing to low-sulphur coal(s) from currently used high-sulphur coal(s). In addition to its potential for sulphur dioxide control, switching fuel has historically been a way to improve combustion properties by a reduction in the sulphur content of the feedstock (Table 8.4) by blending high-sulphur coals (Table 8.5; Figure 8.3) with lower-sulphur coals which allows the coal plant to deal with declining, or variable, coal quality.

As a passing note, the petroleum-refining and natural gas industries have been aware for some time that feedstock blending prior to processing has become a 'way of life' in the wake of declining reserves of the better crude oils and hydrocarbon-rich natural gas streams (Speight, 1991). The coal industry has also been involved in feedstock blending throughout its history, but perhaps to a lesser extent, since the analysis of a coal sample from one part of a seam can differ significantly from that of coal at another part of the seam!

An issue that might arise as a result of coal switching is the impact on overall plant operability caused by burning a type, or rank, of coal for which the plant was not designed. Fuel changes may require lowered temperatures, degraded heat rates, reduced generating capacities, and modifications to the electrostatic precipitator used for particulate emissions control (Lausen and Schioth, 1992); use of a low-sulphur coal, for example, which may have a different ash composition, can produce intolerable slagging and fouling.

Table 8.4 Average sulphur content of coal for utility plant use

Year	Quantity (million tons)	Average sulphur (%)
1978	476	1.80
1979	553	1.73
1980	591	1.60
1981	576	1.50
1982	598	1.51
1983	593	1.47
1990	580[a]	1.45[a]

[a] Estimates.
Source: Gilleland and Swisher (1986).

Table 8.5 Sulphur content of United States coals (by region)

Region	No. samples	Organic S (%)	Pyritic S (%)	Total S (%)
N. Appalachian	227	1.00	2.07	3.01
S. Appalachian	35	0.67	0.37	1.04
E. Midwest	95	1.63	2.29	3.92
W. Midwest	44	1.67	3.58	5.25
Western	44	0.45	0.23	0.68
Alabama	10	0.64	0.69	1.33

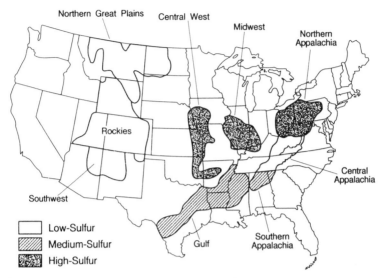

Figure 8.3 Distribution of sulphur content of coals in the United States

From a technical standpoint, coal blending and switching for sulphur dioxide reduction are attractive and broadly applicable.

Fluidized-bed combustion

Coal combustion is an important aspect of coal utilization (Essenhigh, 1981) and developments in combustion technology (Ceely and Daman, 1981) plus investigations of the effects, and production, of pollutants have arisen as major issues (Reid, 1981; Slack, 1981; Pierce *et al.*, 1991). In order to increase the efficiency of coal combustion systems, the concept of fluidized-bed coal combustion has passed through the early development stages and into commercial use.

Fluidized-bed combustion, or fluid-bed combustion (FBC) (Figure 8.4), is a procedure that has become important in coal plants because of its inherently low sulphur dioxide and nitric oxide emissions that are produced during coal combustion. The furnace configuration allows a mixture of fuel and limestone to be suspended and thoroughly mixed in an upward flow of air (Pierce *et al.*, 1991).

In fluidized-bed combustion, the coal is fluidized at 790–925°C (1450–1700°F) by air forced through the bed. The reaction zone of burning coal usually occupies less than 5% of the fluidized bed, but burning is fairly even and rapid due to the turbulence in the bed, and all the bed particles are heated quickly (Figure 8.5). As an extra benefit, boiler tubes submerged in the bed absorb heat from the turbulent solids to convert water (in the tubes) to steam.

The high heat transfer rates also permit lower combustion temperatures resulting in the formation of relatively low levels of nitrogen oxide(s) (see, for example, Amand *et al.*, 1991) and, since the bed temperatures are below coal ash fusion temperatures, a variety of coals can be burned without modification of the original design (Amand and Leckner, 1991).

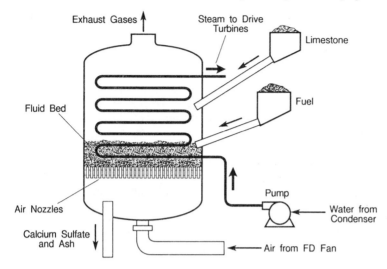

Figure 8.4 Schematic representation of a fluidized-bed combustor

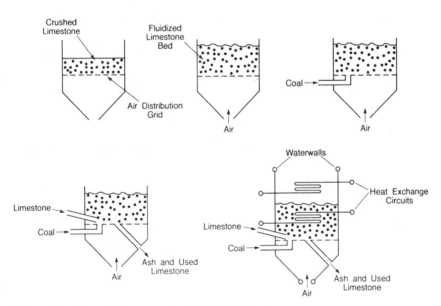

Figure 8.5 General elements of a fluidized-bed boiler

Fluidized-bed combustion reduces emissions by controlling combustion parameters and by injecting a sorbent, or a pollutant absorbent (such as crushed limestone), into the combustion chamber along with the coal. Pulverized coal mixed with crushed limestone is suspended on jets of air (or fluidized) in the combustion chamber. As the coal burns, sulphur is released and combines with the limestone to form a new solid waste product, a mixture of calcium sulphite and calcium sulphate. Some of the solid waste is removed with the bed ash through the bottom of the boiler. Small ash particles, or fly ash, that escape the boiler are captured with dust collectors (cyclones and baghouses). More than 90% of the sulphur released from coal can be captured in this manner.

As limestone is mixed with the coal, most of the sulphur can be captured during the combustion process. The operating temperatures of the fluidized bed are particularly appropriate to the efficiency of the reaction to form calcium sulphite and sulphate. The furnace space above the bed (the 'freeboard') provides additional time to convert any unburned coal that escapes the bed as well as to complete the sulphur capture.

The limestone in the fluidized bed reacts with the sulphur dioxide and nitrogen oxide(s) and more than 90% of the sulphur dioxide and nitrogen oxide(s) can be captured. In addition, combustion can be maintained at about 815°C 1500°F), instead of the much higher temperatures necessary for conventional pulverized-coal boilers (Pierce et al., 1991).

Fluidized-bed combustors can be manufactured to operate either at atmospheric pressure or under pressure. The atmospheric presure type of combustor operates at normal atmospheric pressure while the pressurized type operates at pressures 6–16 times higher than normal atmospheric pressure.

The pressurized fluidized-bed boiler offers potentially higher efficiency, reduced operating costs and less waste products than does the state-of-the-art atmospheric fluidized-bed boiler. A new type of atmospheric fluidized-bed boiler offers circulating (entrained) fuel flow instead of the bubbling bed or fixed bed used in earlier approaches. Circulating fluidized beds allow for finer coal feed, better fuel mixing, higher efficiencies and increased capture of sulphur dioxide.

Examples of advanced combustion technologies include low-NO_x burners, slagging combustors, cyclone combustors, vortex combustors, pulsed combustors, limestone-injection multistage burners, coal- and gas-reburning technologies and fluidized-bed combustors (Khan et al., 1990).

Atmospheric fluidized-bed combustion

This process option (Figure 8.6) offers the capability of burning low-grade coals and waste fuels while meeting environmental regulations. A range of advantages have been claimed: (a) emission control without scrubbers; (b) dry solid waste; (c) furnace design independent of ash content or properties; (d) fuel flexibility with the ability to use low-grade fuels.

Pressurized fluidized-bed combustion

If the furnace is pressurized to a fireside pressure of 90–200 psi (620–1380 kPa) the fluidized-bed boiler system is classified as a 'pressurized fluidized-bed combustion' (PFBC) system (Figure 8.7).

The use of the pressurized fluidized-bed combustor boiler in a combined-cycle system as a means of producing both steam and hot pressurized flue gas to drive separate turbine generators is a highly energy-efficient use of coal for power generation (Marrocco et al., 1991). Operation at elevated pressure reduces the combustor size and enhances the general performance parameters.

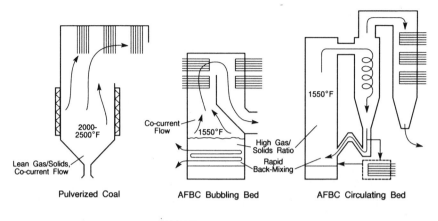

Figure 8.6 Illustration of the working concept of a fluidized-bed combustor

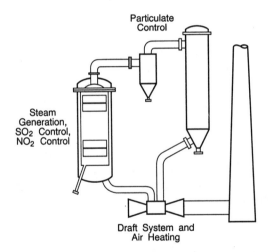

Figure 8.7 Illustration of a pressurized fluidized-bed combustion system

Integrated coal gasification combined cycle (ICGCC)

The integrated coal gasification combined-cycle (ICGCC or IGCC or CGCC) system (Figure 8.8) utilizes the principal components of a coal gasification facility to convert coal into a clean low-to-medium-Btu (kJ) gas that can be consumed in a gas turbine and a boiler to provide the thermal and mechanical energy used for the production of power (Jenkins, 1991; CST, 1991).

In this process, coal is converted into a fuel and/or synthesis gas. The conversion is accomplished by introducing air or oxygen and steam with the coal into the reactor employing the concepts of moving, fluid, or entrained

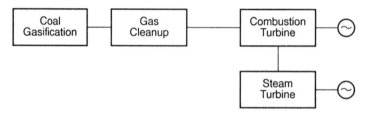

Figure 8.8 General elements of the integrated gasification combined-cycle system

bed. The product mix (i.e. carbon monoxide, carbon dioxide, methane, hydrogen, water, nitrogen, hydrogen sulphide, etc.) is influenced by the type of coal, the use of air or oxygen as well as by the thermodynamics and chemistry of the gasification reactions under the operating conditions.

Compressed air is mixed and burned with the fuel gas and the hot combustion gas is expended through a turbine. A portion of the turbine output drives the air compressor and the remaining energy drives a generator to produce electric power. A waste heat boiler, steam turbine and generator are added to the gas turbine system to form a 'gas turbine combined cycle'.

One of the most important advantages exhibited by an integrated coal gasification combined-cycle power plant over direct coal combustion systems is the ability to control sulphur emissions to any extent necessary. A high level of sulphur removal (above 95%) can be accomplished at a small cost in terms of efficiency, operating parameters and capital cost.

Limestone injection with multistage burners (LIMB)

In its simplest form (Figure 8.9), the process involves the injection of limestone into the boiler where it is calcined to lime, which then absorbs sulphur dioxide. Because of its simplicity, it was one of the first methods tested several decades ago. The results, however, were very poor: at a stoichiometric ratio of 2.0 (which was considered to be about all the limestone the boiler could stand), removal was only of the order of 15–20%.

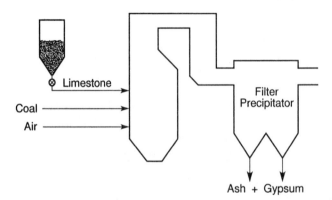

Figure 8.9 Simplified illustration of the limestone injection system

Both spray dryer scrubbing and limestone injection are fairly simple processes and for this reason should be less expensive than limestone scrubbing. However, the key factor is the sulphur content of the coal; as sulphur content of the coal increases, limestone scrubbing appears to be favoured over limestone injection.

Limestone injection with multistage burners (LIMB) is an approach (Figure 8.10) also based on the injection of sorbents into the boiler for the direct capture of sulphur dioxide from the combustion gases.

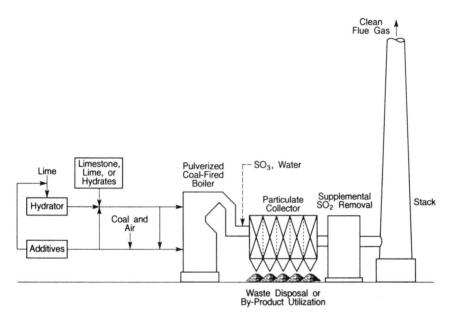

Figure 8.10 Illustration of the integrated LIMB system

The intimate mixing of the sorbent with the coal leads to increased contacting of the active sorbent with the sulphur early in the flame. Furthermore, contacting the calcium and sulphur under fuel-rich conditions leads to the formation of calcium sulphide, which is stable to higher temperatures. And application of a gas-reburning–sorbent-injection concept is predicted to reduce the emissions of sulphur dioxide and nitrogen oxide(s) (Bartok *et al.,* 1991).

There is also the potential for a temperature reduction in the burners with the attractive side option of a reduction in nitrogen oxide emissions. In fact, low-NO_x burners of various designs have been developed which are capable of retrofit applications.

Other control technologies

Additives

The emphasis in recent years has been on the development of additives to improve scrubber performance, both in regard to absorption efficiency and scrubber reliability. Most of the effort has gone into testing organic acids which enhance efficiency by buffering at the gas/liquid interface.

For example, emphasis has been placed on adipic acid, or a by-product mixture containing adipic acid along with other organic acids such as formic acid. Magnesium compounds have been claimed to increase absorption efficiency by as much as 30 percentage points for a relatively small addition. Thiosulphate has been used to inhibit oxidation with the objective of reducing scaling and improving system reliability.

More recently, additives for the control of, or reduction in, the nitrogen oxide(s) emissions in flue gases include chemicals such as ammonia (NH_3) or urea (H_2NCONH_2), i.e. chemicals which release amino (NH_2) radicals at low temperatures (Jodal et al., 1990; Mjornell et al., 1991; Pierce et al., 1991; Pachaly et al., 1992). Hydrogen and the lower-molecular-weight hydrocarbons (methane, ethane and butane) have also been tested (Leckner et al., 1991).

In general, although additives are effective, they have not been adopted widely. The preference appears to be to regard them as a last resort, for use only if the system does not meet expectations.

Particulate removal

Particulate removal (Licht, 1988) is often accomplished by the use of mechanical precipitators (Figure 8.11) large enough to meet the emission standards without any help from scrubbers, although a prescrubber offers valuable assistance.

Baghouses (Figure 8.12) are popular because they remove particulates down to a very low level and may be less costly than precipitators.

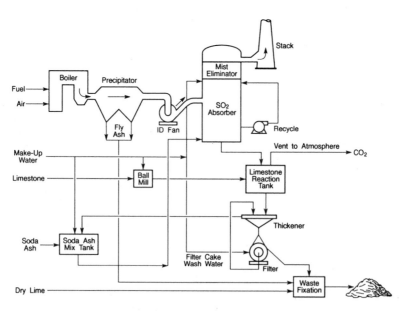

Figure 8.11 Illustration of utility combustor showing the placement of environmental control systems

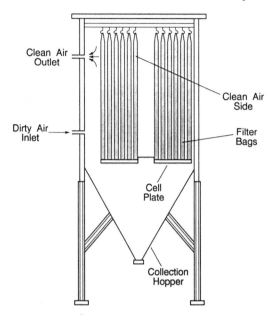

Figure 8.12 Illustration of a fabric filter (baghouse)

Mist elimination

The gas velocities involved in scrubbers generate large amounts of mist which, if not removed, accumulate on downstream surfaces and promote corrosion.

The general practice is to mount the mist eliminator (a zigzag set of baffles) horizontally in the top of a scrubber, although a better performance, but usually at a higher cost, can be attained with a vertically mounted unit (horizontal gas flow) unit.

The usual practice is to mount the mist eliminator in a separate vessel through which the gas passes from the scrubber to the stack. Another arrangement, in which the top of the scrubber is mounted in a 'box-type' arrangement with the baffles making up the sides of the box and the bottom closed, forces the gas to flow horizontally through the sides of the box. Another design offers a horizontal scrubber in which the mist eliminator needs to be in the vertical position because of the horizontal flow of the gas.

The main problem with mist eliminators has been plugging which is caused, primarily, by washing with recycled liquid that contains a high concentration of calcium sulphate and using too much limestone as the absorbent feed. The former can be avoided by using a freshwater wash or, if this causes trouble with the water balance, recycled liquor can be used on a periodic basis by intermittent washing with freshwater to remove the sulphate-laden liquor. Using too much limestone causes mist eliminator plugging when less than 85% is utilized. The limestone is carried up into the unit and forms calcium salts in an area where there is not enough irrigation to keep surfaces free of precipitation. The situation arises mainly because operators use excess limestone as a means of complying with removal regulations.

Flue gas desulphurization

In order to put the concept of flue gas desulphurization into the proper perspective, a short description is necessary at this point. However, it is not the intent to launch into a detailed description of the various processes but to address the means by which the desulphurization of flue gases might be achieved.

Changes in the feedstock to a combustor, such as coal switching or coal cleaning, have the potential to reduce power plant emissions, but they are solutions that are external to the plant itself. To achieve greater removal of sulphur dioxide and nitric oxide, utility operators will need to institute other emission reduction options.

Flue gas cleaning (postcombustion cleaning) involves removing sulphur dioxide, nitrogen oxide(s) and/or particulates from the downstream flue gas after it leaves the boiler (Slack, 1981). However, the primary emphasis seems to have focused on the removal of the oxides of sulphur and nitrogen.

Conventional technology (wet scrubbers) uses lime or limestone to capture sulphur pollutants in the flue gas before it leaves the stack. This technology tends to be plagued by corrosion and pugging; it also produces a wet waste product (sludge), which has high disposal costs. However, the reliability of wet scrubbers has improved significantly, and they have demonstrated their ability to remove more than 90% of the sulphur dioxide.

Advanced postcombustion cleaning technologies encompass two approaches: (a) using the existing flue gas ductwork to inject a sorbent; and (b) inserting one or more separate vessels into the downstream ductwork where pollutant absorbents are added. Using a separate vessel allows a greater residence time for the absorbent to react, but the vessel is larger in size than the ductwork used with in-duct sorbent injection and is costlier to install.

In-duct sorbent injection works inside the ductwork leading from the boiler to the smokestack. Sulphur absorbers (such as hydrated lime) are sprayed into the centre of the duct. By controlling the humidity of the flue gas and the spray pattern of the sorbent, 50–70% of the sulphur dioxide can be removed. Selective additives, such as adipic acid, may remove more than 90% of the sulphur dioxide. The reaction produces dry particles that can be collected downstream. Because the plant's existing ductwork is used, extensive new construction is not needed. This makes in-duct sorbent injection an attractive option for retrofitting smaller, older plants where space requirements might be limited.

When separate vessels are used, one or more process chambers are inserted in the flue gas ductwork, and various sorbents are injected to remove the pollutants. Generally the separate vessels provide a longer residence time for the absorbent to react with the gas, and pollutant capture is greater. Although more costly than in-duct injection, this approach has the potential of capturing more than 90% of the pollutants. Owing to the cost and added size requirements, the use of separate vessels tends to be more suitable to new plant applications or to plants that can accommodate the additional size requirements. Technologies such as the spray dryer and selective catalytic reduction represent approaches that use separate vessels.

Flue gas desulphurization is an established technology for high-efficiency sulphur dioxide removal (EPRI, 1991). Recognition of this, and the fact that flue gas desulphurization is a technology applicable to new and retrofit

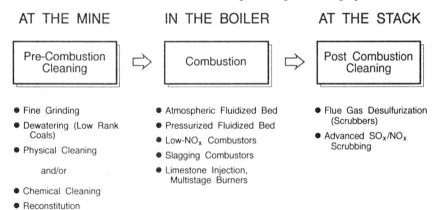

Figure 8.13 Schematic representation of system modifications for emissions control
Source: Gilleland and Swisher (1986)

operations, have caused a surge in research and development efforts aimed at the design of advanced flue gas clean-up systems for the simultaneous control of nitrogen oxides and sulphur oxides (Figure 8.13) (Feibus *et al.*, 1986).

However, flue gas desulphurization technologies are, currently, commercially available and cover a broad spectrum of chemical and physical process types (Doctor *et al.*, 1990). The first flue gas desulphurization installations on coal-fired plants were completed in the early 1970s. A major expansion of the use of flue gas desulphurization technology has taken place since then, with installations currently operating worldwide (Tables 8.6 and 8.7; Figures 8.14 and 8.15) (Couch *et al.*, 1990; Dalton, 1990; Vernon and Soud, 1990; IEA Coal Research, 1991).

Many of the processes described later in this text (Chapter 12) can be used for flue gas desulphurization in so far as, by definition, they are suited to the removal of sulphur-containing species from flue gases. The usual terminology applied to the processes, i.e. gas cleaning, may often be misleading in terms of realistic process definitions and capabilities.

The choice of a flue gas desulphurization process depends upon several factors not the least of which are: (a) coal properties (especially sulphur content); (b) regulations relating to sulphur dioxide emissions (and other waste products); as well as (c) a variety of economic assumptions and cost accounting procedures. Another important consideration in the choice of emission control systems is to avoid producing new pollutants, either by depositing wastes that have a potential to leach, producing contaminating waste water, or by increased carbon dioxide emissions due to the use of limestone and higher-energy use in operating the flue gas treatment plant. In more general terms, the choice will focus on site-specific conditions and issues (Klingspor and Cope, 1987).

At the risk of repetition, but to maintain the current trend of thinking, some of the candidate processes available for flue gas desulphurization will be introduced here on the understanding that the discussion will be limited to the salient features of the process(es) and that more details are supplied elsewhere in this text (Chapter 12).

Table 8.6 Types of flue gas desulphurization units (by MWe) installed in different countries (up to late 1990)

Country	Sorbent injection	Spray dry scrubbers	Limestone gypsum	Other wet processes	Regenerable processes	Not Known	Total
Austria	315	755	690	100			1 860
Canada	300						300
China				700			700
Czechoslovakia					(200)[a]		(200)[a]
Denmark		700				370	1 570
FRG	760	2 920	35 150	800	1 770	470	41 870
Finland	280	260	500				540
France	600						600
Italy					30		30
Japan			13 680	200			13 880
Netherland			2 730				2 730
Sweden	530	550		10			1 090
Turkey			340				340
USA	550	7 600	6 510	57 970	3 010	6 090	81 730
USSR	45						45
Total	3 380	12 785	59 600	59 780	4 810	6 930	147 285

[a] Not in operation.
Source: Kyte (1991).

Table 8.7 Estimates of future installations for flue gas desulphurization

	New GWe	Retrofit GWe	Total GWe
Bulgaria	–	2.3[a]	2.3
Canada	0.6	2.0	2.6
Czechoslovakia	–	2.5[a]	2.5
Denmark	0.4	0.3	0.7
FRG	1.0	4.0[a]	5.0
Finland	0.6	0.4[a]	1.0
Hungary	–	1.0	1.0
India	–	0.5	0.5
Italy	6.0	3.3	9.3
Japan	2.2	–	2.2
Netherlands	1.2	–	1.2
Poland	–	11.0[a]	11.0
Sweden	–	0.1	0.1
Taiwan	2.1	–	2.1
UK	9.5	–	9.5
USA	26.0	50.0[a]	76.0
Yugoslavia	–	2.1[a]	2.1
Total	49.6	79.5	129.1

[a] Not included in the IEA Coal Research database.
Source: Kyte (1991).

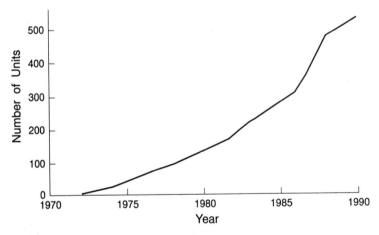

Figure 8.14 Increases in flue gas desulphurization systems on a worldwide basis
Source: Kyte (1991)

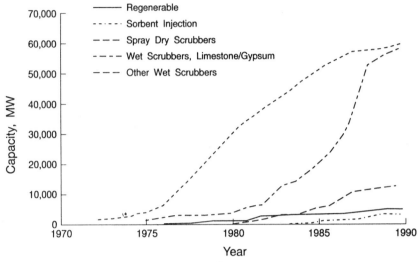

Figure 8.15 Increases in flue gas desulphurization capacity on a worldwide basis
Source: Kyte (1991)

Wet scrubbing

The most commonly used type of flue gas desulphurization procedures are wet scrubbers. The processes include a gas/liquid reaction and result in a wet product (Figure 8.16) (Licht, 1988). The majority of wet scrubbers use a calcium-based slurry as a sorbent, generally of lime or limestone, but sodium-based sorbents can also be used.

In this process, the flue gas is sprayed with a slurry made up of water and an alkaline reagent, usually lime or limestone. The sulphur dioxide present in the flue gas reacts chemically with the reagent in the slurry, forming calcium sulphite and/or calcium sulphate, which is removed and disposed of in the form of a wet sludge:

$$SO_2 + CaO = CaSO_3$$
$$SO_2 + Ca(OH)_2 = CaSO_4 + H_2O$$
$$SO_2 + CaCO_3 = CaSO_3 + CO_2$$

Since the residue of such a process is a wet mix of calcium sulphite and sulphate, an oxidation step can be incorporated to produce gypsum, which can be dried. This removes the need for extensive land area for ponds, landfills or one or the various transportation systems to handle the large volumes of the wet waste.

In addition, scrubbers use large amounts of water, perhaps not quite analogous to the water usage at the oil sands plants in northern Alberta (Speight, 1990) but certainly enough to be of some concern. The institution of such processes in semi-arid/arid regions will make demands upon the water resources that are difficult, if not impossible, to tolerate. In summary, waste management is a major issue for those flue gas desulphurization technologies which involve wet scrubbing. Water management could be a major issue, depending upon regional and other water demands.

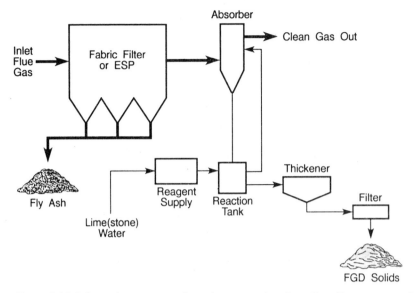

Figure 8.16 Schematic representation of a conventional wet lime/limestone scrubber

The first generation of wet scrubbers were designed to use lime as a sorbent, but limestone has become more common in recent years. Processes producing gypsum are increasingly being favoured. Wet scrubbers are used with a wide variety of coal types and municipal wastes (Donnelly and Felsvang, 1989) on installations up to the largest sizes for coal-fired units. Sulphur removal efficiencies of over 90% are normally achieved.

Dual-alkali wet scrubbers are used less widely. Here a sodium-based solution absorbs sulphur dioxide from the flue gases and the reaction products are regenerated in a second step using lime or limestone for the final sulphur dioxide capture.

Other types of wet scrubber in use on a small number of coal-fired plants employ either aqueous ammonia as a sorbent, giving a fertilizer by-product, or carbide sludge. Another process requires a coastal location to use the natural alkalinity of sea water as a sorbent. In this process large volumes of sea water are needed.

Dry scrubbing

In this process, a dry sorbent (such as pulverized limestone, ($CaCO_3$), hydrated lime ($Ca(OH)_2$) or sodium bicarbonate ($NaHCO_3$)), is fed into the furnace upstream of the particulate collection system (Figure 8.17). The sorbent reacts with the sulphur dioxide whence calcium sulphate or sodium sulphate is produced:

$$CaCO_3 = CaO + CO_2$$
$$Ca(OH)_2 = CaO + H_2O$$
$$CaO + SO_2 = CaSO_3$$
$$2CaSO_3 + O_2 = 2CaSO_4$$
$$2NaHCO_3 = Na_2O + 2CO_2 + H_2O$$

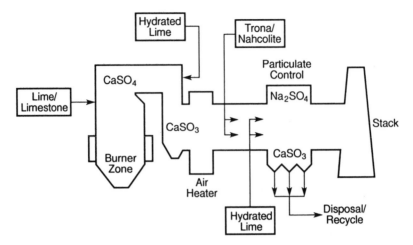

Figure 8.17 Options for dry sorbent injection as a means of emissions control

The initial application of this technology was disappointing because the efficiency of the sulphur dioxide removal was variable and generally low (Gartrell, 1973; Blythe, 1982; Drehmel et al., 1983; Smith and Dhargalkar, 1992).

The reactions above can also proceed, quite effectively, during combustion at reduced temperature, which is a desirable environment for the suppression of nitric oxide (Chughtai and Michelfelder, 1983). By contrast, a conventionally hot, high-heat-release flame zone seems to fuse and seal (sinter) the limestone particles, cutting off their reactivity. The latter reaction fundamentally limited the effectiveness of the earlier attempts at sulphur dioxide reduction by this method.

The dry sorbent injection may be accomplished in several ways. For example, calcium-based sorbents can be injected in the high-, mid- or low-temperature regions of a utility boiler, while sodium-based sorbents are best injected into the low-temperature regions of the boiler (Figure 8.18).

One of the drawbacks of limestone scrubbing is the damage to construction materials by corrosion and erosion. Although the scrubber slurry is not highly acidic (pH 5.0 to 5.5), it has the capability of corroding carbon steel and several alloys used in the construction of equipment. Many systems use a prescrubber before the main scrubber as a method of removing particulate matter and other impurities that are undesirable in gypsum destined for later use.

Other scrubber processes that have been used, but to a limited extent, include indirect limestone (e.g. dual alkali) scrubbing and indirect lime scrubbing. The indirect method (i.e. scrubbing with an absorbent that has a high affinity for sulphur dioxide and then adding lime or limestone to precipitate calcium sulphite or sulphate and regenerate the absorbent at the same time) has seen wide usage. The use of magnesian lime, made by calcining a limestone containing magnesium, was popular at one time. The magnesium makes the lime (called 'thiosorbic') very effective as an absorbent.

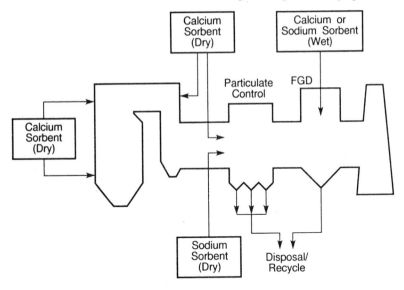

Figure 8.18 Simplified illustration for sulphur dioxide control

The recovery processes are quite attractive, particularly where waste water discharge permits may be difficult to obtain. Although many recovery processes have been developed and some installed, the relatively high costs and difficulties in merchandizing a recovery-type product have proven to be major obstacles in the adoption of such processes.

Dry scrubbing has been developed as an alternative to wet scrubbers especially for smaller-sized installations and where the sulphur content in the coal is usually less than 2% wt/wt. Efficiencies for sulphur dioxide removal of the order of 80–90% are usually recorded. The concept is of special interest because of its suitability for applications where either equipment size or water availability (see above) may be a particular problem.

Regenerable processes

Regenerable processes are processes where the sorbent for sulphur dioxide can be reused following a regeneration step which releases concentrated sulphur dioxide. The sulphur dioxide can then be processed further to liquid sulphur dioxide, or sulphuric acid or elemental sulphur.

The most widely used regenerable flue gas desulphurization system is the Wellman–Lord process (page 327) which uses a gas/liquid reaction with a sodium sulphite solution as the sorbent. The reaction product is regenerated by heat treatment to produce a gas rich in sulphur dioxide that can be further processed to give sulphuric acid or elemental sulphur. Another regenerable process uses magnesium oxide slurry as a sorbent (Chapter 12). As with the Wellman–Lord process, the sorbent is thermally regenerated producing a gas rich in sulphur dioxide. Yet another sorbent uses a sodium-impregnated aluminium substrate (Neal and Woods, 1992) for the reduction of nitrogen oxides and sulphur dioxide in flue gases.

Sulphur oxide/nitrogen oxide processes

There are also combined sulphur oxide/nitrogen oxide (SO_x/NO_x) processes where the flue gas desulphurization is combined with a denitrification step. An example is the activated carbon process where sulphur dioxide is adsorbed on activated carbon (page 255). Ammonia is injected into the flue gases in a mixing chamber before the flue gases enter the second stage. In the second stage nitrogen oxides react catalytically with ammonia to form water and nitrogen. Desorption of the sulphur-laden carbon is achieved by thermal means.

Another combined SO_x/NO_x method involves catalytic absorption in which the first step is to remove nitrogen oxide(s) by selective catalytic reduction (page 278) after which a catalyst is used to convert any sulphur dioxide to sulphur trioxide. The flue gases are cooled and the sulphur trioxide is hydrated to sulphuric acid by water vapour contained in the flue gases (Hjalmarsson, 1990).

References

Amand, L.-E. and Leckner, B. (1991) *Energy and Fuels*, **5**, 809, and references cited therein

Amand, L.-E., Leckner, B. and Andersson, S. (1991) *Energy and Fuels*, **5**, 815, and references cited therein

Argonne National Laboratory (1984) *Coal Cleaning as an Acid Rain Mitigation Strategy: An in-depth examination.* Report No. 84–27.5, Argonne National Laboratory, Argonne, IL

Bartok, W., Folsom, B.A., Payne, R. and Wu, K.T. (1991) *Journal of Environmental Science and Health*, **A26**, 1267

Bielaga, B.A. and Kilbane, J.J. (1990) *Molecular Biological Enhancement of Coal Biodesulfurization.* Report No. DOE/PC88891-T6-DE2003086, United States Department of Commerce, Springfield, VA

Blythe, G.M. (1982) *Dry Limestone Injection Test as a Low Rank Coal Fired Power Plant.* Technical Report No. DCN 82-213-022-06, Radian Corporation, Austin, TX

Ceeley, F.J. and Daman, E.L. (1981) In *Chemistry of Coal Utilization. Second Supplementary Volume*, Wiley, New York, Chapter 20

Chughtai, M.Y. and Michelfelder, S. (1983) *Proc. Symp. on Flue Gas Desulfurization, New Orleans, Louisiana, 1–4 November*

Couch, G., Hessling, M., Hjalmarsson, A-K., Jamieson, E. and Jones, T. (1990) *Coal Prospects in Eastern Europe.* IEA CR/31, IEA Coal Research, London

CST (1991) Florida utility shifting 1,400 MW to IGCC. *Coal and Synfuels Technology*, 23 December, p. 3

Dalton, S. (1990) Status of US flue gas desulphurisation situation in the USA. *Proc. Symp. on Flue Gas Desulphurisation, 19–20 November, Madrid, Spain*

Deurbrouck, A.W. and Hucko, R.E. (1981) In *Chemistry of Coal Utilization. Second Supplementary Volume*, Wiley, New York, Chapter 10

Doctor, R.D., Wilzbach, K.E. and Joseph, T.W. (1990) *Sulfur Emissions Reduction at the Great Plains Coal Gasification Facility: Technical and Economic Evaluations.* Report No. ANL/ESD-1, Argonne National Laboratory, Argonne, IL

Donnelly, J.R. and Felsvang, K.S. (1989) *Proc. 82nd Annual Meeting, Air and Waste Management Association, Anaheim, California, 25–30 June*, Paper 89–109.6

Drehmel, D.C., Martin, G.B. and Abbott, J.H. (1983) *Proc. Symp. on Flue Gas Desulfurization*, United States Environmental Protection Agency (Washington, DC) and Electric Power Research Institute (Research Triangle Park, NC), p. 689

Eliot, R.C. (1978) *Coal Desulfurization Prior to Combustion*, Noyes Data Corp., Park Ridge, NJ

EPRI (1991) *Economic Evaluation of 28 FGD Processes*. Report No. EPRI GS-7193, Electric Power Research Institute, Palo Alto, CA

Essenhigh, R.H. (1981) In *Chemistry of Coal Utilization. Second Supplementary Volume*, Wiley, New York, Chapter 19

Evans, J.J., Dugliss, W.G., Larochelle, M., Paavola, J.O., Sausser, R., Vanhorn, G.R. and Young, A. (1991) *Proceedings. American Power Conference*, **53**, (2), 883

Feibus, H., Voelker, G. and Spadone, S. (1986) In *Acid Rain Control II. The Promise of New Technology*, Southern Illinois University Press, Carbondale, IL, Chapter 3

Frost, J.W. (1991) *Biotic and Abiotic Carbon to Sulfur Bond Cleavage*. Report No. CONF-9106303-DE92003115, United States Department of Commerce, Springfield, VA

Gartrell, F.E. (1973) *Full-scale Desulfurization of Stack Gas by Dry Limestone Injection*. Technical Report No. EPA-650/2-72-019, Vol. 1, Tennessee Valley Authority, Chattanooga, TN

Gilleland, D.S. and Swisher, J.H. (eds.) (1986) *Acid Rain Control II: The Promise of New Technology*, Southern Illinois University Press, Carbondale, ILL

Hall, E.H., Peterson, D.B., Foster, J.F., Kiang, K.D. and Ellzey, V.W. (1975) *Fuels Technology: A State of the Art Review*. Report No. EPA-650/2-75-034, Batelle Laboratories, Columbus, OH

Hart, A.B. (1981) In *Energy and Chemistry* (ed. R. Thompson), Royal Society of Chemistry, London

Hessley, R.K. (1990) In *Fuel Science and Technology Handbook* (ed. J.G. Speight), Marcel Dekker, New York

Hjalmarsson, A-K. (1990) NO_x *Control Technologies for Coal Combustion*, 1EACR/24, IEA Coal Research, London

IEA Coal Research (1991) *IEA Coal Research FGD and NO_x Control Installations Data-base*, IEA Coal Research, London

Jenkins, S.C. (1991) *Proceedings. American Power Conference*, **53**, (1), 732

Jodal, M., Nielsen, C., Hulgaard, T. and Dam-Johansen, K. (1990) *Proc. 23rd Symp. (Int.) on Combustion*, The Combustion Institute, Pittsburgh, PA, p. 237

Khan, S.R., Desai, M.S. and Gawin, A.F. (1990) *Proceedings. American Power Conference*, **52**, 595

Klingspor, J. and Cope, D. (1987) *FGD Handbook, Flue Gas Desulphurisation Systems*, ICEAS/BS, IEA Coal Research, London

Kyte, W.S. (ed.) (1991) *Desulphurization 2: Technologies and Strategies for Reducing Sulphur Emissions*, Hemisphere Publishing Corporation, London, England

Lausen, P. and Schioth, M.R. (1992) *Power Generation Technology*, p. 103

Leckner, B., Karlsson, M., Johansen, K., Weinell, E., Kilpinen, P. and Hupa, M. (1991). *Ind. Eng. Chem. Research*, **30**, 2396

Licht, W. (1988) *Air Pollution Control Engineering*, Marcel Dekker, New York

Marrocco, M., Hollback, J.E. and Stogran, H.K. (1991) *Proceedings. American Power Conference*, **53**, (2), 1041

Meyers, R.A. (ed.) (1981) *Coal Handbook*, Marcel Dekker, New York

Meyers, R.A., Hamersma, J.W., Land, J.S. and Kraft, M.L. (1972) *Science*, **117**, 1187

Miller, K.W. (1991) *Evaluation of Sulfur-reducing Microorganisms for Organic Desulfurization*. Report No. DOE/PC90176-DE92004188, United States Department of Commerce, Springfield, VA

Mjornell, M., Leckner, B., Karlsson, M. and Lyngfelt, A. (1991) *Proc. 11th Int. Conf. on Fluidized Bed Combustion*, American Society for Mechanical Engineers, p. 655

Mohnen, V.A. (1988) *Scientific American*, **259** (2), 30

Neal, L.G. and Woods, M.C. (1992) *Power Generation Technology*, p. 89

Pachaly, R., Hofmann, J.E. and Sun, W.H. (1992) *Power Generation Technology*, p. 79

Pierce, J.L., Duckett, E.J., Haug, R.T. and Albertson, D.M. (1991) *Proceedings. American Power Conference*, **53**, (1), 164

Powell, E.M. and Ulmer, R.C. (1974) *Combustion*, **45**, 23

Reid, W.T. (1981) In *Chemistry of Coal Utilization. Second Supplementary Volume*. Wiley, New York, Chapter 21

Rupinskas, R.L. and Hiller, P.A. (1991) *Proceedings. American Power Conference*, **53**, (2), 867

Slack, A.V. (1981) In *Chemistry of Coal Utilization. Second Supplementary Volume*, Wiley, New York, Chapter 22

Smith, P.V. and Dhargalkar, P.H. (1992) *Power Generation Technology*, p. 97

Speight, J.G. (1983) *The Chemistry and Technology of Coal*, Marcel Dekker, New York

Speight, J.G. (1990) In *Fuel Science and Technology Handbook* (ed. J.G. Speight), Marcel Dekker, New York

Speight, J.G. (1991) *The Chemistry and Technology of Petroleum*, 2nd edn, Marcel Dekker, New York

United States Department of Energy (1990) *Gas Research Program: Implementation Plan*. DOE/FE-0187P, United States Department of Energy, Washington, DC, April

United States Department of Energy (1991) *Clean Coal Technology Demonstration Program*. DOE/FE-0219P, United States Department of Energy, Washington, DC, February

United States General Accounting Office (1990) *Energy Policy: Developing Strategies for Energy Policies in the* 1990s. Report to Congressional Committees, GAO/RCED-90-85, United States General Accounting Office, Washington, DC, June

Vernon, J. and Soud, H. (1990) *FGD Installations on Coal-fired Plants*. IEA CR/22, IEA Coal Research, London

Worthy, W. (1975) *Chemical Engineering News*, **53**, (27), 24

Part Two
Gas cleaning

9

Processing – general concepts

Introduction

Gas processing, although generally simple in chemical and/or physical principles, is often confusing because of the frequent changes in terminology and, often, lack of cross-referencing.

For the purposes of this text, and in an attempt to alleviate some of the confusion that arises from uncertainties in the terminology, gas cleaning (gas processing) can be conveniently subdivided into two sections: (a) the first section (immediately below and Chapter 10) deals with the general proceing concepts; and (b) specific processes (Chapter 12). Because of the nature of gas-processing operations, a general description of the types of equipment used is also given in this text (Chapter 11).

Thus, although gas processing employs different process types there is always an overlap between the various concepts. Therefore, in any text such as this there is a need, for the convenience of the reader, for repetition and this is achieved by subdivision of the subject categories. When necessary, cross-referencing is employed so that the reader should not miss any particular aspect of the processing operations.

There are several process options available for the preliminary purification of gases (Mott, 1972) that will not be covered here, being considered outside the scope of this text. Nevertheless, the reader should be aware that gas, like petroleum (Speight, 1991), needs preliminary treatment to remove any impurities before being subjected to gas cleaning/processing proper. Protection of the chemical agents from other species is recommended for continued efficiency of the gas-cleaning process(es).

The selection of a particular process type (Table 9.1) for a gas-cleaning operation is not a simple choice. Many factors have to be considered, not least of which is the constitution of the gas stream that requires treatment. Indeed, process selectivity indicates the preference with which the process will remove one acid gas component relative to another. For example, some processes remove both hydrogen sulphide and carbon dioxide whilst others are designed to remove hydrogen sulphide only (Table 9.2).

It is important to consider the process selectivity for, say, hydrogen sulphide removal compared with carbon dioxide removal, which will ensure minimal concentrations of these components in the product – hence the need to consider the ration of carbon dioxide to hydrogen sulphide in the gas stream.

There are many variables in gas treating and the precise area of application

Table 9.1 General summary of gas-cleaning processes

Sorbent	Nature of interaction	Regeneration	Examples
Liquid	Absorption + chemical reaction	Yes	Many processes for the removal of CO_2 and H_2S from various gases, with solvents like water + MEA, DEA, DIPA, etc.
			Agents improving physical solubility may be added (Sulfinol process); H_2S may be recovered as such or oxidized to S
Liquid + solid	Absorption + chemical	Varies	Some slurry wash processes for flue gas desulphurization
Liquid	Physical adsorption	Yes	CO_2 and/or H_2S from hydrocarbon gases; solvents: N-methyl pyrrolidone, propylene carbonate, methanol
Solid	Physical adsorption	Yes	Purification of natural gas (H_2S, CO_2); with molecular sieves
		Yes	Gas-drying operations (cyclic regenerative); molecular sieves
		Varies	Odour removal from waste gases (active carbon)
Solid	Chemical reaction	No	H_2S from process gases, with ZnO
		Yes	SO_2 from flue gases, with CuO/Al_2O_3

Table 9.2 Gas-cleaning processes for hydrogen sulphide and carbon dioxide removal (see Chapter 12)

Process	Sorbent	Removes
Amine	Monoethanolamine, 15% in water	CO_2, H_2S
Econamine	Diglycolamine, 50–70% in water	CO_2, H_2S
Alkazid	Solution M or DIK (potassium salt of dimethylamine acetic acid), 25% in water	H_2S, small amount of CO_2
Benfield, Catacarb	Hot potassium carbonate, 20–30% in water (also contains catalyst)	CO_2, H_2S; selective to H_2S
Purisol	n-Methyl-2-pyrrolidone	H_2S, CO_2
Fluor	Propylene carbonate	H_2S, CO_2
Selexol	Dimethyl ether polyethylene glycol	H_2S, CO_2
Rectisol	Methanol	H_2S, CO_2
Sulfinol	Tetrahydrothiophene dioxide (sulfolane) plus diisopropanol amine	H_2S, CO_2; selective to H_2S
Giammarco–Vetrocoke	K_3AsO_3 activated with arsenic	H_2S
Stretford	Water solution of Na_2CO_3 and anthraquinone disulphonic acid with activator of sodium metavanadate	H_2S
Activated Carbon	Carbon	H_2S
Iron Sponge	Iron oxide	H_2S
Adip	Alkanolamine solution	H_2S, some COS, CO_2, and mercaptans
SNPA–DEA	Diethanolamine solution	H_2S, CO_2
Takahax	Sodium, 1,4-napthaquinone, 2-sulphonate	H_2S

of a given process is difficult to define although there are several factors that need to be considered: (a) the types and concentrations of contaminants in the gas; (b) the degree of contaminant removal desired; (c) the selectivity of acid gas removal required; (d) the temperature, pressure, volume and composition of the gas to be processed; (e) the carbon dioxide to hydrogen sulphide ratio in the gas; (f) the desirability of sulphur recovery due to process eonomics or environmental issues.

In addition to hydrogen sulphide and carbon dioxide, gas streams may contain other contaminants such as sulphur dioxide, mercaptans and carbonyl sulphide. The presence of these impurities may eliminate some of the sweetening processes since some processes will remove large amounts of acid gas but not to a sufficiently low concentration. On the other hand, there are those processes that are not designed to remove (or are incapable of removing) large amounts of acid gases but are capable of removing the acid gas impurities to very low levels when the acid gases are in low-to-medium concentrations in the gas stream.

A variety of processes are commercially available for the removal of acid gas from gas streams (Figure 9.1) (Probstein and Hicks, 1990) and generally fall into one of several categories. However, several factors control the choice of an acid gas removal process and these are: (a) gas flow rate; (b) concentration of acid gases in the gas stream; and (c) the necessity to remove carbon dioxide as well as hydrogen sulphide.

Gas purification by some form of sorption by a liquid or solid sorbent is one of the most widely applied operations in the chemical and process industries (Table 9.1). Some of the process options have the potential for sorbent regeneration but, in a few cases, the process is applied in a non-regenerative manner. The interaction between sorbate and sorbent may either by physical in nature or consist of physical sorption followed by chemical reaction. Other

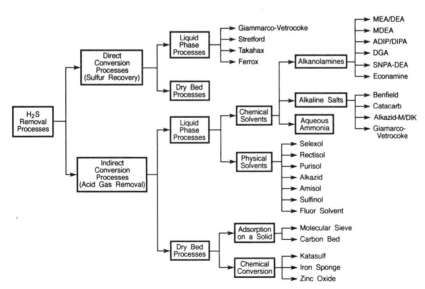

Figure 9.1 Examples of processes used for the removal of hydrogen sulphide
Source: Probstein and Hicks (1990)

gas stream treatments use the principle of chemical conversion of the contaminants with the production of 'harmless' (non-contaminant) products or the conversion to substances which can be removed much more readily than the impurities from which they are derived.

The most important application of the sorption concept, other than gas drying, is the removal of carbon dioxide and/or hydrogen sulphide from gas streams such as natural gas, synthesis gas, primary products from the gasification of coal or heavy fuel oil, and gas streams obtained from the desulphurization and refining of petroleum fractions (Table 9.3).

Many different methods have been developed for the removal of carbon dioxide and hydrogen sulphide, some of which are briefly discussed below. Concentrates of hydrogen sulphide obtained as by-products of gas desulphurization are often converted by partial oxidation to elemental sulphur (Claus process) (Chapter 12).

For large gas volumes containing high concentrations of carbon dioxide mixed with hydrogen sulphide, a likely but by no means unique sequence of treatments is possible (Figure 9.2) (Probstein and Hicks, 1990). Most of the carbon dioxide and hydrogen sulphide are removed in a regenerable liquid absorbent which is continuously circulated, and the final traces of hydrogen sulphide which, for example in a refining scenario, might poison processing catalysts are removed in a solid adsorbent which can be regenerated or discarded. Off-gas from the solvent section of the process may be treated in a Claus unit for the recovery of sulphur. The final clean-up of the Claus plant off-gas, usually referred to as 'tail gas', can be by a direct conversion process (see Chapters 10 and 12).

Table 9.3 Comparative summaries of various acid gas removal processes (see Chapters 10 and 12)

	Chemical adsorption		
Feature	Amine processes	Carbonate processes	Physical absorption
Absorbents	MEA, DEA, DGA, MDEA	K_2CO_3, $K_2CO_3 + MEA$ $K_2CO_3 + DEA$, $K_2CO_3 +$ arsenic trioxide	Selexol, Purisol, Rectisol
Operating pressure (psi)	Insensitive to pressure	> 200	250–1000
Operating temp. (°F)	100–400	200–250	Ambient temperature
Recovery of absorbents	Reboiled stripping	Stripping	Flashing, reboiled, or steam stripping
Utility cost	High	Medium	Low–medium
Selectivity H_2S, CO_2	Selective for some amines (MDEA)	May be selective	Selective for H_2S
Effect of O_2 in the feed	Formation of degradation products	None	Sulphur precipitation at low temperature
COS and CS_2 removal	MEA: not removed DEA: slightly removed DGA: removed	Converted to CO_2 and H_2S and removed	Removed
Operating problems	Solution degradation; foaming; corrosion	Column instability; erosion; corrosion	Absorption of heavy hydrocarbons

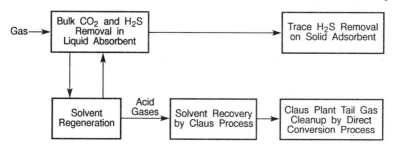

Figure 9.2 Example of a process sequence for acid gas removal from gas streams
Source: Probstein and Hicks (1990)

Table 9.4 Factors affecting gas absorption

Zone	Factor
Gas	Gas velocity
	Molecular weights and sizes of host and solute gases
	Temperature
	Concentration gradient
Solvation	Solubility in host liquid
Liquid	Molecular weights and sizes of host liquid and solute gas
	Temperature
	Viscosity of liquid
	Liquid surface velocity
	Surface renewal rate
	Concentration gradient of solute gas
	Concentration gradient of neutralizing reagent

However, the actual choice of a gas-processing scheme will depend on a number of factors (Table 9.4) and these include the presence of other impurities, gas pressure, solvent efficiency and energy requirements.

Absorption

Absorption is the separation of solute gases from gaseous mixtures by transfer into a liquid solvent. This recovery is achieved by contacting the gas stream with a liquid that offers specific or selective solubility for the gas(es) to be recovered. The process is applied extensively in industry to purify process streams or recover valuable components of the stream as well as to remove toxic or noxious components (pollutants) from an effluent gas stream.

It is noteworthy at this point that absorption is achieved by dissolution (a physical phenomenon) or by reaction (a chemical phenomenon) (Table 9.1). Chemical adsorption processes adsorb sulphur dioxide on to a carbon surface where it is oxidized (by oxygen in the flue gas) and absorbs moisture to give sulphuric acid impregnated into and on the adsorbent. Various methods are used to remove the acid from the adsorbent. These processes are attractive because regeneration of the adsorbent is generally simpler and more convenient than the regeneration of most chemical absorbents.

The absorption process can use liquid or solid absorbents (Barboteau and Dalaud, 1972; Ward, 1972) although the definition of the process is open to question because of the participation of chemical effects. To the chemical purist, many of the absorption processes that have been classed as 'physical' for several decades (or at least since the early days of the gas-processing industry) would fall into the category of chemical processes. But it is not the intent of this text to add another layer to the strata of confusion. What is, or has been, will remain!

Desorption, or stripping, is essentially a reversal of the absorption process in which the material moves from the liquid to the gaseous phase.

Wet absorption is one of the most popular, if not the most popular, options for the removal of sulphur dioxide from gas streams. A variety of processes exist, many of which are on-stream or are very close to being commercial in so far as the process, or a close configuration of it, has been commercialized (Cortelyou, 1969).

The absorption of acid gases in liquid absorbents, usually organic or inorganic alkaline solutions, has been the subject of a considerable volume of both research and industrial development (Zenz, 1979). Most of this work has been directed towards the recovery of sulphur dioxide from the flue gases of power plants burning sulphur-containing coal or oil, and the many processes which have resulted are covered elsewhere in this volume.

Nevertheless, the absorption process involves the following steps: (a) diffusion of the solute gas molecules through the host gas to the liquid boundary layer; (b) solvation of the solute gas in the hot liquid; and (c) diffusion of the solute gas based on concentration gradient. The removal of the solute gas from the liquid boundary layer is often accomplished by adding neutralizing agents to the host liquid to change the molecular form of the solute gas (absorption accompanied by chemical reaction).

Several factors (Table 9.4) affect the rate of absorption or the kinetics of transfer from the host gas. The transfer of gas into the liquid phase tends to become the controlling factor in the rate of absorption, except for situations of very high gas solubility. The transfer systems require the effective contact of liquid and gas, with maximum surface contact and surface renewal, and minimum consumption of energy.

As currently practised, acid gas removal processes involve the selective absorption of the contaminants into a liquid which is passed countercurrent to the gas. Thence, the absorbent is stripped of the gas components (regeneration) and recycled to the absorber (Figure 9.3). The process design will vary and, in practice, may employ multiple absorption columns and multiple regeneration columns.

Before entering the absorber, the crude gas must be free of all heavy hydrocarbons and separators fitted with baffles are placed in series to collect particles of water and hydrocarbons (Figure 9.4). The performance of the absorber is affected by the efficiency of the separators because the hydrocarbons cause foaming which leads to the instability of the wash and an increased cost of treatment, and consequently heavy losses of absorbing solution.

As noted previously above, liquid absorption processes, which usually employ temperatures below $50\,^{\circ}C$ ($120\,^{\circ}F$), are classified either as physical solvent processes or chemical solvent processes. The former employ an organic solvent and absorption is enhanced by low temperatures, or high pressure, or both. Regeneration of the solvent is often accomplished readily

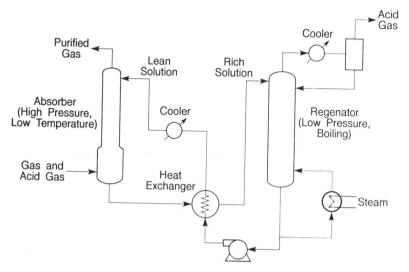

Figure 9.3 Illustration of a gas-cleaning (absorber) system

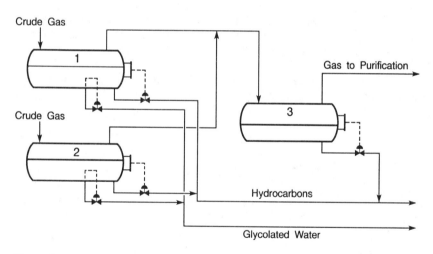

Figure 9.4 A preliminary gas-treating (water and hydrocarbon) removal system

(Staton *et al.*, 1985). In the latter (i.e. chemical solvent processes), absorption of the acid gases is achieved mainly by use of alkaline solutions such as amines or carbonates (Kohl and Riesenfeld, 1979). Regeneration (desorption) can be brought about by use of reduced pressure and/or high temperature whereby the acid gases are stripped from the solvent.

The most common way of removing carbon dioxide and hydrogen sulphide from gas streams is by washing with solvents which are selective for these impurities. Other examples of the application of gas absorption systems include the removal of mercaptans from gas streams, the recovery of carbon monoxide in petrochemical plants, the removal of sulphur oxides from power plant stack gases, and the recovery of silicon tetrafluoride and hydrogen fluoride from fertilizer production stack gases.

The distribution of the acid gas components over the gas and liquid phases is, to a first approximation, governed by Henry's law if only physical interactions occur:

$$x_i = p_i/H_i$$

where x is the mole fraction, p the partial pressure and H the Henry coefficient; subscript i refers to component i of the gas.

If the separation is based solely on differences in H_i, the rate of dissolution is governed by mass transfer from the gas to liquid phase. It is possible to obtain lower equilibrium partial pressures by applying solvents in which the acid components react, e.g. with a base. At moderate degrees of conversion lower equilibrium pressures of impurities can be obtained.

The separation may be either by physical dissolution or by physical dissolution followed by chemical reaction (Tables 9.5 and 9.6). In some processes hydrogen sulphide is removed selectively by oxidizing absorbed hydrogen sulphide to elemental sulphur by aerating the solution.

The two types of absorption processes also behave differently with respect to the operating temperature although the basic process concepts may appear to be similar, if not the same (Figures 9.5 and 9.6).

In a physical absorption system, the amount of gas dissolved at equilibrium increases with decreasing temperature, with the result that the process becomes more efficient. In a chemical absorption system, however, a temperature decrease may result in the less complete removal of contaminants when the rate of dissolution is largely determined by the rate at which the reaction proceeds in the liquid phase.

Table 9.5 Simplified classification system for acid gas removal processes (see also Chapters 10 and 12)

Chemical absorption (chemical solvent processes)	Physical absorption (physical solvent processes)
Alkanolamines:	
MEA	Selexol
SNPA: DEA (DEA)	Rectisol
UCAP (TEA)	Sulfinol[a]
Selectamine (MDEA)	
Econamine (DGA)	
ADIP (DIPA)	
Alkaline salt solutions:	
Hot potassium carbonate	
Catacarb	
Benfield	
Giammarco–Vetrocoke	
Non-regenerable:	
Caustic	

[a] A combined physical/chemical solvent process.

Table 9.6 Physical absorption processes for gas cleaning

Process	Absorber solvent
Fluor Solvent	Propylene carbonate
Purisol	N-methyl pyrrolidone
Rectisol	Methanol (below 0°F, −18°C)
Selexol	Polyethylene glycol dimethyl ether
Sulfinol	Sulfolane (tetrahydrothiophene dioxide) mixed with an alkanolamine and water

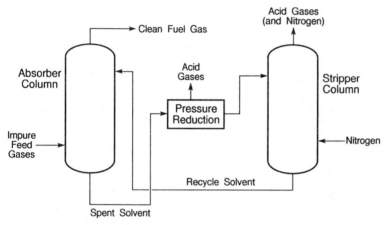

Figure 9.5 Gas cleaning by physical absorption

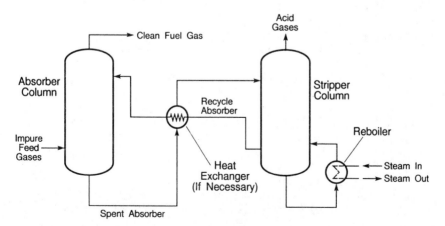

Figure 9.6 Gas cleaning by chemical absorption

The rate of dissolution of carbon dioxide in an aqueous liquid consists of the following steps:

$$CO_2(gas) = CO_2(liquid)$$
$$CO_2(liquid) + OH^- = HCO_3^-$$

followed by further reactions, if these can occur.

The second step is generally considered to be so slow that it limits the process rate under unfavourable conditions. This indicates that the absorption can be speeded up by raising the temperature. Although this causes the concentration of carbon dioxide in the liquid phase to decrease, the effect is overshadowed by the faster attack on carbon dioxide by hydroxyl ions, resulting in higher rates. A disadvantage is that the equilibrium pressure of carbon dioxide increases with increasing temperatures, other conditions remaining the same.

A second way of overcoming limitations due to chemical kinetics is to add a solvent in which the physical solubility of carbon dioxide is much higher. This causes an increase in carbon dioxide (liquid) and a proportional increase in the rate of bicarbonate formation, and the equilibrium partial pressure of carbon dioxide increases.

There are a number of chemical absorption processes applied in industry (Table 9.6). One of these, the Sulfinol process, uses a solvent showing high physical solubility for carbon dioxide and carbonyl sulphide. This speeds up the dissolution of the carbon dioxide and also raises the rate at which the carbonyl sulphide is hydrolysed. These advantages are to some extent offset by the higher solubility of other gas components in the absorption liquid. In other words, the Sulfinol process can be considered as a hybrid between a physical and a chemical absorption process.

When applying chemical absorption processes to gases obtained by the partial oxidation of heavy fuels or by coal pyrolysis or gasification, care should be taken that hydrogen cyanide (HCN) is not present in the feed to the absorber because it would hydrolyse to formic acid and thus permanently neutralize the base. This implies that hydrogen cyanide should be removed beforehand, e.g. by water wash or by catalytic conversion.

There are many reports in the literature indicating that additives may be added to speed up carbon dioxide absorption in chemical absorption processes. Examples are methanol, ethanol, glycols, glycerol, sucrose and dextrose. It is difficult to establish whether these materials act by increasing the physical solubility of carbon dioxide by catalysis of the reaction of carbon dioxide to the bicarbonate ion or by both mechanisms. In many cases the increased rate of dissolution seems to be due to both effects.

Absorption of hydrogen sulphide is invariably fast because mass transfer to the solution is followed by rapid dissociation into a proton (H^+) and a bisulphide ion (HS^-). This is utilized in some chemical processes to obtain partial selectivity for hydrogen sulphide in the presence of carbon dioxide. The temperature must be kept relatively low to ensure that carbon dioxide absorption is limited to a small fraction.

This type of separation is possible with secondary and tertiary amine-containing solvents and is based on kinetic and not on equilibrium phenomena. Absorption of carbon dioxide and hydrogen sulphide in solutions is accompanied by a thermal effect. With physical absorption, where the forces of interaction between solute and solvent are weak, the heat of dissolution is usually small, i.e. less than 5 kcal/mole (20 kJ/mol). This indicates that regeneration of the fat solution by raising the temperature is not very effective and that the driving force for regeneration should be obtained mainly by lowering the pressure. Thus, physical absorption is most economic when gases containing high partial pressures of acid components must be treated to products of at best moderate purity. Another consequence is that the amount of heat required for regeneration is relatively small.

In the case of absorption by chemical reaction, however, the exothermic heat of absorption is often much higher, exceeding 25 kcal/mole (100 kJ/mol) in many processes. In such cases regeneration can be effected by raising the temperature, lowering the pressure, or by both methods. The heat requirements for regeneration of chemically active solvents are much higher than for solvents showing physical interactions only. This does not mean that chemical treatment invariably requires more energy than physical absorption, because the amount of solvent circulating in the process is often much smaller for chemical absorption and the energy required for this circulation may offset the energy advantage of physical absorption processes as compared with chemical absorption.

All gas purification plants in which physical absorption is applied have similar basic structures. Any differences are due to solvent properties, product quality requirements, the desired degree of heat economy, which is often greater as plant size increases, and similar process requirements. Another important factor is the extent to which selectivity between carbon dioxide and hydrogen is needed.

Solvents used for gas-cleaning processes should have a high capacity for acid gas, a low tendency to dissolve valuable feed components such as hydrogen and low-molecular-weight hydrocarbons, low vapour pressure at operating temperatures to minimize solvent losses, low viscosity, thermal stability, an absence of reactivity towards gas components, a low tendency towards fouling of equipment and corrosion, as well as acceptable cost. Poly(ethylene glycol dimethyl ether) and N-methyl pyrrolidone are reasonably selective towards hydrogen sulphide in the presence of carbon dioxide.

The simplest systems for the removal of carbon dioxide and hydrogen sulphide contain a single absorption column of the packed or plate type for countercurrent gas/liquid contact at high pressures (up to 1250 psi, 9×10^3 kPa) and a desorption section where the dissolved gases are recovered by two-stage expansion. The gases from the first expansion stage may contain enough valuable components to warrant returning them to the absorber after recompression. If the unit has a large capacity, it is usual to recover part of the energy by reducing the pressure in an expansion turbine.

Among processes that have been applied to the purification of impure gas streams, and which are illustrative of physical solvent procedures, are the Rectisol and Selexol processes (Chapter 12). The Benfield hot potassium carbonate process (illustrative of a chemical solvent process) (Chapter 12) has been used to clean up gases derived from the partial oxidation of petroleum and coal, as might occur in a petroleum refinery. This note is made remembering the comment in regard to the chemical purist (see above)!

Use of liquids for the absorption process offers the attractive option of selective recovery through fractionation in so far as the more volatile components of gas streams can be removed by a combination of absorption and pressure 'flashing' (Figure 9.7) or even partial stripping (Figure 9.8). In

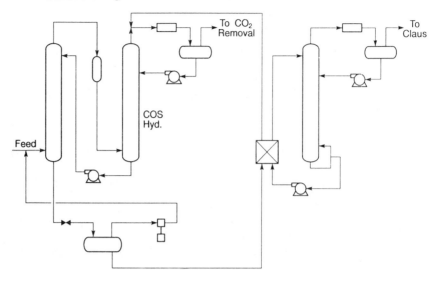

Figure 9.7 Hydrogen sulphide removal with flash–recycle

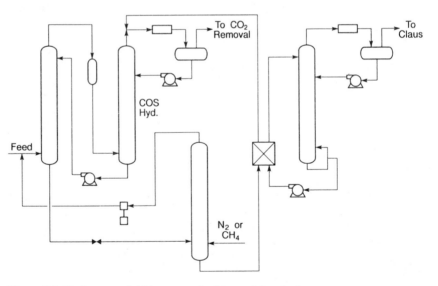

Figure 9.8 Hydrogen sulphide removal with partial stripping

fact, the most common and economical way to remove components of higher volatility ('lights' or 'light ends') is to recycle the gases given off in flashing. Solutes that have been absorbed under pressure come out of solution as the pressure is relieved.

Thus, as the pressure is reduced, first to atmospheric and then to vacuum, all the solutes come off and the solvent is regenerated. Cessation of the procedure at intermediate pressures will leave the less volatile components in solution.

Although the process is not 'clean' in so far as the lighter components will contain some of the less volatile components, these latter species can be removed in an absorber when the flash gas is recompressed and added to the fresh feed.

There are also processes where the gas purification involves absorption of one or more of the constituents into a solid, i.e. the procedure involves the transfer of a component from the gas phase to the solid phase whereby the absorbed material diffuses throughout the absorbent. There is also the option of chemical reaction between the gas contaminant and the solid.

Amine washing

The washing of a gas with amines is identical, from the point of view of the principle involved, with washing it with potassium carbonate (Kohl and Riesenfeld, 1979). The amine reacts chemically with the acid gases with the liberation of an appreciable amount of heat and it is necessary to compensate for the absorption of heat.

Amine washing involves the use of a range of amine derivatives which are chosen on the basis of gas composition and the degree of contaminant removal required (Chapter 10). Amine derivatives such as ethanolamine (mono-ethanolamine, MEA), diethanolamine (DEA), triethanolamine (TEA), methyldiethanolamine (MDEA), diisopropanolamine (DIPA) and dig-lycolamine (DGA) have been used in commercial applications (Tables 9.3 and 9.5) (Kohl and Riesenfeld, 1979; Maddox et al., 1985; Polasek and Bullin, 1985; Jou et al., 1985; Pitsinigos and Lygeros, 1989).

The chemistry is often complex (Chapters 10 and 12) but can be represented by simple equations for low partial pressures of the acid gases:

$$2RNH_2 + H_2S = (RNH_3)_2S$$
$$2RHN_2 + CO_2 + H_2O = (RNH_3)_2CO_3$$

At high acid gas partial pressures, the reactions will lead to the formation of other products:

$$(RNH_3)_2S + H_2S = 2RNH_3HS$$
$$(RNH_3)_2CO_3 + H_2O = 2RNH_3HCO_3$$

The reaction of hydrogen sulphide with the amine is extremely fast, the absorption of hydrogen sulphide being limited by mass transfer; this is not so for carbon dioxide.

Regeneration of the solution leads to near complete desorption of carbon dioxide and hydrogen sulphide. A comparison between monoethanolamine, diethanolamine and diisopropanolamine shows that monoethanolamine is the cheapest of the three but shows the highest heat of reaction and corrosion; the reverse is true for diisopropanolamine.

In the process (Figure 9.9), the raw gas stream, after passage through the separator, enters the base of the absorber where it is washed in countercurrent contact with an amine solution. The purified gas leaves the head of the column and then passes into a separator which enables any entrained amine to be recovered. The regenerated amine is drawn into a tank and fed with the aid of a pump to the top of the absorber. The charged amine leaves the base of the column and passes into a degassing vessel.

The gases (hydrogen sulphide, carbon dioxide and hydrocarbons if present)

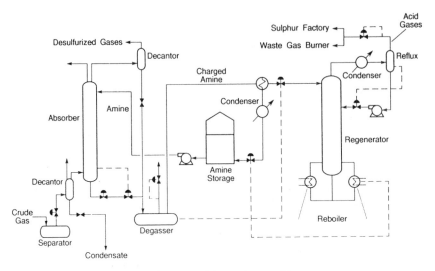

Figure 9.9 Hydrogen sulphide removal by amine washing

are subjected to a reduction of pressure in the fuel gas collector. The amine, now at reduced pressure, is reheated by a battery of heat exchangers; the hydrogen sulphide and the carbon dioxide (and a small percentage of hydrocarbons) pass into the head of the column and are cooled in two coolers which condense the water. The latter separates out and is fed back as reflux; the acid gases are directed to the sulphur recovery unit.

There are several physicochemical constraints which can affect the performance of amine-washing operations leading to losses of the amines: (a) losses by entrainment; (b) losses through the vapour pressure of the gases; and (c) losses by degradation.

Losses by entrainment take place when foaming occurs in the absorber or the regenerator. Heavy hydrocarbons are the principal cause of foaming. For this reason it appears necessary to check the correct operation of the separators placed at the inlets of the units. The installation of a degasser, an intermediate vessel between the absorption and regeneration stages, allows the release and effective separation of the dissolved hydrocarbons. By this means, foaming losses are reduced or the addition of an anti-foaming agent can be an effective means of reducing foaming.

Losses due to vapour pressure effects can be reduced by washing the purified gas with water or glycol on a number of plates at the head of the absorber. Use may also be made of adsorbents (e.g. bauxite, an ore having aluminium oxide (Al_2O_3) as the main constituent) capable of being regenerated, otherwise (in the amine processes) it is necessary to use amines with a very low vapour pressure (triethanolamine and methyldiethanolamine, for example).

Finally, that high-boiling amines may suffer losses may appear to be contradictory since the amines are stable at the regeneration temperatures; on the other hand, the amines are subject to chemical degradation. The presence of oxygen reacting with the hydrogen sulphide, for example, can give free sulphur which will react with the hot amine to form dithiocarbonates or other non-regenerable products.

In addition, the gas stream to be treated may also contain some acidic compounds (such as formic acid (HCO_2H), acetic acid (CH_3CO_2H), carbon disulphide (CS_2), carbonyl sulphide (COS)) which react with the amines to produce non-regenerable materials. These must be eliminated from the absorbing solution since they adversely affect the absorption, increase the viscosity and may possibly activate foaming.

In order to avoid an increase in the concentration of these residues, it is necessary to remove a small quantity of regenerated solution from the bottom of the regenerator and to feed it to a small associated plant (Figure 9.10) where the amine is purified.

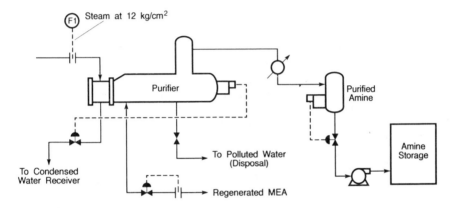

Figure 9.10 Ancillary purge purification system for an amine-treating operation

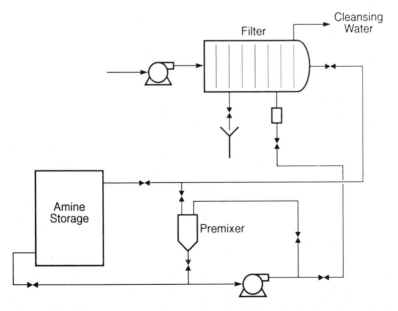

Figure 9.11 Filtration system for an amine-treating plant

Purification of the amine from any solid particles (which can initiate foaming) can, if necessary, be achieved by filtration (Figure 9.11). The filtration medium will also be an adsorbent for hydrocarbons (carbon or activated earth) and allow effective filtration of the amine as well as efficient separation of the condensates from the crude gas stream.

Foaming can be reduced by the addition of anti-foaming products which modify the surface tension, or, if the foaming persists, by increasing the temperature of the amine. It is advantageous to reduce the absorption by raising the temperature, rather than by condensing the hydrocarbons in the absorber, because there is a very considerable increase in the partial pressures of the hydrocarbons when the crude gas contains a high proportion of constituents to be absorbed.

Carbonate washing

Carbonate washing is a mild alkali process for the removal of acid gases (such as carbon dioxide and hydrogen sulphide) from gas streams (Kohl and Riesenfeld, 1979) and uses the principle that the rate of absorption of carbon dioxide by potassium carbonate increases with temperature. It has been demonstrated that the process works best near the temperature of reversibility of the reactions:

$$K_2CO_3 + CO_2 + H_2O = 2KHCO_3$$
$$K_2CO_3 + H_2S = KHS + KHCO_3$$

Thus, it is necessary to provide a minimum amount of steam to regenerate the solution. At high pressure, the release of pressure suffices for the regeneration, the action of hydrogen sulphide and carbon dioxide on the carbonate being almost non-exothermic.

The heat of reaction is lower than that noted for alkanolamine absorption and the reaction between dissolved carbon dioxide and the solution components is slow at low temperatures. Therefore, the absorbers of the first generation of hot carbonate processes operated close to $100\,°C$ ($212\,°F$) and the regeneration is achieved, mainly, by pressure reduction, with only a slight-to-moderate rise in temperature. Since complete regeneration, necessary to obtain low partial pressure of acid gases above the regenerated solution, consumes much energy when applied to the entire volume of solution, split stream regeneration was often applied. In such plants, the fully regenerated part of the solution was introduced at the top of the absorber to obtain high gas purity.

More modern processes usually operate with an activator (see the previous discussion on process principles earlier in this chapter). This permits either operation at lower temperatures or a reduction in size of the absorber when the temperature is kept high.

In the process (Figure 9.12), the feed gas passes through a separator and then enters the base of the absorber, and the purified gas leaves the head of the column and passes to a separator which allows any entrained absorbent solution to be recovered. The regenerated carbonate is returned to the head of the absorber; the rich carbonate is subjected to reduced pressure and, after passage through heat exchangers, goes to the top of the generator.

The acid gases and the water vapour pass through a cooler and a reflux vessel, where the water separates out from them; the acid gases are led to the

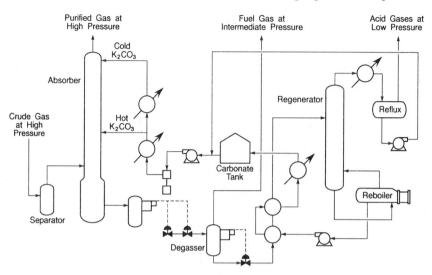

Figure 9.12 A carbonate-treating system for gas cleaning

sulphur recovery unit and the water is refluxed to the head of the regenerator. A reboiler completes the regeneration of the carbonate and makes up the heat losses of the plant.

The hydrogen sulphide content of some gases may be too high for single-stage removal and it may be preferable to use the amine process despite the disadvantage of the volatility of the amine.

Water washing

Washing with water, in terms of the outcome, is analogous to washing with carbonate (Kohl and Riesenfeld, 1979) and it is also possible to carry out the desorption step by pressure reduction. The absorption is purely physical and there is also a relatively high absorption of hydrocarbons, which are liberated at the same time as the acid gases.

The crude gas, at ambient temperature (but above 20 °C, 68 °F) in order to avoid the formation of hydrates (which occurs when gases contain specific amounts of hydrogen sulphide and carbon dioxide), enters at the bottom of the main washing column and comes into direct contact with the water fed to the top of the tower (Figure 9.13). The washing takes place in countercurrent flow on plates, which may be either stepped plates or plates of the conventional type; Raschig rings (Chapter 11) may be used with an almost equal efficiency.

The washing takes place under pressure (about 1000 psi, 7×10^3 kPa) and the gas, after washing, deposits the entrained water in a separator from where it is reintroduced into circulation. The charged water from the bottom of the absorber undergoes a reduction in pressure (to about 200 psi, 1.4×10^3 kPa). During this release of pressure, fuel gas rich in hydrogen sulphide and carbon dioxide is produced. The mixture of these gases is washed in a secondary washing column supplied with water in countercurrent to absorb the maximum amount of acid gases. The residual gas from the top of the second

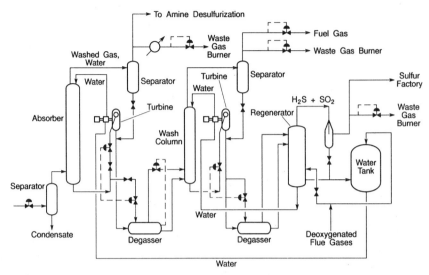

Figure 9.13 A water wash system for gas cleaning

washer is passed into an amine-washing unit to refine the gases (for combustion purposes) whilst recovering the hydrogen sulphide.

The water leaving the secondary washing column and that brought to a reduced pressure are degassed in a third column to recover any other dissolved gases by reducing the pressure even further (water temperature influences the efficiency of gas washing) (Figure 9.14). This regenerated water is then discharged into a reservoir whence it is taken to feed both the main and secondary washing columns. An increase in the throughput of water permits a higher recovery of acid gases but involves a higher reabsorption of hydrocarbons (Figures 9.15–9.18).

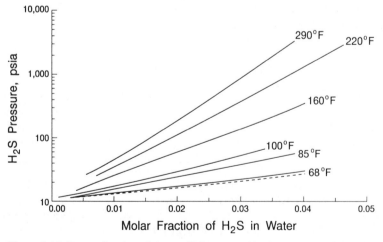

Figure 9.14 Determination of the coefficient $K = Y/X$ for the hydrogen sulphide–water system

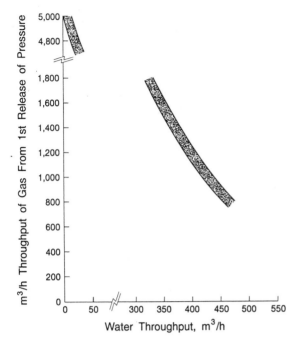

Figure 9.15 Relationship of gas cleaned with water throughput

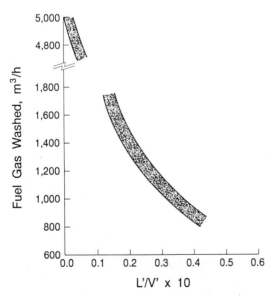

Figure 9.16 Variation of amount of gas cleaned as a function of $L'/V' \times 10$: L' = throughput of water, m^3/h; V' = throughput of gas washed, m^3/h ($L/V \times 10$)

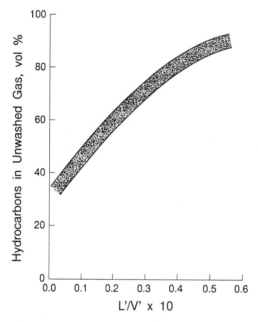

Figure 9.17 Percentage volume of hydrocarbons in the gas washed as a function of $L'/V' \times 10$: L' = throughput of water, m^3/h; V' = throughput of gas washed, m^3/h ($L/V \times 10$)

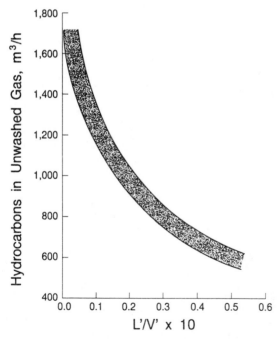

Figure 9.18 Volume of hydrocarbons in the gas washed as a function of $L'/V' \times 10$: L' = throughput of water, m^3/h; V' = throughput of gas washed, m^3/h

One of the characteristics of this process configuration is the severe working conditions: the high percentage of hydrogen sulphide and carbon dioxide in the gas, and the higher pressure, giving high concentrations of hydrogen sulphide and carbon dioxide in the water. The mixture of hydrogen sulphide and carbon dioxide, dissolved in water, is more corrosive than the water–hydrogen sulphide and water–carbon dioxide systems.

Higher efficiencies are also claimed when a wetted-wick column (Table 9.7; Figure 9.19) is used for the absorption of carbon dioxide in water owing to the greater area of contact (Lee, 1991).

Table 9.7 Comparison of a wetted-wick column with a packed tower (see Chapter 11)

	Packed column advantage/disadvantage	Wetted-wick column advantage/disadvantage
Flooding	yes – disadvantage	no – advantage
Excess flow of liquid at wall	yes – disadvantage	no – advantage
Channelling	yes – disadvantage	no – advantage
Backmixing	yes – disadvantage	no – advantage
Wetted surface	small – disadvantage	large – advantage
Pressure drop	large – disadvantage	small – advantage
Liquid stagnation	much – disadvantage	little – advantage
Energy efficiency	low – disadvantage	high – advantage
Liquid flow rate	high – advantage	low – disadvantage
Scaleup design	easy – advantage	difficult – advantage
Regular maintenance	easy – advantage	difficult – disadvantage

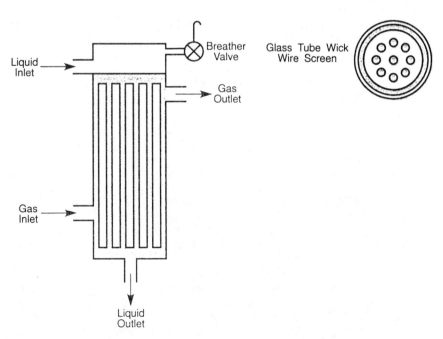

Figure 9.19 Schematic representation of a wetted-wick absorption column

Adsorption

Adsorption is defined as the concentration of one or more components of a gaseous or liquid mixture at the surface of a solid adsorbent (Kovach, 1979).

Adsorption processes utilize the concept of contaminant removal by concentration on the surface of a solid material; the method has been in commercial practice for several decades (Mantell, 1951). The commercial adsorbents are generally granular solids which have been prepared to have a large surface area per unit weight. These materials are frequently used in fixed beds for the purification and dehydration of the natural gas. In most fixed-bed adsorption systems, the adsorption is exothermic and the bed temperature can be raised to extreme process limits if the evolution of heat is not controlled.

In the thermal-swing adsorption concept, adsorption is carried out at near-ambient temperature followed by thermal regeneration using a portion of the cleaned gas or steam. On the other hand, pressure-swing adsorption processes are employed for the separation of bulk gas mixture (Richter, 1987). The adsorption is carried out at an elevated pressure level to give a product stream enriched in the more weakly adsorbed component. After the column is saturated with the stronger adsorbing component of the gas stream, it is regenerated by depressurization and purging with a portion of the product gas. During this time, the second column comes on-stream and, thus, pressure-swing adsorbers are built in pairs. The cycle is repeated after raising the pressure to the adsorption level. Numerous combinations of these cyclic steps in conjunction with the internal recycling of gases have been patented. A typical application is for the recovery of hydrogen from the purge gas of ammonia or methanol synthesis loops.

Adsorption, as practised in the industry, involves the removal of impurities from gas streams by concentration on the surface of a solid material (Fulker, 1972). The quantity of material adsorbed is proportional to the surface area of the solid and, consequently, adsorbents are usually granular solids with a large surface area per unit mass.

If the forces of interaction are physical in character, such as van der Waals' forces, the term physical adsorption is used. The rate of this type of adsorption is so high that equilibrium is usually reached almost instantaneously, the attainment of equilibrium being governed solely by diffusional limitations.

When a chemical interaction between an adsorbate and an adsorbent occurs, the phenomenon is called chemisorption, i.e. the chemical reaction occurring at and limited to the surface of the solid.

Chemisorption shows many characteristics of chemical reactions: it may be irreversible, and its rate is sometimes so low that equilibrium is reached only after a considerable period of time. Since the forces of interaction are usually much stronger in chemisorption, the heat released during physical adsorption (up to 10 kcal/mole, 40 kJ/mol) is much lower than for chemisorption, where thermal energies up to 50 kcal/mole (200 kJ/mol) and higher have been observed.

The exothermic nature of many adsorption processes also implies that the amount adsorbed at equilibrium decreases with increasing temperature. Adsorbents applied in physical adsorption processes include activated alumina (Al_2O_3), bauxite (an alumina ore), silica gel (SiO_2), active carbon and molecular sieves (Kohl and Riesenfeld, 1979). The quantity adsorbed by these materials depends on the internal surface area, which determines to a large extent the capacity of the adsorbent. However, molecular size is also important, particularly with molecular sieves (Hersh, 1961; Kovach, 1979).

Molecular sieves are crystalline aluminosilicates which contain relatively large cavities that are accessible to adsorbate species through relatively narrow pore mouth dimensions (Table 9.8) (Kohl and Riesenfeld, 1979). The molecular sieves have high capacities for small, polar molecules like water, hydrogen sulphide and carbon dioxide but do not adsorb larger molecules, which cannot enter the cavities in the crystal structure. Thus, separations are possible based on differences in molecular polarity as well as in size. These solids exhibit a wide spectrum of pore structures, surface polarity and chemistry which makes them specifically selective for the separation of many different gas mixtures (van den Berg and de Jong, 1980).

Table 9.8 Examples of the properties of a (Linde) molecular sieve

Pore mouth width (nm)	0.4	
Molecules adsorbed	$H_2O,H_2S,CO_2,NH_3,SO_2,C_2H_4,C_2H_6,C_3H_6$	
Molecules excluded	C_3H_8 and higher hydrocarbons	
Capacity at 25°C (kg/100 kg)	at $p_{eq}=2.67$ kPa	at $p_{eq}=0.13$ kPa
H_2O	22	17
H_2S	9.5	4.5
CO_2	11	3.5

One of the disadvantages of using sulphur-containing fuels in the utility industry is the occurrence of sulphur dioxide in the combustion products. However, there are many processes for the removal of sulphur dioxide from utility stack gases. They may be categorized into four broad groups: (a) dry absorption processes, (b) wet absorption processes, (c) catalytic oxidation processes, and (d) wet and dry adsorption processes. The wet absorption processes appear to be the most popular; however, processes in each category are available commercially.

The use of solid adsorbents to recover and recycle sulphur dioxide from acid plant tail gas has the potential to effect almost complete removal of the pollutants concerned, certainly to much lower levels than have been achieved thus far by any liquid absorption process, even using strong alkali.

Such processes can employ either a fixed-bed system (Figure 9.20) in which two absorbent beds are used alternately, one being regenerated while the other is collecting, or a moving-bed system in which the resin bed moves first through an adsorbing zone and then through a regeneration zone. However, the ease with which the process can be applied depends on matching the time required for adsorption and regeneration.

One of the most common uses for the adsorption process in the natural gas industry is dehydration (Chapter 10) and adsorption can be achieved using many different solid adsorbents. However, the great majority of dehydration adsorbents are based on silica, alumina (including bauxite), carbon and molecular sieves. The silica- and alumina-based materials are used primarily for dehydration whilst the carbon and molecular sieves have expanded usage in the adsorption of organic materials also.

These processes are best applied to gas streams that have only moderate concentrations of hydrogen sulphide and where the carbon dioxide is not

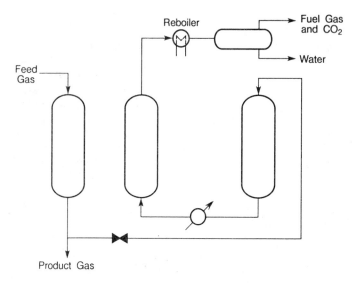

Figure 9.20 Schematic representation of a system for the removal of carbon dioxide, hydrogen sulphide and water by adsorption

required to be removed from the gas stream. These process types are not as widely used as the liquid absorption processes but they do offer advantages such as (a) simplicity, (b) high selectivity for hydrogen sulphide, and (c) process efficiency.

In a general sense (molecular sieves being the exception), the equipment used for the adsorption process is essentially the same for each adsorbent and the adsorbents are interchangeable.

In the simplest configuration (Figure 9.21), the dehydration operation for the treatment of a gas stream will consist of two contact reactors filled with the adsorbent but with only one reactor on-stream at any given time (the in-parallel twin-reactor concept used in the delayed coking operation; see, for example, Speight (1991) and references cited therein). This allows one bed to be regenerated whilst not in use. Regeneration can be accomplished by the passage of a hot gas through the bed. The regenerated bed is then brought on-stream and the used bed is regenerated.

The beds are mostly operated in a cyclic regenerative manner and the process consists of the following phases:

(a) *Adsorption.* Impurities are removed from the feed gas at low temperatures, but if the concentration of impurity is high the amount of heat released is also high and cooling must be applied. If only small amounts of contaminants are to be removed, adiabatic adsorption is possible.
(b) *Regeneration.* Regeneration/reactivation is achieved by raising the temperature as a clean purge gas is passed through the adsorbent. Part of the product gas or an auxiliary gas may be used for this purpose. A reduction of the pressure helps to speed up the regeneration.
(c) *Cooling.* The hot regenerated adsorbent bed is cooled by passing through cold regeneration gas. This not only prepares the bed for another adsorption phase but also serves to recover part of the heat.

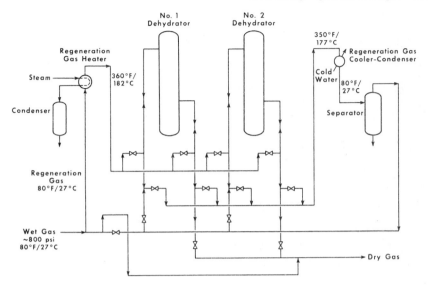

No. 1
Dehydrator

No. 2
Dehydrator

350°F/
177°C

Regeneration
Gas Heater

Regeneration Gas
Cooler-Condenser

Steam

360°F/
182°C

Cold
Water

80°F/
27°C

Condenser

Separator

Regeneration
Gas
80°F/27°C

Wet Gas
~800 psi
80°F/27°C

Dry Gas

Figure 9.21 Schematic representation of a gas dehydration system

The complete cycle is repeated at periodic intervals depending upon the nature of the natural gas passing through the unit and is, because of the twin-reactor concept, a continuous process. There has been some development of adsorption–dehydration units in which the adsorbent moves from the adsorption zone to the regeneration zone in a similar manner to the movement of catalyst during petroleum processing (Speight, 1991).

Among the advantages of fixed-bed adsorption separations are the following: (a) high product purity can be attained; (b) on-stream life of the adsorbent, particularly with molecular sieves, can be up to 5 years; (c) within limits, fixed-bed processes are insensitive to variations in pressure, temperature, feed rate and contaminant concentration in the feed; (d) there appears to be a diminished incidence of corrosion problems. However, a disadvantage is that a separate process may be required for treating acid gases obtained as a by-product; this is particularly important as there are cases where the combination of fixed-bed adsorption with contaminant aftertreatment is less economic than the application of a single wash with a liquid agent.

Adsorption is usually measured in terms of 'isotherms'. The simplest model for a type 1 (Figure 9.22) isotherm of a pure gas is the Langmuir equation

$$n = mKP(1 + KP)^{-1}$$

where

$$K = K_0 \exp(q/RT)$$

The two-parameter (m,K) Langmuir isotherm is applicable to most homogeneous adsorbents for which the heat of adsorption (q) is independent of coverage (n). Using this approach, it is possible to develop adsorption isotherms for the removal of gases (hydrogen sulphide is used as the example) from process gas streams (Table 9.9).

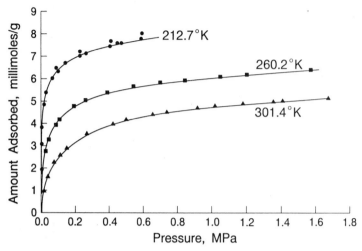

Figure 9.22 Examples of adsorption isotherms (ethylene on activated carbon)

Most commercial adsorbents are heterogeneous, indicated by the variation of q with n. An approach to account for heterogeneity is to assume that the adsorbent consists of homogeneous patches and that the overall amount adsorbed can be obtained by integration of the contributions of each patch:

$$n(P) = \int n_H(q)\mathrm{d}q$$

where n_H is the homogeneous isotherm (such as the Langmuir isotherm) on a patch defined by the energy q, and (q) is the probability density function of q on the adsorbent surface.

One such heterogeneous isotherm uses the constants m, b and t which are functions of temperature:

$$n = mP(b + P^1)^{-1/t}$$

It is also worthy of note that many commercial processes for the purification of gas streams, such as the removal of hydrogen sulphide, are designed primarily for operation at low temperatures and the use of gas-cooling equipment (Figure 9.23) is necessary. However, gases leaving the various reactors have temperatures ranging from 400 to 1000 °C (750 to 1830°F) or up to 1500°C (2730°F) for the extremely high-temperature gasifiers. Thus, there is the potential for the development of high-temperature processes which employ a solid absorbent that is not volatile, under process conditions, to absorb, and react with, the hydrogen sulphide (Table 9.10). Such processes are particularly attractive for use with the combined-cycle power systems.

Chemical conversion

In chemical conversion processes, the contaminants in gas streams are converted to compounds which are not objectionable or which can be removed from the gas stream with greater ease than the original contaminant. For example, a number of processes have been developed which remove

Table 9.9 Summary of the characteristics of gas-cleaning processes (see also Chapter 10)

	Chemical adsorption	Physical adsorption	Chemical absorption	Physical absorption
Equilibrium relation:	Almost irreversible adsorption	Very fast, reversible physical adsorption	Equilibrium reaction	Physical equilibrium (Henry's law)
Maximum load:	Limited by stoichiometry	Limited by surface area of adsorbent	Limited by stoichiometry	No clear limit observed
Product purity:	Extremely high	High	High	High only when low temperatures are applied
H_2S/CO_2 selectivity:	Very selective	Almost non-selective	Partial selectivity obtainable	Partial selectivity obtainable
H_2S/hydrocarbon selectivity:	Very selective	Almost non-selective	Selective	Partially selective
Energy requirements: for desorption	High (if regenerative)	Moderately high	High	Low
for other operations	low	Rather low	Low	Moderate to high
total	–	Moderately high	Rather high	Moderate
Feed requirements:	Low partial pressure of contaminants	Not feasible with very high partial pressure of contaminants	No limitations	High partial pressure of contaminants required
Mode of operation:	Mostly batch: sometimes cyclic regenerative	Cyclic regenerative	Continuous	Continuous
Main applications:	Guard beds	Aftertreatment	Generally applicable	Bulk removal of contaminants

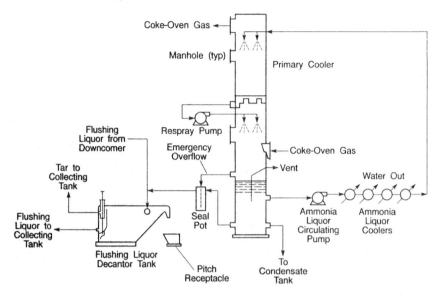

Figure 9.23 Schematic representation of a gas-cooling system

Table 9.10 Examples of processes for high-temperature gas cleaning (see Chapter 12)

Process	Absorber
Battelle Molten Salt	Calcium carbonate (in a carbonate melt)
Conoco	Calcium carbonate and magnesium oxide
Iron Oxide	Iron oxide (oxidizer)

hydrogen sulphide and sulphur dioxide from gas streams by absorption in an alkaline solution (Tables 9.11 and 9.12).

As noted above, there is always the issue of the classification of these processes, i.e. absorption, adsorption or chemical? It is considered appropriate, but more for convenience than to raise any form of intellectual scientific debate, that these processes be considered as 'chemical conversion' processes.

For example, the Giammarco–Vetrocoke process (Chapter 12) uses solutions of potassium carbonate to absorb hydrogen sulphide after which it is oxidized to sulphur by the catalytic action of the arsenic compounds:

$$Na_3AsO_3 + 3H_2S = Na_3AsS_3 + 3H_2O$$
$$Na_3AsS_3 + 3Na_3AsSO_4 = 3Na_3AsO_3S + Na_3AsO_3$$
$$Na_3AsO_3S = Na_3AsO_3 + S$$
$$2Na_3AsO_3 + O_2 = 2Na_3AsSO_4$$

The Stretford process (Chapter 12) is a low-temperature ($<40°C$, $105°F$) process in which hydrogen sulphide is absorbed in an alkali solution to form sodium bisulphide

$$Na_2CO_3 + H_2S = NaHS + NaHCO_3$$

Table 9.11 Examples of gas-cleaning processes which use weak alkali (see Chapter 12)

Process	Absorber in aqueous solution
Adip	Alkanolamine (diisopropanolamine, DIPA)
Alkazid M	Potassium methylamine propionate
Alkazid DIK	Potassium dimethylamine acetate
Benfield ⎫ Catacarb ⎭	Potassium carbonate and 'additives'
Econamine	Alkanolamine (diglycolamine, DGA)
MDEA	Alkanolamine (methyldiethanolamine, MDEA)
SNPA–DEA and DEA	Alkanolamine (diethanolamine, DEA)
Sulfiban	Alkanolamine (monoethanolamine, MEA)
Vacuum Carbonate	Sodium carbonate

Table 9.12 Examples of the chemical reactions involved in the removal of sulphur dioxide from gas streams

Process	Chemical reactions[a]
Limestone slurry scrubbing	$CaCO_3 + SO_2 \rightarrow CaSO_3 + CO_2$
Lime slurry[b] scrubbing	$CaCO_3 + heat \rightarrow CaO + CO_2$
	$CaO + SO_2 + 2H_2O \rightarrow CaSO_3.2H_2O$
Magnesia slurry scrubbing with regeneration	$Mg(OH)_2 + SO_2 \rightarrow MgSO_3 + 2H_2O$
	$MgSO_3 \rightarrow MgO + SO_2$
Sodium base scrubbing	$Na_2SO_3 + H_2O + SO_2 \rightarrow 2NaHSO_3$
	$2NaHSO_3 \rightarrow Na_2SO_3 + H_2O + SO_2$
Catalytic oxidation ion	$2SO_2 + O_2 \rightarrow 2SO_3$
	$H_2O + SO_3 \rightarrow H_2SO_4$
Sodium citrate scrubbing	$SO_2 + OH \rightarrow HSO_3^-$
	$2H_2S + SO_2 \rightarrow 3S + 2H_2O$

[a] All of these processes (except for the catalytic oxidation process) depend on the absorption of sulphur dioxide by an acid–base reaction. The first two processes are 'throwaway' processes in so far as they produce considerable quantities of waste material. The other processes include sulphur product recovery.
[b] May require as much as 200 lb calcium oxide per ton of coal.

which is, in turn, converted to sulphur using sodium vanadate:

$$NaHS + NaVO_3 = NaHVO_3 + S$$
$$NaHVO_3 + ADA = NaVO_3 + (reduced)ADA$$

Regeneration of the reduced vanadate is accomplished by use of an organic acid (anthraquinone disulphonic acid, ADA) which, in the presence of air, brings about oxidation of the reduced vanadate.

On the other hand, there are those processes which utilize a solid material and are often classed as adsorption processes but which, in reality, cause separation of the undesirable constituents by chemical changes. In spite of the ambiguity which exists in terms of the classification of these processes, and recognition of this ambiguity is essential for the gas-processing aficionado, it is considered more appropriate that they be classified under this subheading.

Such processes are the iron oxide and other processes employing metal oxides (Chapter 12). As a historical note, the use of iron oxide as a gas purification agent is not new, having been used since the middle of the nineteenth century for the 'purification' of coke-oven gas.

The 'dry' purification process, in which hydrogen sulphide is removed from gas streams by hydrated iron oxide ($Fe_2O_3.H_2O$) with continuous regeneration of the resultant iron sulphide by oxygen (air), remains the most commonly used process for this purpose:

$$Fe_2O_3 + 3H_2S = Fe_2S_3 + 3H_2O$$

There are several known forms of ferric oxide (Table 9.13) and, with only one exception (which appears only to exist in acid conditions), they all react completely with hydrogen sulphide to give ferric sulphide, usually in the hydrated form which is more correctly represented by the following equations:

$$2Fe_2O_3.H_2O + 6H_2S = 2Fe_2S_3.H_2O + 6H_2O$$
$$2Fe_2O_3 + 6H_2S = 2Fe_2S_3.H_2O + 4H_2O$$

Generally, the best results for the removal of hydrogen sulphide from gas streams have been achieved by using the hydrated form of ferric oxide, $Fe_2O_3.H_2O$. The less efficient results obtained from anhydrous ferric oxide were attributed both to the tendency of these oxides to produce the anhydrous sulphide which is slower to reoxidize than the hydrate and to the induction period of their reaction with hydrogen sulphide which renders them unsuitable for the complete removal of hydrogen sulphide. Dehydration of this sulphide, which takes place readily at temperatures above 40–50 °C (140–122 °F), has an adverse effect on the process and higher temperatures can result in the decomposition of the ferric sulphide with more permanent loss of activity. The sulphur dioxide can subsequently be reduced to sulphur by reduction over carbon or by capture by calcium (Chapters 10 and 12).

However, the process is much more complex than is indicated by these simple equations. For example, iron (ferrous) sulphide (FeS) is also formed, particularly at higher temperatures. Any oxygen present reacts to give

Table 9.13 Description of the different forms of ferric oxide

αFe_2O_3	Bright red, with low magnetic susceptibility; very stable on heating
$\alpha Fe_2O_3.H_2O$	Yellow brown, with low susceptibility; changes to αFe_2O_3 on heating at 250–300 °C (482–572 °F)
$\beta Fe_2O_3.H_2O$	Bright yellow with low susceptibility; changes to αFe_2O_3 on heating above 100 °C (212 °F)
γFe_2O_3	Chocolate brown; intensely magnetic; on heating at 300 °C (572 °F) changes to αFe_2O_3 with loss of magnetic properties, the conversion being complete in about 24 h
$\gamma Fe_2O_3.H_2O$	Orange with low susceptibility; on heating at 300 °C changes to γFe_2O_3 with great increase in magnetic properties, but on prolonged heating at this temperature slowly converted to αFe_2O_3 with a fall of susceptibility as for γFe_2O_3
$\delta Fe_2O_3.H_2O$	Dark red–brown, intensely magnetic; on heating at 150 °C (302 °F) is converted completely to αFe_2O_3 with a great fall in susceptibility
Amorphous Fe_2O_3	Red–brown, low susceptibility hardly changed on heating; at 100 °C becomes αFe_2O_3 even under boiling water, when colour becomes bright red

elemental sulphur:

$$2H_2S + O_2 = 2S + 2H_2O$$

If too much sulphur is formed the process rate diminishes because of diffusion limitations. The same effect is noted when the oxide is overheated and loses water or changes in crystal structure. Thus, careful temperature control is necessary.

Roasting the iron sulphide in air returns the sulphide mainly to ferric oxide:

$$Fe_2O_3 + 3H_2S = Fe_2S_3 + 3H_2O$$
$$3Fe_2S_3 + 13O_2 = 2Fe_3O_4 + 9SO_2$$
$$4Fe_3O_4 + O_2 = 6Fe_2O_3$$

Iron oxide will also remove residual suspended tar, a distinct advantage when the process is used for the treatment of town gas (Chapter 5).

A limited number of regenerations are possible; the sulphide is converted to oxide by the addition of small quantities of oxygen to recirculating inert gas:

$$2Fe_2S_3 + 3O_2 = 2Fe_2S_3 + 6S$$

If the oxygen concentration is too high during the regeneration process, deactivation by overheating and the formation of sulphur dioxide occur. This process can also be applied using fluidized beds at higher pressure.

Iron oxide and copper oxide are also excellent examples of those materials that can be used when it is necessary, or preferable, to treat hot gases after they have left a reactor.

Another adsorbent, zinc oxide, is used extensively for the removal of sulphur compounds (hydrogen sulphide (H_2S), carbonyl sulphide (COS), carbon disulphide (CS_2) and mercaptans (R-SH)) from gas streams prior to their passage with steam over nickel-based catalysts for the production of synthesis gas. Natural gas and/or methane/ethane mixtures are treated at about 420°C (790°F) whilst gas streams which contain butane, or higher hydrocarbons, should be desulphurized at lower temperatures to avoid thermal cracking of the butane and deposition of soot on the catalyst.

Generally, zinc oxide is a better sorbent than iron oxide for hydrogen sulphide; the concentration of hydrogen sulphide in gas streams can be reduced to less than 1 ppm at temperatures of about 300°C (570°F). In fact, thermodynamic calculations indicate that at 300°C (570°F), atmospheric pressure and a water concentration of 5 mole%, removal of hydrogen sulphide from hydrogen is possible to a concentration of about 10 ppb.

The chemical reactions involved in sulphur compound removal by zinc oxide are varied by the presence of hydrogen which renders the process more effective since zinc can operate as a hydrogenation catalyst. However, steam should not be added to the hydrocarbon gases before sulphur compound removal over zinc oxide because of the reversible reaction

$$ZnS + H_2O = ZnO + H_2S$$

The better performance of zinc oxide is specifically related to the difficulty of reducing the oxide to zinc. However, a major drawback of zinc oxide is that *in situ* regeneration by oxidation of zinc sulphide to zinc oxide is not possible because the active surface diminishes appreciably by sintering. Also, much of the mechanical strength is lost, and the consequent formation of fines results in a high pressure drop over beds of regenerated sorbents. This low mechanical strength also prevents the use of zinc oxide in regenerative moving-bed or fluid-bed applications. As a result, zinc oxide is applied only to remove small amounts of hydrogen sulphide in guard beds, e.g. to protect catalysts used in downstream operations such as copper/zinc oxide low-temperature shift or methanol production catalysts.

Processes for the removal of contaminants from gas streams that utilize the concept of chemical conversion of the contaminants usually accomplish the objectives by heterogeneous catalysis using solid catalysts in fixed-bed catalytic reactors.

The catalytic conversion processes that are used to purify gas differ from other process concepts in so far as the contaminants are (a) removed from the gas stream by a physical process, and (b) converted to compounds that are not objectionable. Therefore, the products may remain in the gas stream or they can be removed from the gas stream with greater ease than the original contaminants.

In this type of process, in which the reactants and the catalysts are in different phases (heterogeneous catalysis), efficient contact between the catalyst and the reactants is essential. Indeed, the reaction may be regarded as a sequence of steps in which each step is critical to the smooth operation of the process and which are: (a) transfer of the reactants from the gas stream to the catalyst surface; (b) adsorption on to the surface; (c) chemical reaction on the surface of the catalyst; (d) desorption of the product from the catalyst surface; and (e) incorporation of the product into the gaseous phase (Hougen and Watson, 1947). Thus, mass transfer of the reactants and the products are very important aspects of the process operation. In addition, the frequent occurrence of catalyst poisoning is another factor that can affect the efficient operation of the process. In order to diminish catalyst loss due to poisoning, the use of catalyst guard beds (see also Speight, 1981) is highly recommended.

There are many other process concepts that are used, or have been proposed, for the removal of sulphur dioxide from gas streams. For example, to mention only a few of the options, ferrous oxide has been suggested as an adsorbent, as has phosphate rock (Cortelyou, 1969); manganese dioxide in a dry gas-phase regeneration system (Brennan, 1967), rather than the wet system used in the Mitsubishi DAP–Mn process (Chapter 12); lime which removes the sulphur as calcium sulphide and also has the potential to remove vanadium metal (*Chemical Engineering*, 1971); copper oxide which removes the sulphur dioxide by high-temperature adsorption (Buckingham and Homan, 1971; Dautzenberg *et al.*, 1971).

Perhaps one of the simpler approaches to the recovery of sulphur dioxide from stack gases involves the catalytic oxidation of sulphur dioxide to sulphur trioxide in the stack gas stream. The sulphur trioxide can be recovered, without absorbent recycling, and corrosion-resistant construction materials must be used in the catalytic oxidation process equipment where sulphur trioxide reacts with moisture to form sulphuric acid.

The oxidation of sulphur dioxide to sulphur trioxide

$$2SO_2 + O_2 = 2SO_3$$

is an exothermic reversible reaction, the equilibrium mixture containing a greater proportion of sulphur trioxide at lower temperatures. The reactor used to accomplish the oxidation of the dioxide to the trioxide is a multistage catalytic reactor with means for reducing the temperature of the reactants

Table 9.14 Examples of catalysts used for the reduction or oxidation of organic sulphur compounds in gases

	Nickel subsulphide[a]	Nickel subsulphide[b]	Copper or nickel thiomolybdate	Molybdenum disulphide or cobalt thiomolybdate	Zinc oxide	Zinc oxide and copper
Support	Fireclay	China clay	Activated alumina	Bauxite	No support	–
Gas treated	Coal gas	Coal gas	Coal gas	Refinery gas and vaporized light distillate with hydrogen		Hydrogen with low CO content
Reaction temperature	420°C (756°F)	300/380°C (572/716°F)	300/360°C (572/680°F)	380°C (716°F)	400°C (752°F)	250/350°C (482/662°F)
Example of reaction pressure (atmospheres absolute)	1	1	1	15	15	100
Approximate period before regeneration or replacement (days)	30	100	25	300 (Varies with sulphur content of treated gas)	200	200
Typical space velocity, approx.	500	1000	800	500/750	200/300	>1000
Results achieved	Approx. 90% removal of organic sulphur compounds other than thiophene			Approx. 100% reduction of organic sulphur compounds including thiophene; hydrogenation of unsaturated hydrocarbons	Hydrogenation of COS, CS_2 and mercaptans	Reduction to ppm of sulphur compounds including thiophene and organic sulphides
Final form of sulphur	H_2S	SO_2 mainly H_2S trace	H_2S	H_2S	ZnS	Sulphided catalyst

[a] Used as reduction catalyst.
[b] Used as oxidation catalyst.

between stages. Catalysts, such as those based on vanadium pentoxide as the active agent, promote the reaction at temperatures between 400 and 595°C (750 and 1150°F).

Early plants generally used a two-stage oxidation sequence but most modern plants have a multistage operation requiring four or more stages of catalyst.

A substantial reduction of the sulphur dioxide content can also be achieved if the bulk of the sulphur trioxide formed in the early stages of the converter is removed and the remaining gas allowed to reach a new equilibrium composition. This is the basis of the process commonly known as double absorption or double catalysis (Donovan and Stuber, 1967; Browder, 1971). By removing the sulphur trioxide between stages of the converter, a considerable reduction in the sulphur dioxide concentration in the equilibrium mixture results. Plants of this type usually contain two or three stages of catalyst in the primary reactor and one or two in the secondary, the two reactors commonly being built into one vessel.

Other catalysts, or novel sorbents (Hedges and Diffenbach, 1990), are also available for the adsorptive removal of sulphur compounds from gas streams (Table 9.14).

References

Barbouteau, L. and Dalaud, R. (1972) In *Gas Purification Processes for Air Pollution Control* (ed. G. Nonhebel), Butterworths, London, Chapter 7

Brennan, P.J. (1967) *Chemical Engineering*, 9 October, p. 114

Browder, T.J. (1971) *Chemical Engineering Progress*, **67**, (5), 45

Buckingham, P.A. and Homan, H.R. (1971) *Hydrocarbon Processing*, **50** (8), 121

Chemical Engineering (1971) 8 February, p. 78

Cortelyou, C.G. (1969) *Chemical Engineering Progress*, **65**, (9), 69

Dautzenberg, F.M., Nader, J.E. and van Ginneken, A.J.J. (1971) *Chemical Engineering Progress*, **67**, (8), 86

Donovan, J.R. and Stuber, P.J. (1967) *Journal of Metals*, **19**, (11), 45

Fulker, R.D. (1972) In *Gas Purification Processes for Air Pollution Control* (ed. G. Nonhebel), Butterworths, London, Chapter 9

Hedges, S.W. and Diffenbach, R.A. (1990) *Proc. 25th Intersociety Energy Conversion Engineering Conference, Piscataway, NJ*, **6**, p. 185

Hersh, C.K. (1961) *Molecular Sieves*, Reinhold, New York

Hougen, O.A. and Watson, K.M. (1947) *Chemical Process Principles*, Vol. 3, Wiley, New York

Jou, F.Y., Otto, F.D. and Mather, A.E. (1985) In *Acid and Sour Gas Treating Processes* (ed. S.A. Newman), Gulf Publishing Company, Houston, TX, Chapter 10

Kohl, A.L. and Riesenfeld, F.C. (1979) *Gas Purification*, 3rd edn, Gulf Publishing Company, Houston, TX

Kovach, J.L. (1979) In *Handbook of Separation Techniques for Chemical Engineers* (ed. P.A. Schweitzer), McGraw-Hill, New York, Section 3.1

Lee, K.R. (1991) *International Chemical Engineering*, **31**, 379

Maddox, R.N., Bhairi, A., Mains, G.J. and Shariat, A. (1985) In *Acid and Sour Gas Treating Processes* (ed. S.A. Newman), Gulf Publishing Company, Houston, TX, Chapter 8

Mantell, C.L. (1951) *Adsorption*, McGraw-Hill, New York

Mott, R.A. (1972) In *Gas Purification Processes for Air Pollution Control* (ed. G. Nonhebel), Butterworths, London, Chapter 4

Pitsinigos, V.D. and Lygeros, A.I. (1989) *Hydrocarbon Processing*, **58**, (4), 43

Polasek, J. and Bullin, J. (1985) In *Acid and Sour Gas Treating Processes* (ed. S.A. Newman), Gulf Publishing Company, Houston, TX, Chapter 7

Probstein, R.F. and Hicks, R.E. (1990) *Synthetic Fuels*, pH Press, Cambridge, MA

Richter, E. (1987) Erdol und Kohle. *Erdgas-Petrochemie*, **40**, 432

Speight, J.G. (1981) *The Desulfurization of Heavy Oils and Residua*, Marcel Dekker, New York

Speight, J.G. (1991) *The Chemistry and Technology of Petroleum*, 2nd edn, Marcel Dekker, New York

Staton, J.S., Rousseau, R.W. and Ferrell, J.K. (1985) In *Acid and Sour Gas Treating Processes* (ed. S.A. Newman), Gulf Publishing Company, Houston, TX, Chapter 5

van den Berg, P.J. and de Jong, W.A. (1980) *Introduction to Chemical Process Technology*, Delft University Press, Delft, Chapter VI

Ward, E.R. (1972) In *Gas Purification Processes for Air Pollution Control* (ed. G. Nonhebel), Butterworths, London, Chapter 8

Zenz, F.A. (1979) In *Handbook of Separation Techniques for Chemical Engineers* (ed. P.A. Schweitzer), McGraw-Hill, New York, Section 3.2

10

General process classification

Introduction

The processes that have been developed to accomplish gas purification vary from a simple once-through wash operation to complex multistep systems with options for recycling the gases. In many cases, the process complexities arise because of the need for recovery of the materials used to remove the contaminants or even recovery of the contaminants in the original, or altered, form.

The purpose of the preliminary purification of process gases in their crude state is the removal of materials such as liquid vapours (i.e. water, tars and aromatics such as benzenes and/or naphthalenes) and mechanically carried solid particles which are either process products and/or dust. Furthermore, in some processes, preliminary purification might also include the removal of ammonia gas. For the purpose of this section, 'preliminary purification' does not include the removal of hydrogen sulphide, carbon dioxide, sulphur dioxide, and the like.

For example, cleaning of town gas is the means by which the crude 'tarry' gases from retorts or coke ovens are (first), in a preliminary step, freed from tarry matter and condensable aromatics (such as naphthalene) and (second) purified by the removal of materials such as hydrogen sulphide, other sulphur compounds, and any other unwanted components that will adversely affect the use of the gas (Chapter 5).

The gas, once cleaned to specifications, is distributed through a pipeline system to the domestic or industrial consumer. On the other hand, a process gas might be needed as a fuel gas or as a chemical feedstock. In both cases, just as for the town gas, cleaning is necessary because of environmental or process considerations.

The preliminary purification of crude gases is achieved by cooling and/or washing (with water or oil) and electrostatic precipitation. However, gas purification proper can be readily divided into the following categories: (a) removal of gaseous impurities; and (b) removal of particulate impurities.

· There is also a need for subdivision of these two categories as dictated by needs and process capabilities: (a) coarse cleaning whereby substantial amounts of unwanted impurities are removed in the simplest, most convenient, manner; (b) fine cleaning for the removal of residual impurities to a degree sufficient for the majority of normal chemical plant operations, such as catalysis or preparation of normal commercial products, or cleaning to a degree sufficient to discharge an effluent gas to the atmosphere through a chimney (Chapter 2); (c) ultra-fine cleaning where the extra step (and expense!) is only justified by the nature of the subsequent operations or the need to produce a particularly pure product.

To make matters even more complicated, a further subdivision of the processes, which applies particularly to the removal of gaseous impurities, is by process character in so far as there are processes which rely upon chemical and physical properties/characteristics of the gas stream to enhance separation of the constituents.

Each subcategory, or possibility, must be considered in terms of technical efficiency, maintenance of performance and cost.

All of these processes, in one form or another, can be found in industrial practice and are worthy of comment in this chapter. But the prime purpose of the chapter is to add an additional dimension to the process concepts that were introduced earlier (Chapter 9).

Petroleum gas cleaning

The most common way of removing sulphur from petroleum is by hydrodesulphurization (HDS) typically at 1000 psi (7×10^3 kPa) and 370–425°C (700–800°F) over a catalyst in an atmosphere of hydrogen (Speight, 1981). The desulphurized overhead is usually blended back with the bottoms to give a desulphurized product having the desired sulphur content.

During hydrodesulphurization of petroleum feedstocks, the majority of the hydrogen sulphide (H_2S) is removed with the overhead of the 'light-ends fractionator' in the light-ends treatment portion of the process. The hydrogen sulphide is usually absorbed from sour gas by an amine solution (Chapter 9) and the products are dependent upon the partial pressure of the hydrogen sulphide.

For example, at low partial pressures the process chemistry might be represented in the following manner:

$$2RNH_2 + H_2S = (RNH_3)_2S$$
$$2RHN_2 + CO_2 + H_2O = (RNH_3)_2CO_3$$

whereas at high partial pressures the process chemistry might be represented somewhat differently:

$$(RNH_3)_2S + H_2S = 2RNH_3HS$$
$$(RNH_3)_2CO_3 + H_2O = 2RNH_3HCO_3$$

The acid gas produced from the regeneration step can then be processed in a Claus sulphur plant.

Other refining steps, such as hydrocracking and catalytic cracking, will accomplish some desulphurization (Speight, 1981, 1991). However, hydrodesulphurization of fuel oil is required if low sulphur contents (e.g. below 0.3 wt%) are to be obtained.

Natural gas cleaning

Natural gas used for fuel may also contain a minimal (perhaps acceptable for detection purposes) amount of hydrogen sulphide (or other sulphur compound) per unit volume. Consequently, very little sulphur dioxide pollution results from burning 'sweet' natural gas. When sour natural gas is produced, it is usually processed in a gas-treating plant which removes liquefied petroleum gas (LPG) liquids, sour components and moisture from the gas (Speight, 1990).

Most gas-treating plants utilize an amine system to remove sour (sulphurous or acid) components from the gas (Chapter 9). Amines such as ethanolamine (monoethanolamine (MEA), diethanolamine (DEA), triethanolamine (TEA), diglycolamine (DGA), diisopropanolamine (DIPA) and sulfinol, an aqueous solution of sulfolane and diisopropanolamine, may be utilized to remove hydrogen sulphide and carbon dioxide from gas streams:

$$2RNH_2 + H_2S = (RNH_3)_2S$$
$$2RHN_2 + CO_2 + H_2O = (RNH_3)_2CO_3$$

At high acid gas partial pressures, the reactions will lead to the formation of other products:

$$(RNH_3)_2S + H_2S = 2RNH_3HS$$
$$(RNH_3)_2CO_3 + H_2O = 2RNH_3HCO_3$$

Potassium (or sodium) carbonate is also utilized to remove carbon dioxide and hydrogen sulphide, as well as other components, from gas streams:

$$K_2CO_3 + CO_2 + H_2O = 2KHCO_3$$
$$K_2CO_3 + H_2S = KHS + KHCO_3$$

Each gas of these gas-sweetening systems has an area of optimum application (Goar, 1971a,b; Speight, 1990).

The sour 'acid gas' produced from the regeneration step of the gas-sweetening system is normally processed in a modified Claus sulphur plant for the purpose of recovering elemental sulphur from the hydrogen sulphide.

Coal gas cleaning

The 'average' (a most inaccurate term to say the least since it bears very little relationship to the chemistry of the combustion of the different sulphur forms) sulphur content of coal burned to generate electricity is generally assumed to be of the order of 2.5% wt/wt. Many coals have a much higher sulphur content (see also Chapter 1) (Berkowitz, 1979; Speight, 1983; Hessley et al., 1986; Hessley, 1990) and, because of a variety of geographical, economic and political issues, such coals are (or have to be!) used for power generation.

Organic sulphur comprises 50–60% of the total sulphur present in coal; it is an integral part of the coal structure and cannot be removed by mechanical means (see also Chapter 8). Pyritic sulphur accounts for most of the remaining sulphur in coal. Gravity separation techniques can readily remove pyritic sulphur from coal if the pyrite particles in the coal are fairly large. The coal industry has used these techniques for many years. Many American coals permit the removal of about half of the pyritic sulphur in this way. The pyrite in some coals, however, is too fine to permit separation by these methods.

Any gases, such as hydrogen sulphide and/or carbon dioxide, that are the products of coal processing can be removed, just as for natural gas cleaning (above), by application of an amine-washing procedure (Benson, 1981) (see also Chapter 9):

$$2RNH_2 + H_2S = (RNH_3)_2S$$
$$2RHN_2 + CO_2 + H_2O = (RNH_3)_2CO_3$$
$$(RNH_3)_2S + H_2S = 2RNH_3HS$$
$$(RNH_3)_2CO_3 + H_2O = 2RNH_3HCO_3$$

There are also 'solvent extraction' methods for producing low-sulphur and low-ash 'coal' but hydrotreatment of the coal extract is also required. In these methods, the organic material is extracted from the inorganic material in coal. A study has indicated that solvent-refined coal will probably not penetrate the power generation industry on a large scale for several years to come.

Acid gas removal

The removal of acid gases from gas streams can generally be classified into two categories: (a) chemical absorption processes and (b) physical absorption processes. There are several such processes which fit into these categories (Tables 10.1 and 10.2; Figure 10.1). The features of the individual processes vary so much (Table 10.3), and there are so many variables, that only the more important processes will be addressed in this text (van den Berg and de Jong, 1980; Bodle and Heubler, 1981).

For the purposes of this text acid gases are considered to be hydrogen sulphide and carbon dioxide whilst sulphur oxide(s) and nitrogen oxide(s) are considered elsewhere (in the next two sections).

In more general and simple process terms, acid gas removal is most often achieved by contact of the gas with an alkaline solution (Figure 10.2), used in a variety of processes which also suffer a variety of limitations (Table 10.3) but offer variation in the relative selectivity for hydrogen sulphide, carbon dioxide and hydrocarbons (Table 10.4) (Bodle and Heubler, 1981; Wesch, 1992).

Most of the treating agents rely upon physical absorption and chemical reaction. When only carbon dioxide is to be removed in large quantities, or when only partial removal is necessary, a hot carbonate solution or one of the

Table 10.1 Simple classification system for acid gas removal processes

Chemical absorption (chemical solvent processes)	Physical absorption (physical solvent processes)
Alkanolamines:	
MEA	Selexol
SNPA: DEA (DEA)	Rectisol
UCAP (TEA)	Sulfinol[a]
Selectamine (MDEA)	
Econamine (DGA)	
ADIP (DIPA)	
Alkaline salt solutions:	
Hot potassium carbonate	
Catacarb	
Benfield	
Giammarco–Vetrocoke	
Non-regenerable:	
Caustic	

[a] A combined physical/chemical solvent process.

Table 10.2 Examples of processes for acid gas removal from gas streams (see Chapter 12)

Process	Sorbent	Removes
Amine	Monoethanolamine, 15% in water	CO_2, H_2S
Econamine	Diglycolamine, 50–70% in water	CO_2, H_2S
Alkazid	Solution M or DIK (potassium salt of dimethylamine acetic acid), 25% in water	H_2S, small amount of CO_2
Benfield, Catacarb	Hot potassium carbonate, 20–30% in water (also contains catalyst)	CO_2, H_2S; selective to H_2S
Purisol (Lurgi)	n-Methyl-2-pyrrolidone	H_2S, CO_2
Fluor	Propylene carbonate	H_2S, CO_2
Selexol (Allied)	Dimethyl ether polyethylene glycol	H_2S, CO_2
Rectisol (Lurgi)	Methanol	H_2S, CO_2
Sulfinol (Shell)	Tetrahydrothiophene dioxide (sulfolane) plus diisoprapanol amine	H_2S, CO_2; selective to H_2S
Giammarco–Vetrocoke	K_3AsO_3 activated with arsenic	H_2S
Stretford	Water solution of Na_2CO_3 and anthraquinone disulphonic acid with activator of sodium metavanadate	H_2S
Activated Carbon	Carbon	H_2S
Iron Sponge	Iron oxide	H_2S
Adip	Alkanolamine solution	H_2S, some COS, CO_2, and mercaptans
SNPA–DEA	Diethanolamine solution	H_2S, CO_2
Takahax	Sodium, 1,4-napthaquinone, 2-sulphonate	H_2S

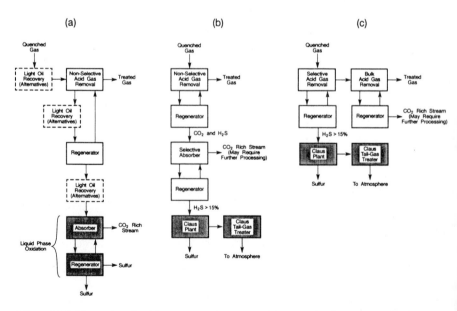

Figure 10.1 Examples of acid gas removal systems: (a) non-selective removal followed by liquid-phase oxidation; (b) non-selective removal followed by selective removal of carbon dioxide; (c) selective carbon dioxide removal

Table 10.3 Comparative summaries of various acid gas removal processes (see Chapters 10 and 12)

Feature	Chemical adsorption		Physical absorption
	Amine processes	*Carbonate processes*	
Absorbents	MEA, DEA, DGA, MDEA	K_2CO_3, $K_2CO_3 + MEA$ $K_2CO_3 + DEA$, $K_2CO_3 + arsenic$ trioxide	Selexol, Purisol, Rectisol
Operating pressure (psi)	Insensitive to pressure	> 200	250–1000
Operating temp. (°F)	100–400	200–250	Ambient temperature
Recovery of absorbents	Reboiled stripping	Stripping	Flashing, reboiled, or steam stripping
Utility cost	High	Medium	Low–medium
Selectivity, H_2S, CO_2	Selective for some amines (MDEA)	May be selective	Selective for H_2S
Effect of O_2 in the feed	Formation of degradation products	None	Sulphur precipitation at low temperature
COS and CS_2 removal	MEA: not removed DEA: slightly removed DGA: removed	Converted to CO_2 and H_2S and removed	Removed
Operating problems	Solution degradation; foaming; corrosion	Column instability; erosion; corrosion	Absorption of heavy hydrocarbons

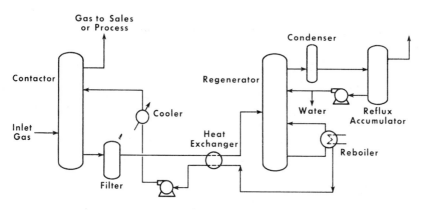

Figure 10.2 Schematic representation of an alkali treatment system for acid gas removal from gas streams

physical solvents is the most economical selection. The Sulfinol solvent, a mixture of an aqueous amine (chemical solvent) with sulfolane (physical solvent), is reported to be particularly advantageous (Taylor *et al.*, 1991).

Hydrogen sulphide may be removed solely by the use of several processes (Table 10.5) (Grosick and Kovacic, 1981). Most sulphur removal processes concentrate on removing the hydrogen sulphide because over the years it has been considered the greatest health hazard and also because it is corrosive,

Table 10.4 More detailed summary of acid gas removal processes (see Chapter 12)

Solvents	General operating range: CO_2 partial pressure	Selectivity[a]	
		H_2S/CO_2	CO_2/HC
Amines:			
MEA	Low	1	8
DEA	Low	1	8
DGA	Low	1	8
MDEA	Low	3	8
TEA	Low	3	8
DIPA	Low	2	8
Hot carbonate:			
Catacarb	Moderate	2	9
Benfield	Moderate	6	9
Alkazid	Moderate	3	9
Giammarco–Vetrocoke	Moderate	9	9
Physical solvents			
Methanol	High	7	2
DMEPED	High	9	3
Propylene carbonate	High	4	3
N-methylpyrrolidone	High	9	3
Mixed solvents:			
Sulfolane + DIPA (Sulfinol)	Moderate	2	3
Methanol + DGA (Amisol)	Moderate	1	3

[a] Arbitrary number scale 1–10: 1 = poor selectivity; 10 = excellent selectivity.

Table 10.5 General characteristics of some acid gas removal processes (see Chapter 12)

Process	H_2S removal efficiency	Organic sulphur removal	Form of recovered sulphur	Regeneration medium
Absorption–desorption:				
Vacuum carbonate				
Single stage	90–93%	None	Rich H_2S acid gas	Steam
Two stage	98%	None	Rich H_2S acid gas	under vacuum
Sulfiban	90–98%	Partial	Lean H_2S acid gas	Steam
Ammonia processes	90% normal, 98% maximum	None	Acid gas containing $H_2S + NH_3 + CO_2 + RCN$	Steam
Oxidation type:				
Stretford	99% +	None	Elemental sulphur	Air
Takahax	99% +	None	Ammonium sulphate or elemental sulphur	Air
Giammarco–Vetrocoke	< 1 ppm	None	Elemental sulphur	Air

particularly when water is present. With an increasing emphasis on eliminating or minimizing sulphur discharge to the atmosphere, attention in the newer and more effective processes is turning to the removal of other sulphur compounds from gas. Generally, specifications dictate a very low hydrogen sulphide content to be transmitted by pipeline.

A number of processes are available for the removal of hydrogen sulphide from gas streams. These processes can be categorized as those based on physical absorption, adsorption by a solid, or chemical reaction. Physical absorption processes suffer from the fact that they frequently encounter difficulty in reaching the low concentrations of hydrogen sulphide required in the sweetened gas stream. However, there are processes which, with proper attention and care to regeneration cycles, can meet this specification.

Solid-bed adsorption processes suffer from that fact that they are generally restricted to low concentrations of hydrogen sulphide in the entering sour gas stream. The development of a short-cycle adsorption unit for the removal of hydrogen sulphide might help remove part of this low-concentration restriction for the solid-bed adsorption processes. In general, chemical processes are able to meet the regulated hydrogen sulphide levels with little difficulty. However, they suffer from the fact that a material that will react satisfactorily with hydrogen sulphide will also usually react with carbon dioxide.

The most well-known hydrogen sulphide removal process is based on the reaction of hydrogen sulphide with iron oxide (often also called the Iron Sponge process or the Dry Box method; Figure 10.3) (see Chapters 9 and 12) in which the gas is passed through a bed of wood chips impregnated with iron oxide:

$$Fe_2O_3 + 3H_2S = Fe_2S_3 + 3H_2O$$

after which the bed is regenerated by passing air through it:

$$2Fe_2S_3 + 3O_2 = 2Fe_2O_3 + 6S$$

The bed is maintained in a moist state by circulation of water or a solution of soda ash.

The method is suitable only for small-to-moderate quantities of hydrogen

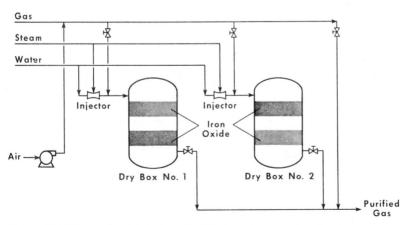

Figure 10.3 Illustration of the Iron Oxide (Iron Sponge, Dry Box) process

sulphide. Approximately 90% of the hydrogen sulphide can be removed per bed but bed clogging by elemental sulphur occurs and the bed must be discarded. The use of several beds in series is not usually economical.

The removal of larger amounts of hydrogen sulphide from gas streams requires continuous processes, such as the Ferrox process (Figure 10.4) or the Stretford process (Figure 10.5) (Chapter 12).

The Ferrox process is based on the same chemistry as the Iron Oxide process except that it is fluid and continuous. The Stretford process employs a solution containing vanadium salts and anthraquinone disulphonic acid (Maddox, 1974).

Most hydrogen sulphide removal processes involve fairly simple chemistry (Table 10.6) with the potential for regeneration with 'return' of the hydrogen sulphide. However, if the quantity involved does not justify installation of a

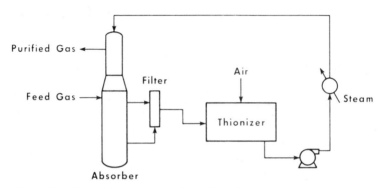

Figure 10.4 Schematic representation of the Ferrox process

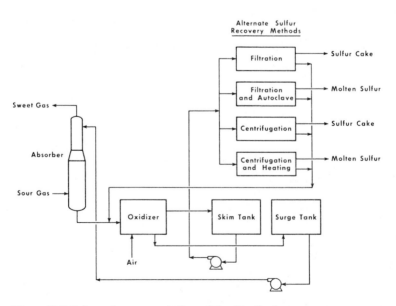

Figure 10.5 Schematic representation of the Stretford process

Table 10.6 Examples of the chemical reactions involved in the removal of hydrogen sulphide from gas streams

Name	Reaction	Regeneration
Caustic soda	$2NaOH + H_2S \rightarrow NaS + 2H_2O$	None
Lime	$Ca(OH)_2 + H_2S \rightarrow CaS + 2H_2O$	None
Iron oxide	$Fe_2O_3 + 3H_2S \rightarrow Fe_2S_3 + 3H_2O$	Partly by air
Seaboard	$Na_2CO_3 + H_2S \rightleftharpoons NaHCO_3 + NaHS$	Air blowing
Thylox	$Na_4As_2S_5O_2 + H_2S \rightarrow Na_4As_2S_6O + H_2O$	Air blowing
	$Na_4As_2S_6O + \frac{1}{2}O_2 \rightarrow Na_4As_2S_5O_2 + S$	
Girbotol	$2RNH_2 + H_2S \rightleftharpoons (RNH_3)_2S$	Steaming
Phosphate	$K_3PO_4 + H_2S \rightleftharpoons KHS + K_2HPO_4$	Steaming
Phenolate	$NaOC_6H_5 + H_2S \rightleftharpoons NaHS + C_6H_5OH$	Steaming
Carbonate	$Na_2CO_3 + H_2S \rightleftharpoons NaHCO_3 + NaHS$	Steaming

sulphur recovery plant, usually a Claus plant (Figure 10.6), it will be necessary to select a process which produces elemental sulphur directly:

$$2H_2S + 3O_2 = 2SO_2 + 2H_2O$$
$$2H_2S + SO_2 = 3S + 2H_2O$$

or

$$2H_2S + O_2 = 2S + 2H_2O$$

The conversion can be achieved by reacting the hydrogen sulphide gas directly with air in a burner reactor if the gas can be burnt with a stable flame.

Other equilibria which should be taken into account are the formation of sulphur dimer, hexamer and octamer as well as the dissociation of hydrogen sulphide:

$$H_2S = H_2 + S$$

Carbonyl sulphide and carbon disulphide may be formed, especially when the gas is burned with less than the stoichiometric amount of air in the presence of hydrocarbon impurities or large amounts of carbon dioxide.

Equilibrium data on the reaction between hydrogen sulphide and sulphur dioxide indicate that the equilibrium conversion is almost complete (100%) at relatively low temperatures and diminishes at first at higher temperatures, in accordance with the exothermic nature of the reaction. A further rise in temperature causes the equilibrium conversion to increase again. This is a consequence of the dissociation of the polymeric sulphur into monatomic sulphur.

Catalysis by alumina is necessary to obtain good equilibrium conversions: the thermal Claus reaction is fast only above 500°C (930°F) (Dowling et al., 1990; Chou et al., 1991). There is also a lower temperature limit which is not caused by low rates but by sulphur condensation in the catalyst pores and consequent deactivation of the catalyst. The lower limit at which satisfactory operation is still possible depends on the pore size and size distribution of the catalyst; with alumina-based catalysts having wide pores, the conversion proceeds satisfactorily at about 200°C (390°F) (Lagas et al., 1989; Luistra and d'Haene, 1989).

In all Claus process configurations several conversion steps in adiabatic, i.e. cheap, reactors are used, with intermittent and final condensation of the sulphur produced. There are two main process forms, depending on the concentration of hydrogen sulphide and other sulphur compounds in the gas to be converted, i.e. the straight-through and the split-flow oxidation process.

The straight-through process (Figure 10.6) is applicable when the gas stream contains more than 50% v/v hydrogen sulphide. Feed gases of this type can be burnt with the stoichiometric amount of air to give sulphur.

The combustion reactor is followed by a combined waste heat boiler and sulphur condenser from which liquid sulphur and steam are obtained. The gases are then reheated by in-line fuel combustion to the temperature of the first catalytic converter, which is usually kept at about 350°C (660°F) to decompose any carbonyl sulphide and carbon disulphide formed in the combustion step. A second catalytic converter, operating at as low a temperature as possible, is also employed to obtain high final conversions.

Caution is necessary to avoid condensation of an aqueous phase in the system because of the extreme corrosivity of the liquid phase produced. Another operating issue concerns mist formation, a phenomenon which

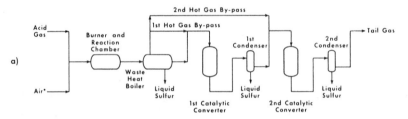

* Sufficient Air is Added to Burn 1/3 of Total H₂S to SO₂ and All Hydrocarbon to CO₂.

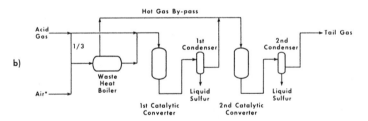

* Sufficient Air is Added to Burn All H₂S to SO₂ and All Hydrocarbon to CO₂ in 1/3 of Acid Gas.

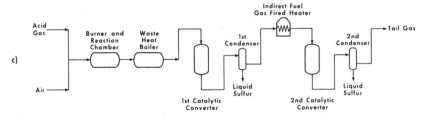

Figure 10.6 Schematic representation of three Claus process configurations

occurs very readily when condensing sulphur, and a series of demisters are necessary to prevent this. Residual sulphur is converted to sulphur dioxide by incineration of the tail gas from the process (Chapter 12) to prevent the emission of other sulphur compounds and to dilute the effluent to reduce ground-level sulphur dioxide concentrations.

The split-flow configuration (Figure 10.6) is used whenever the hydrogen sulphide content of the feed is in the range 15–50% v/v. Such gases cannot be burned with a stable flame with the stoichiometric quantity of air needed to give sulphur, but stoichiometric combustion to sulphur dioxide is possible. The feed is now split into two flows, usually with a volume ratio of 2:1, and the smaller glow is converted to sulphur dioxide.

The gases are recombined, reheated if necessary, and then fed to a catalytic converter which operates at low temperatures as a deterrent to the formation of carbonyl sulphide. Usually, only two such converters are necessary to give high sulphur yields.

Molecular sieves and membranes have been undergoing development for the removal of hydrogen sulphide and carbon dioxide from gas streams, especially when the amount of the acid gas(es) is low (Benson, 1981; Chiu, 1990; Winnick, 1991) (see also Chapter 12). The most appropriate use of the sieves and the membranes would be use of the sieve selectively to remove hydrogen sulphide, without removing much of the carbon dioxide, and/or use of membranes permeable to hydrogen sulphide but not to carbon dioxide.

Removal of sulphur-containing gases

Since the time when English kings recognized that the burning of coal could produce noxious fumes (Chapter 1) there has been a series of attempts (not always continuous) to mitigate the amounts of noxious gases entering the atmosphere, the least of which have been attempts to reduce the amount of sulphur oxide(s) (particularly sulphur dioxide) released to the environment.

Historically, the first method for removing sulphur dioxide from flue gases consisted of simple water scrubbing of the flue gas to absorb the sulphur dioxide into solution (Plumley, 1971); the method was first used in London during the 1930s. Since then, various regulatory organizations in many countries (the United States, Canada, Japan, Europe in general, and others) have set standards for sulphur dioxide emissions (Table 10.9) which must be met immediately or in the very near future.

Sulphur dioxide represents a high percentage of the sulphur oxide pollutants generated in combustion. The removal of the sulphur dioxide from the combustion gases before they are released to the stack is essential and a considerable number of procedures exist for flue gas desulphurization (FGD). These procedures may be classified as wet or dry (Tables 10.7 and 10.8) depending on whether a water mixture is used to absorb the sulphur dioxide or whether the acceptor is dry.

The procedures can be classified further as regenerative or non-regenerative depending upon whether the chemical used to remove the sulphur dioxide can be regenerated and used again or whether it passes to disposal.

In the wet regenerative processes either elemental sulphur or sulphuric acid is recovered. In the wet non-regenerative lime and/or limestone procedures, a calcium sulphite or calcium sulphate sludge is produced and disposed of. Wet

Table 10.7 Examples of 'wet' processes for gas treating

Process	Absorber
Non-regenerative:	
Lime/Limestone	Lime or limestone slurry
Regenerative:	
Bureau of Mines Citrate	Citrate solution
Chemico	Magnesia slurry
Chiyoda Two-Stage	Dilute sulphuric acid
Double (or Dual) Alkali	Alkali or sulphite solution
Molten Salt	Molten carbonates
Stone and Webster/Ionics	Sodium hydroxide solution
TVA Ammonia	Ammonia solution
Wellman–Lord SO_2 Recovery	Sulphite solution

Table 10.8 Examples of 'dry' processes for gas treating

Process	Absorber
Carbon Adsorption	Activated carbon
Cat-Ox	Oxidation
Copper Oxide	Oxidation
Limestone Injection	Limestone

Table 10.9 Current national emission standards for sulphur (mg SO_2/m^3) (IEA, 1991)

Country	New plants	Existing plants
Austria	200–400	200–400
Belgium[c]	400–2000(250)[d]	
Canada[a]	740	
Denmark[c]	400–2000	
Finland	400–660	660
France[c]	400–2000	
FRG[c]	400–2000	400–2500
Greece[c]	400–2000	
Ireland[c]	400–2000	
Italy[c]	400–2000	400–2000
Japan	[b]	[b]
Luxembourg[c]	400–2000	
Netherlands[c]	200–700	400–700
Poland	540	2700–4000 (1800–3000)
Portugal[c]	400–2000	
Spain[c]	2400–9000	2400–9000
Sweden	290	290–570
Switzerland	400–2000	400–2000
Taiwan	2145–4000	2145–4000
Turkey	400–2000	
UK[c]	400–2 000	
USA	740–1480[e]	[e]

[a] Guidelines.
[b] Set on a plant-by-plant basis according to nationally defined formulae.
[c] EC countries.
[d] From 1995.
[e] Clean Air Act Amendments, 1990.

scrubbing processes contact the flue gas with a solution or slurry for the removal of sulphur dioxide.

In a wet limestone process (Chapter 12), the flue gas contacts a limestone slurry in a scrubber:

$$CaCO_3 + SO_2 = CaSO_3 + CO_2$$
$$2CaSO_3 + O_2 = CaSO_4$$

Both the sulphite and the sulphate absorb water of hydration to form the sulphite/sulphate sludge.

In a wet lime process (Chapter 12), a gas stream containing the sulphur dioxide is reacted with a wet lime slurry to form the sulphite:

$$Ca(OH)_2 + SO_2 = CaSO_3 + H_2O$$

with subsequent conversion, by oxidation, to the sulphate.

In the dual-alkali non-regenerative procedure (Chapter 12), the flue gas is scrubbed with a soluble alkali such as sodium sulphite which is subsequently regenerated with lime to form insoluble calcium sulphite:

$$Na_2SO_3 + SO_2 + H_2O = 2NaHSO_3$$
$$2NaHSO_3 + Ca(OH)_2 = Na_2SO_3 + 2H_2O + CaSO_3$$

after which disposal of the calcium sulphite slurry occurs and the spent absorbent, sodium bisulphite, is regenerated by thermal means:

$$2NaHSO_3 = Na_2SO_3 + SO_2 + H_2O$$

In all of these procedures, water leaves the system as vapour in the flue gas.

In either the lime/limestone or sodium sulphite/lime scrubbing processes the hydrated calcium sulphite and calcium sulphate can be of some environmental concern when the issue of disposal arises. More than anything else, this has promoted efforts to develop alternative dry scrubbing procedures for the removal of sulphur dioxide. And the dry systems have the additional advantage of reducing the pumping requirements necessary for the wet systems.

Thus there is a tendency to advocate the use of dry lime scrubbing systems thereby producing a waste stream that can be handled by 'conventional' fly ash removal procedures. There are also dry processes, such as the metal oxide processes, in which sulphur dioxide can be removed from gas streams by reaction with a metal oxide (Chapters 9 and 12). These processes, which are able to operate at high temperatures (about $400\,°C$, $750\,°F$), are suitable for hot gas desulphurization without an energy-wasteful cooling step. The metal oxides can usually be regenerated by aerial oxidation to convert any metal sulphide(s) back to the oxide(s) or by use of a mixture of hydrogen and steam.

Membranes have found increasing industrial use in the past two decades (Porter, 1990; Hsieh, 1991; Ho and Sirkar, 1992) and have also been suggested as being appropriate for the separation of hydrogen sulphide and sulphur dioxide; they are also applicable to higher temperature conditions (McKee et al., 1991; Shaver et al., 1991; Winnick, 1991).

The removal of carbonyl sulphide from gas streams, especially those that are destined for the manufacture of synthesis gas, has also been investigated using the principle of hydrogenation:

$$COS + H_2 = CO + H_2S$$
$$COS + H_2 = CO_2 + H_2S$$

Removal of nitrogen-containing gases

The occurrence of nitrogen in natural gas can be a major issue if the quantity is sufficient to lower the heating value. Thus several plants for the removal of nitrogen from the natural gas have been built, but it must be recognized that nitrogen removal requires liquefaction and fractionation of the entire gas stream, which may affect process economics. In many cases, the nitrogen-containing natural gas is blended with a gas having a higher heating value and sold at a reduced price depending upon the thermal value (Btu/ft^3, kJ/m^3).

Of equal interest is the occurrence of nitrogen compounds in gases produced by coal combustion. These compounds, the oxides, originate from the organically bound nitrogen in the coal:

$$[2N]_{coal} + O_2 = 2NO$$
$$2NO + O_2 = 2NO_2$$

Nitrogen oxides are formed during burning by oxidation, at the high temperatures, of the nitrogen in the fuel and in the air. This has given rise to the terminology 'thermal NO_x' and 'fuel NO_x' as a means of distinguishing between the two sources of nitrogen oxides. But, be that as it may, nitrogen oxides from whatever source are pollutants that must be removed from gas streams.

Indeed, many coal-fired boilers are being built with burners designed to reduce the formation of nitrogen oxide by delaying fuel/air mixing, or distributed fuel addition, thereby establishing fuel-rich combustion zones within the burner whereby the reduced oxygen level maintains a low level of production of nitrogen oxides (Slack, 1981; Wendt and Mereb, 1990). Other procedures employ ammonia to reduce the nitrogen oxides by injecting the ammonia and oxygen into the postcombustion zone:

$$4NO + 4NH_3 + O_2 = 4N_2 + 6H_2O$$

Vanadium oxide/aluminium oxide and iron/chromium, as well as the systems based on iron oxide itself, have also been reported to be successful for the removal of nitrogen oxides from gas streams.

Nitrogen compounds must be absorbed in several chemical and related processes, the most important of which is the absorption of nitrogen peroxide in water for the manufacture of nitric acid. Absorption of nitrous gases also takes place in the lead-chamber process used for the production of sulphuric acid, in the metallurgical industries where metals are treated with nitric acid, and in the purification of several tail gases.

At this point, it is also worth while to give some consideration to the production of nitric acid from nitric oxide (using simple chemistry), as might be envisaged in the formation of nitrous and nitric acids in an industrial setting or even in the atmosphere; this latter phenomenon would result in the deposition of acid rain.

Thus, nitric oxide is oxidized to nitrogen peroxide:

$$2NO + O_2 = 2NO_2$$

and dimerization of nitrogen peroxide gives nitrogen tetroxide:

$$2NO_2 = N_2O_4$$

after which the combination of nitrogen peroxide and nitric oxide gives nitrogen trioxide:

$$NO + NO_2 = N_2O_3$$

In the presence of water vapour nitrogen trioxide can be hydrated to nitrous acid:

$$N_2O_3 + H_2O = 2HNO_2$$

In the liquid phase the gross reaction equation of the formation of nitric acid is

$$3NO_2 + H_2O = 2HNO_3 + NO$$

Liquids removal

Gas liquids

The recovery of liquid hydrocarbons can be justified either because it is necessary to make the gas saleable or because economics dictates this course of action. The justification for building a liquids recovery (or a liquids removal) plant depends on the price differential between the enriched gas containing the higher-molecular-weight hydrocarbons and lean gas with the added value of the extracted liquid.

If saleability of the gas is the only reason for processing, the removal of liquids can be achieved as a field operation using either refrigeration processes (see next subsection), adsorption methods (see next subsection and Chapter 9), or a crude oil enrichment technique (see next section).

Most of the plants for liquids separation recover a substantial portion of the propane and essentially all of the butanes and higher-molecular-weight hydrocarbons. Whilst these products can be conveniently transported using the more conventional methods (pipeline, road and sea) (Chapter 6), ethane recovery depends on the availability of a product pipeline, although small amounts can be moved by road or rail when mixed with higher-molecular-weight hydrocarbons.

Water removal

Water is a common impurity in gas streams, especially natural gas, and its removal is necessary to prevent its condensation and the formation of ice or gas hydrates. In the liquid phase water will cause corrosion or erosion problems in pipelines and equipment, particularly when carbon dioxide and hydrogen sulphide are present in the gas. The simplest method of water removal (refrigeration or cryogenic separation) is to cool the natural gas to a temperature at least equal to or (more preferentially) below the dewpoint.

In a majority of cases, cooling alone is insufficient and, for the most part, impractical for use in field operations. Gases may be dried by absorption of the water in spray chambers using organic liquids such as glycerine or aqueous solutions of salts such as lithium chloride, and by use of columns with countercurrent flow of sulphuric acid or phosphoric acid or organic liquids.

Desiccants are classified as: solid adsorbents which remove water vapour by the phenomena of surface adsorption and capillary condensation (silica gel and activated alumina); solid absorbents which remove water vapour by chemical reaction (fused anhydrous calcium sulphate, lime and magnesium perchlorate); deliquescent absorbents which remove water vapour by chemical reaction and dissolution (calcium chloride and potassium hydroxide); or liquid absorbents which remove water vapour by absorption (sulphuric acid, lithium chloride solutions and ethylene glycol).

Silica gel and alumina have a high capacity for water adsorption (up to 8% by weight). Bauxite, an alumina ore consisting of about 60% wt/wt crude alumina, will adsorb up to 6% by weight water and molecular sieves will adsorb up to 15% by weight water. Silica is usually selected for the dehydration of sour natural gas because of its high tolerance to hydrogen sulphide and to protect molecular sieve beds from plugging by sulphur. Alumina 'guard' beds, which will serve as protectors by the act of attrition (see also Chapter 9 and Speight (1981)), may be placed ahead of the molecular sieves to remove the sulphur compounds.

Deliquescent salts and hydrates are often used as concentrated solutions because of the difficulties in handling, replacing and regenerating the wet corrosive solids. The degree of drying possible with solutions is much less than with the corresponding solids. But where only moderately low humidities are required and large volumes of air are dried, solutions are satisfactory.

Organic liquids suitable for water removal processes are hygroscopic liquids such as diethylene or triethylene glycol (Hicks and Senules, 1991). Ethylene glycol can be directly injected into the gas stream in refrigeration-type plants (see Figure 10.7). Solid adsorbents (e.g. alumina, silica gel and molecular sieves) may also be used for water removal.

The use of a hygroscopic fluid, such as ethylene glycol, for natural gas dehydration is a relatively simple operation (Figure 10.7). The overhead stream from the regenerator is cooled with air fins at the top of the column or

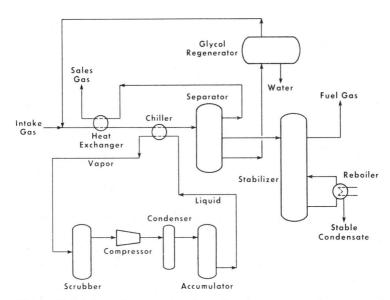

Figure 10.7 Schematic representation of a water removal system using glycol treatment

by an internal coil through which the feed flows. The countercurrent vapour–liquid contact between the gas and the glycol produces a dewpoint of the outlet stream that is a function of the contact temperature and the residual water content of the stripped or lean glycol.

The regeneration (stripping) of the glycol is, in most cases, limited by temperature; diethylene and triethylene glycol will decompose at, or prior to, their respective boiling points. Techniques such as the stripping of hot triethylene glycol with dry gas (e.g. heavy hydrocarbon vapours, the Drizo process; see page 287) or vacuum distillation are recommended.

Liquid desiccants (concentrated acids and organic liquids) are generally liquid at all stages of a drying process. Soluble desiccants (calcium chloride and sodium hydroxide) include those solids which are deliquescent in the presence of high concentrations of water vapour.

Some adsorbent processes for water removal (Figure 10.8) employ a two-bed adsorbent treater: while one bed is removing water from the gas, the other is undergoing alternate heating and cooling. On occasion a three-bed system is used: one bed is adsorbing, one is being heated, and one is being cooled. An additional advantage of the three-bed system is the facile conversion to a two-bed system so that the third bed can be maintained or replaced, thus ensuring continuity of the operations and reducing the risk of a costly plant shutdown.

Gases may also be dried by compression to a partial pressure of water vapour greater than the saturation pressure to effect condensation of liquid water; cooling below the dewpoint of the gas with surface condensers or cold-water sprays; and compression cooling, in which liquid desiccants are used in continuous processes in spray chambers and packed solid desiccants are generally used in an intermittent operation that requires periodic interruption for the regeneration of the spent desiccant.

The mechanical methods of drying gases – compression and cooling and

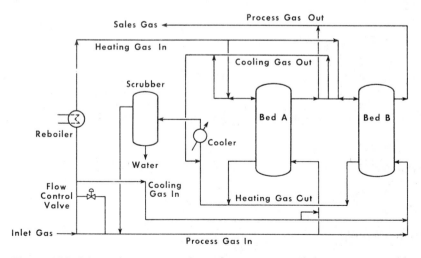

Figure 10.8 Schematic representation of water removal by treatment with an adsorbent

refrigeration – are used in large-scale operations and generally are more expensive methods than those using desiccants. Such mechanical methods are used when compression or cooling of the gas is required.

Enrichment

The purpose of crude enrichment is to produce natural gas for sale and an enriched tank oil. The tank oil contains more light hydrocarbon liquids than natural petroleum and the residue gas is drier (leaner; i.e. has lesser amounts of the higher-molecular-weight hydrocarbons). Therefore, the process concept is essentially the separation of hydrocarbon liquids from the methane to produce a lean, dry gas.

Crude oil enrichment is used where there is no separate market for light hydrocarbon liquids, or where the increase in API gravity of the crude will provide a substantial increase in the price per unit volume as well as volume of the stock tank oil.

A very convenient method of enrichment involves manipulation of the number and the operating pressures of the gas–oil separators (traps). However, it must be recognized that alteration or manipulation of the separator pressure will affect the gas compression operation as well as influence other processing steps.

One method of removing light ends involves the use of pressure reduction (Figure 10.9) in which the stripping of light ends is achieved at low pressure, after which the pressure of the stripped crude oil is elevated so that the oil will act as an absorbent. The crude oil, which becomes enriched by this procedure, is then reduced to atmospheric pressure in stages or using fractionation (rectification).

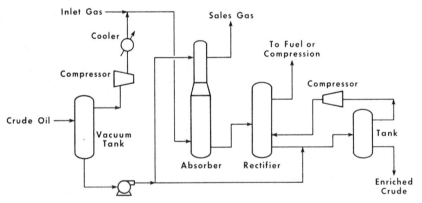

Figure 10.9 Schematic representation of a pressure reduction system for light-ends removal

Fractionation

Fractionation processes are very similar to other processes which are classed as 'liquids removal' processes but often appear to be more specific in terms of

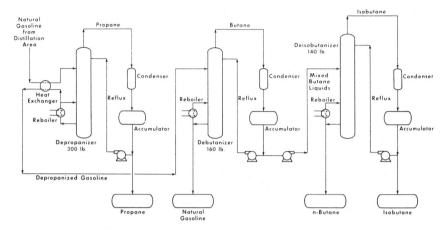

Figure 10.10 Schematic representation of a gas fractionation system

objectives. Hence there is a need to place the fractionation processes into a separate category. The fractionation processes are those processes which are used (a) to remove the more significant product stream first or (b) to remove any unwanted light ends from the heavier liquid products.

In the general practice of natural gas processing, the first unit will be a de-ethanizer, followed by a depropanizer, then by a debutanizer and, finally, a butane fractionator (Figure 10.10). Thus each column can operate at a successively lower pressure thereby allowing the different gas streams to flow from column to column by virtue of the pressure gradient and without necessarily the use of pumps.

Solvent recovery

Solvent recovery can be conveniently achieved by use of the adsorption concept, especially with carbon as the adsorbent.

In such a process using a fairly standard two-adsorber system, the solvent-laden gas stream is passed through an adsorbent (e.g. carbon) in one of two adsorbers, whilst the second adsorber, which has previously been on-stream, is undergoing regeneration, usually with (live) steam.

The mixture of steam and solvent passes through the distillate pipe into the condenser. The condensed mixture of steam and solvent is, in the case of non-miscible solvents, automatically separated into two layers in a separator. The recovered solvent is collected in the tank from which it may be pumped for reuse. When solvents that are miscible or partially miscible with water are used, such as acetone, ethyl acetate, etc., the condensate or water layer from the separator is distilled so that the water-soluble constituents can be recovered in a dry state for further use.

After the carbon has been dried and cooled by means of cold air, the adsorber is again put on-stream. In this way the carbon can be used almost indefinitely, at least theoretically. When non-miscible solvents are used the recovery plant may be simplified by drying and cooling with solvent-laden air.

References

Benson, H.E. (1981) In *Chemistry of Coal Utilization. Second Supplementary Volume* (ed. M.A. Elliott), Wiley, New York, Chapter 25

Berkowitz, N. (1979) *Introduction to Coal Technology*, Academic Press, New York

Bodle, W.W. and Huebler, J. (1981) In *Coal Handbook* (ed. R.A. Meyers), Marcel Dekker, New York, Chapter 10

Chiu, C.-H. (1990) *Hydrocarbon Processing*, **69**, (1), 69

Chou J.S., Chen, D.H., Walker, R.E. and Maddox, R.N. (1991) *Hydrocarbon Processing*, **70**, (4), 38

Dowling, N.I., Hyne, J.B. and Brown, D.M. (1990) *Industrial and Engineering Chemistry Research*, **29**, 2332, and references cited therein

Goar, B.G. (1971a) *Oil and Gas Journal*, p. 75 (12 July)

Goar, B.G. (1971b) *Oil and Gas Journal*, p. 84 (12 July)

Grosick, H.A. and Kovacic, J.E. (1981) In *Chemistry of Coal Utilization. Second Supplementary Volume* (ed. M.A. Elliott), Wiley, New York, Chapter 18

Hessley, R.K. (1990) In *Fuel Science and Technology Handbook* (ed. J.G. Speight), Marcel Dekker, New York

Hessley, R.K., Reasoner, J.W. and Riley, J.T. (1986) *Coal Science*, Wiley, New York

Hicks, R.L. and Senules, E.A. (1991) *Hydrocarbon Processing*, **70**, (4), 55

Ho, W.S.W. and Sirkar, K. (1992) *Membrane Handbook*, Van Nostrand Reinhold, New York

Hsieh, H.P. (1991) *Catalysis Reviews – Science and Engineering*, **33**, (1&2), 1

Lagas, J.A., Borboom, J. and Heijkoop, G. (1989) *Hydrocarbon Processing*, **68**, (4), 40

Luistra, E.A. and d'Haene, P.E. (1989) *Hydrocarbon Processing*, **68**, (7), 53

McKee, R.L., Changela, M.K. and Reading, G.J. (1991) *Hydrocarbon Processing*, **70**, (4), 63

Maddox, R.N. (1974) *Gas and Liquid Sweetening*, Campbell, Norman, OK

Plumley, A.L. (1971) *Combustion*, p. 36 (October)

Porter, M.C. (1990) *Handbook of Industrial Membrane Technology*, Noyes, Park Ridge, NJ

Shaver, K.G., Poffenbarger, G.L. and Grotewold, D.R. (1991) *Hydrocarbon Processing*, **70**, (6), 77

Slack, A.V. (1981) In *Chemistry of Coal Utilization. Second Supplementary Volume* (ed. M.A. Elliott), Wiley, New York, Chapter 22

Speight, J.G. (1981) *The Desulfurization of Heavy Oils and Residua*, Marcel Dekker, New York

Speight, J.G. (1983) *The Chemistry and Technology of Coal*, Marcel Dekker, New York

Speight, J.G. (1990) *Fuel Science and Technology Handbook*, Marcel Dekker, New York

Speight, J.G. (1991) *The Chemistry and Technology of Petroleum*, 2nd edn, Marcel Dekker, New York

Taylor, N.A., Hugill, J.A., van Kessel, M.M. and Verburg, R.P.J. (1991) *Oil and Gas Journal*, **89**, (33), 57

van den Berg, P.J. and de Jong, W.A. (1980) *Introduction to Chemical Process Technology*, Delft University Press, Delft

Wendt, J.O.L. and Mereb, J.B. (1990) *Nitrogen Oxide Abatement by Distributed Fuel Addition*. Report No. DE92005212, Office of Scientific and Technical Information, Oak Ridge, TN

Wesch, I.M. (1992) *Oil and Gas Journal*, **90**, (8), 58

Winnick, J. (1991) *High-temperature Membranes for H_2S and SO_2 Separations*. Report No. DE92003115, Office of Scientific and Technical Information, Oak Ridge, TN

11

Processing equipment

Introduction

Throughout this text, mention is often made of the types of equipment that are used to effect the separation of contaminants from gas streams. Therefore, it is appropriate at this point to present a simple outline of the types of equipment that are employed in gas-cleaning operations.

There are several sources of reference related to the theory, design and construction of equipment (Sherwood, 1937; Mantell, 1951; Schweitzer, 1979; Green and Maloney, 1984; Avallone and Baumeister, 1987). It is not the intent of this chapter to make any attempt to supplant such major works. Rather, it is the intent of this chapter to present an overview of the types of equipment (Table 11.1) that are used for gas-cleaning operations in order that the reader might gain a better understanding of the parameters involved in the industrial operations.

Absorption equipment

Equilibrium conditions for absorption processes vary depending upon whether the system is based upon physical absorption, chemical absorption, or on a combination of both physical and chemical interaction (Figure 11.1). Thus, the purpose for which absorption systems are to be used is to produce from a gas stream the lowest, possible concentration of contaminants in the product gas and certainly the most legal according to regulations (Slack, 1981).

Thus, there are several requirements which an absorption system should satisfy and these are as follows: (a) homogenization of the gas phase should be avoided; (b) the liquid should pass through the apparatus countercurrent to the gas stream; (c) any heat of the reaction should be removed as it is evolved; (d) a large gas–liquid interface; (e) a large volume of the gas phase (Fair et al., 1984). Consequently, several variants of the system are available (e.g. Figures 11.2–11.5).

Examples of the equipment used for this service are: absorption towers, of which there are several varieties including spray towers (high surface, no surface renewal); tray towers (high surface renewal); and packed towers (high surface, high surface renewal) and venturis (high surface, no surface renewal). However, it must be noted here that reference is often made in certain technical

Table 11.1 Examples of contaminants produced by industrial processes and the types of control equipment used for gas cleaning

Industry	Air contaminants emitted	Control technologies
Aluminium reduction	Particulates, CO, SO_2, hydrocarbons and fluorides	Cyclone, electrostatic precipitators, scrubbers, baghouse
Cement plants	Particulates, CO, SO_2, NO_2	Electrostatic precipitators and baghouses, scrubber
Coal preparation	Dust, smoke, particulates, sulphur oxides, H_2S	Cyclones and venturi
Coke plants	Particulates, SO_2, CO, H_2S, methane, ammonia, H_2, phenols, hydrogen cyanide, N_2, benzene, xylene	Scrubbers and badhouses, electrostatic precipitators
Fertilizer industry (chemical)	PH_3, P_2O_5 HF, SiF_4	Scrubber
	NH_3, NH_4Cl, SiF_4, HF, NO_2	Cyclone, electrostatic precipitator, baghouse
Kraft pulp mills	Mercaptans, H_2S, H_2, organic sulphides and disulphides	Caustic scrubbing, static precipitator, packed tower and cyclone, venturi scrubbers
Municipal and industrial incinerators	Particulates, CO, SO_2, ammonia, organic acids, aldehydes, NO_2, hydrocarbons, HCl	Settling chambers, baghouses
Non-ferrous smelters; Copper	SO_2, particulates	Settling chambers, cyclones or scrubbers and electrostatic precipitators
Lead	SO_2, CO, particulates	Cyclones and baghouse or precipitators, settling chambers
Zinc	Particulates and SO_2, CO	Cyclone scrubber, electrostatic precipitator, baghouse
Paint and varnish manufacturing	Acrolein, other aldehydes and fatty acids, phthalic anhydride (sublimed); ketones, fatty acids, formic acids, acetic acid, glycerine, acrolein, other aldehydes, phenols and terpenes; from tall oils, hydrogen sulphide, alkyl sulphide, butyl mercaptan, and thiofene; olefins, branched-chain aromatics and ketone solvents	Scrubbers
Rendering plants	SO_2, mercaptans, ammonia	Condenser, scrubber
Steel mills	CO, particulates, SO_2, CO_2, NO_2	Cyclone, scrubber, electrostatic precipitator or venturi scrubber, settling chamber

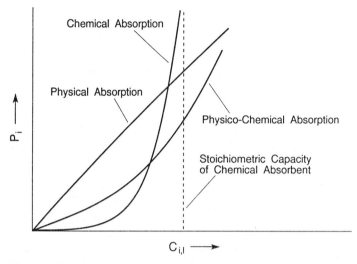

Figure 11.1 Equilibria relationships for gas cleaning by absorption in liquids

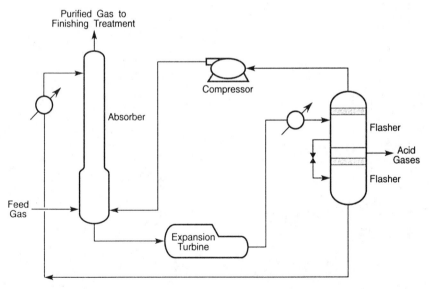

Figure 11.2 Schematic representation of a physical absorption system with two-stage expansion regeneration

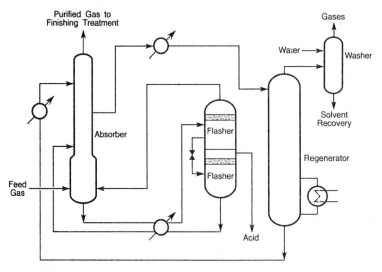

Figure 11.3 Schematic representation of a physical absorption system with two-stage expansion and thermal regeneration

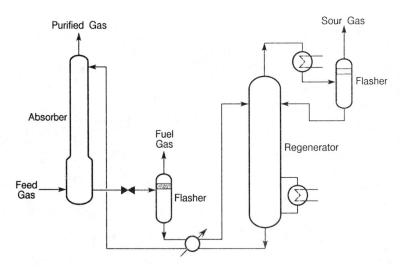

Figure 11.4 Schematic representation of a chemical absorption system

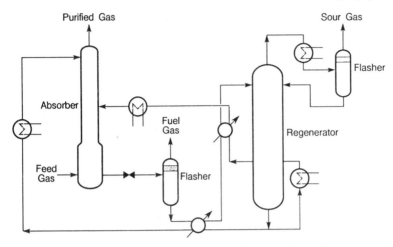

Figure 11.5 Schematic representation of a chemical absorption system with split-stream regeneration

publications to 'absorption scrubbers' as well as to 'absorption towers', and for the purposes of this text a very general definition is in order.

A tower (see the next subsection) is a system whereby the gas comes into contact with a liquid of some depth. With the exception of the 'bubble tower', a tower may often be full, near to capacity. Towers, as defined in this text, also contain packings that are usually absent from scrubbing equipment. On the other hand, a scrubber (see the next section) is defined as a means by which a gas is washed, and often takes the form of a spray of liquid with which the gas comes into contact.

Towers

The removal of unwanted gaseous contaminants in gas absorption towers is effected by bringing the gas into contact with a liquid in which the impurities alone are soluble. The usual procedure is for the solvent to be introduced at the top of the tower and the gas at the bottom, so that countercurrent contact between the gas and the liquid is made in the tower.

Absorption is a diffusional process: the molecules of the gas being absorbed have to pass by diffusion through the gas phase, cross the interface between gas and liquid, and then pass into the liquid phase. The molecules can move either by molecular diffusion, which is a slow process, or by eddy diffusion, a much faster process in which appreciable amounts of the gas or liquid move as a whole. The difference between the concentration of the gas in the gas phase and that in the liquid phase is the cause of the molecular movement. The concentration of the gas in the gas phase is usually expressed as a partial pressure, the units of which are different from those used for concentrations in the liquid phase.

According to Henry's law, the relationship between gas partial pressure and concentration can be described conveniently as

$$c_1 = Hp_1$$

where c_1 = concentration of soluble gas in liquid

H = solubility coefficient

p_1 = partial pressure of soluble gas in equilibrium with liquid of concentration c_1

Thus, the cause of the molecular movement from gas to liquid can be expressed either as a difference of partial pressures (gas-phase basis) or a difference of concentrations (liquid-phase basis). In an absorption tower, the composition of the gas and the liquid vary as they pass through it, and consequently the movement between the gas and liquid phases will also vary.

Efficient absorption is the basic tool for exploiting natural selectivity, but it will not prevent the co-absorption of other components: carbon dioxide, for example, will almost always be co-absorbed with hydrogen sulphide. The desorption process which usually follows absorption does, however, offer a means of making further separation between the components of the gas streams.

A practical manner in which staging can be introduced into the desorption process involves stripping or enrichment (Chapter 10). Instead of flash tanks, a tower can be used to carry out the desorption of excess carbon dioxide from the rich ('fat') liquid; the tower will be operated at a pressure somewhat lower than the absorption pressure. Rich liquid is first flashed at the top of the tower or in a separate vessel and passes countercurently down the tower to a flow of stripping gas. Indeed, the use of staging in the desorption process, as might be experienced with a series of towers (Figure 11.6) rather than with a single tower, is a highly effective method of treating gas.

The rate of absorption of a gas in a liquid is given by

$$W = K_{gm} A (pm)$$

where W = rate of absorption (lb/h)

K_{gm} = mean overall coefficient on gas-phase basis

A = surface area (ft^2)

pm = mean driving force on gas-phase basis (atm)

The aim of the tower design is to obtain the highest possible gas absorption rate or to increase the values of K_{gm} and A as much as possible. The value of

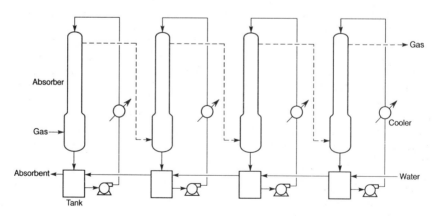

Figure 11.6 Illustration of an absorption system using a series of packed towers

K_{gm} is determined by the resistances of the gas and liquid phases and is, simply, the conductance of the two phases.

Therefore, the reciprocal of K_{gm} is resistance, and is the sum of the resistances of the two phases:

$$1/K_{gm} = 1/k_{gm} + 1/k_{lm}$$

where k_{gm} = mean gas-phase coefficient (lb/h ft^2 atm)
$\quad\quad k_{lm}$ = mean liquid-phase coefficient (lb/h ft^2(lb/ft^3))

In order to increase the value of K_{gm}, the resistances of the phases need to be reduced. This can be achieved by increasing the turbulence in the system since molecular diffusion alone occurs in streamline flow, while in turbulent flow this is replaced by the much more rapid eddy diffusion.

Thus, absorption towers should be designed either to increase turbulence, or to increase surface area, or both.

Selection

Until recently, the type of tower almost invariably selected for gas absorption operations was the packed tower, in which the tower shell was filled with a packing. In addition, plate towers, formerly used almost exclusively in distillation, find wide application in gas absorption processes. In these plate towers, the tower shell, instead of being filled with packing, is fitted with a number of separate plates of various types spaced at intervals up the tower.

The usual construction is a plate tower with bubble caps which contains the whole absorption system. In this type of absorption tower, the homogenization of gas and liquid is usually restricted to one stage only. Absorption towers of this construction normally measure 6.5 ft (2 m) in diameter and 49–50 ft (15 m) in height and contain 30–40 plates.

In the second type of absorption system, each absorption stage corresponding with the liquid layer on a plate, plus the gas space over it in the plate tower, consists of a horizontal cylinder filled with liquid to about one-fourth of its height. The gas and liquid pass in countercurrent through a cascade of these cylinders.

The packed column absorption tower contains a commercially available geometric shape, ranging from 1 to 4 in (2.5 to 6 cm), in forms that provide either large surfaces or the capability for surface renewal, or both.

Absorption systems have been constructed for operation at atmospheric pressure as well as at elevated pressures. In atmospheric pressure systems, the gas leaving the tower is usually passed through additional absorption towers through which an appropriate solution is circulated to effect a reduction of the noxious content of the gas product.

In general, the following items need to be considered in the design of an absorption system:

(a) *Column diameter*. At a given gas feed rate the tower diameter is determined by the admissible gas velocity. This gas velocity is usually kept below the value derived from loading correlations so as to provide sufficient residence time for the 'reaction' (absorption) to occur.

(b) *Column height*. The height should be adapted to the diameter of the column. The maximum value of the dimensions will often depend on the needs of the system, i.e. practical circumstances.

(c) *Packing*. The dimensions of the packing elements should correspond to the diameter of the tower. The type of packing material is determined by the demand of a large free space.

(d) *Liquid velocity*. The velocity of the circulating liquid is determined by the requirement of a sufficient degree of wetting of the packing material. Moreover, the liquid velocity should be high enough to restrict the temperature rise per pass to a few degrees.

The first selection criterion to be given consideration in the choice of an absorption tower is that between packed and plate towers. This depends, for example, on the corrosive character of the substances being handled. With corrosive substances the packed tower is at an advantage, since cheap ceramic packings can be used and only the tower shell needs to be made of expensive corrosion-resistant materials. With plate towers, however, the entire tower would have to be made of such corrosion-resistant materials. Formerly, plate towers gave a much higher pressure drop than packed towers, and the latter were therefore chosen where gas-pumping costs were important. At the present time, however, plates with a lower pressure drop are available, though still higher than for packed towers.

Spray towers

Gas absorption equipment (often called scrubbers and ranging in capacity from $100\,\text{ft}^3/\text{min}$ ($4.7 \times 10^{-4}\,\text{m}^3/\text{s}$) to $2\,000\,000\,\text{ft}^3/\text{min}$ ($940\,\text{m}^3/\text{s}$)) generally employs water (or solutions of reagents in water) as the host liquid, although hydrocarbon liquids are used for specific applications in the chemical industry. The overall efficiency of the operation can be high, especially in the case of absorption accompanied by chemical reaction as evidenced by its use for air pollution control in the steel, fertilizer, glass, and pulp and paper industries.

The spray tower (Figure 11.7) is most commonly used when transfer requirements are low. Spray towers with spray nozzles are designed for effective droplet size distribution.

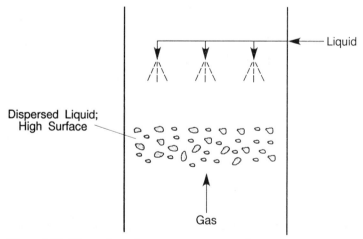

Figure 11.7 Illustration of a spray tower operation

Tray towers

The tray tower (Figure 11.8) is used to a lesser extent than the spray tower (see the previous subsection) owing to a lower efficiency in low-concentration gas systems. Tray towers have elutriated sieve trays with baffles to produce aeration and spray droplets by utilizing the pressure drop for the passage of gases through the tray perforations into the liquid passing over the tray. The baffles, each located above a perforation, provide an impingement surface to produce the turbulence required.

Other aspects of such a tower design include gas humidification sprays at the gas inlet, and entrainment separators for the liquid droplets present in the effluent gases. Most of the dust removal here actually takes place in the region of the spraying droplets above the tray, after the gases have been brought to a high local velocity induced by the pressure drop established as the dust-laden gases are forced through the tray perforations.

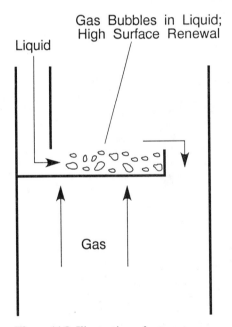

Figure 11.8 Illustration of a tray tower operation

Packings

The packed tower (Figure 11.9) is the predominant system used in gas absorption because of its efficiency and additional capability of being able to handle particulates in gas streams. It is often used in conjunction with a solids recovery unit, such as a venturi (see the final section).

The simplest type of packing consists of lumps of coke or quartz, and these are still used, especially where corrosion is a problem. More widely used, however, are the Raschig rings (Figure 11.10) with a height equal to their diameter; these can be stacked in the tower, if of diameter 3 in (7.5 cm) or over,

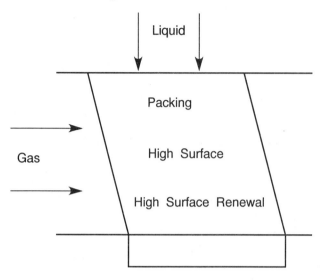

Figure 11.9 Illustration of a packed tower operation

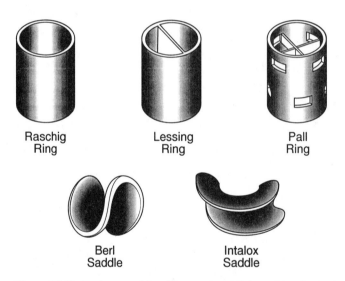

Figure 11.10 Various packing (ring and saddle) used in absorption towers

or used as a random packing. Stacking is useful in that it reduces the tendency to foul.

Random rings have the advantage that they tend to promote even liquid distribution, but it is doubtful if the inside surfaces of the rings are wetted, so that the surface area available for absorption is reduced. It has also been found that random rings tend to lie with their axes horizontal, thus restricting the gas flow and increasing the pressure drop. It is because of these difficulties that Pall rings were devised. Raschig rings are available in a variety of ceramics, metals and alloys; the choice of material depends upon the corrosion characteristics of the system.

Lessing rings, which are actually Raschig rings with a central partition, are used, among other things, for base support courses for beds of random rings.

The Pall ring is, in effect, a Raschig ring with a slotted wall, parts of the wall being pushed in towards the axis of the ring, where they meet. With such a slotted wall, passages are available for the gas to flow through, even when the ring is horizontal, so that the pressure drop is lower than for Raschig rings.

As well as the slots, a Pall ring also provides openings through which the liquid can flow on to the inside wall of the ring, and thus the area available for absorption is increased. A ceramic packing which gives similar improved performance to the Pall ring, but which is of good mechanical strength, is provided by the Intalox saddle which is an improvement on the earlier Berl saddle, in that it is cheaper to produce. In addition, the design is such as to avoid the 'nesting' to which the Berl saddle is subject.

Grid packings made of wood are widely used in cooling towers, but grids of other materials, such as carbon, metal, glass or plastics, can be used by absorption. They have a low pressure drop and permit high throughputs, but good initial liquid distribution is necessary as the grids are not self-distributing. Wooden grids can be plain, or have the bottom edges serrated to prevent excessive drop from one layer to another; grids in the other materials, except carbon, are plain.

Glass grid packing deserves special mention because of its possible use in a specific gas purification process: the removal of moisture from sulphur dioxide before its conversion to sulphur trioxide. The absorbent used is strong sulphuric acid, so that corrosion is a difficulty. Ordinary, cheap, rough-cast slates are suggested where the temperature does not exceed 150°C (302°F); above this temperature and up to 300°C (572°F) the more expensive toughened glass could be used.

A completely different type of packing consists of an assembly of expanded-metal elements, built up to form a honeycomb structure of cells, each cell having a diamond-shaped cross-section. The aim of the expanded metal is to produce films of liquid which are broken up into spray by the gas passing through the mesh. This form of packing, known as structured packing (Kean et al., 1991), is now common in various columns.

Plates

A large number of different types of plate have been developed in recent years, but mainly for distillation purposes. Only those which, as far as is known, have been used in absorption towers will be mentioned here.

The bubble-cap plate (Figure 11.11) is the best known and oldest type. In operation, liquid flows down from plate to plate by means of downcomers, passing usually from side to side of each plate in turn in so doing. The gas is made to pass up through risers into a number of bubble caps, often with serrated edges, from which it is then passed out into the liquid in the form of bubbles. The bubble-cap plate has a high efficiency but a low throughput compared with some of the newer types.

The sieve plate, which is coming increasingly into use, consists merely of a plate perforated with a large number of holes, usually 3/16 in (5 mm) in diameter. This gives a more even bubbling action than with other types of

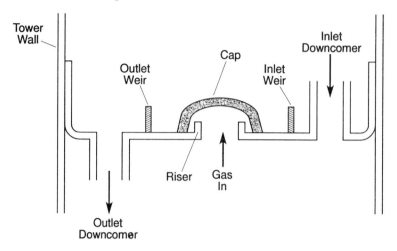

Figure 11.11 Schematic representation of a bubble-cap plate

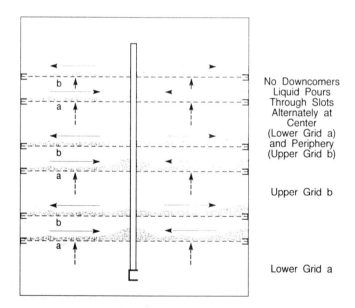

Figure 11.12 Illustration of a Kittel plate operation

plate, and higher efficiencies as a result. Since the holes are of such low diameter, a small amount of erosion can appreciably alter the characteristics of the plate. For this reason, sieve plates are usually made of stainless steel.

The standard Kittel plate (Figure 11.12) consists of two grids, a few inches apart, made from expanded metal and fabricated in such a way that the gas passing through the grid is given a spiral motion which in turn is imparted to the liquid. On the upper grid, liquid flow is towards the wall; on the lower, inwards to the centre.

Polygonal Kittel plates (Figure 11.13) are used where liquid rates are large, e.g. in carbon dioxide absorption towers. These are fitted with alternate annular and central downcomers, single sheets of mesh being used for the plate across which the liquid flows. The spiral motion is produced not only by the gas flow, but also by means of vanes fitted in the downcomers.

The 'shower tray', like the polygonal Kittel plate, is suitable for use where liquid rates are very high, e.g. where the liquid and gas volume flow rates are of the same order. The shower tray consists of a perforated plate of free area and weir height such that all the liquid washes through the plate, while the gas passes up through what would normally be the downcomer. This arrangement allows the liquid to entrain gas with it into the liquid on the plate below, thus creating a surface area available for absorption by the formation of bubbles. In this way, use is made of the potential energy of the liquid in surface formation.

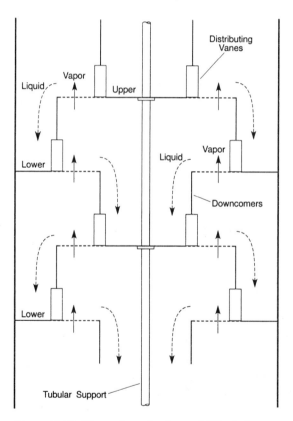

Figure 11.13 Illustration of polygonal Kittel plate operation

Scrubbers

The term 'scrubbers' as employed here can also be used as the term for an absorption tower and care must be taken to qualify the use of the equipment.

For example, the spray tower might also, under various circumstances, be referred to as a scrubber.

Commercial separations that involve a contacting action between contaminant-containing or dust-laden gases and liquids to bring the solids into a slurry form for removal must utilize methods of low energy requirements. And the energy consumption is reflected in the presure drop required to operate the various types of scrubbers (Maas, 1979; Slack, 1981). Those of the lowest pressure drop are limited to a particle size separation above 10 μm, while others at progressively higher pressure-drop characteristics will give a particle separation down to as low as 0.5 μm. Favourable gas-contacting methods are therefore hoped to be found for a design among the following types of scrubbing devices, alone or in some combination.

Of the general types of scrubbing device outlined here, only a limited number may find use in the larger operations, especially those operations which may have flow rates in excess of 500 000 ft^3/min (230 m^3/s).

Selection

There are several types of commercial scrubbers available that have been used for gas cleaning (Figure 11.14).

This means that it will be necessary to make a choice between the simpler and less expensive types of scrubbers, which are less prone to plugging, and the types of scrubbers which promote mass transfer more effectively but which may be more expensive and more prone to plugging. In some installations, the scrubbers must remove part or all of the dust in addition to the gaseous contaminants, e.g. hydrogen sulphide and sulphur dioxide. This requires a scrubber with some pressure drop and even more resistance to plugging.

Unlike most scrubber installations, those attached to sources of sulphur dioxide emission, such as power plants and smelters, are subject to widely varying gas flows. This can be compensated for by the installation of several scrubbers on-stream and shutting off one or more if the gas produced decreases and the flow diminishes.

The most difficult problem in many scrubbing operations is the removal of mist from the gas and at the same time avoiding scaling of the mist eliminator surfaces. Since chevrons (zigzag baffles) are relatively inexpensive, have a low pressure drop, and take up little room, they are used in most lime–limestone scrubbing installations. The usual arrangement is to install the chevron at the top of the scrubber in a horizontal position, perpendicular to the gas flow (see Figure 11.14). Usually, three or more 'direction changers' (individual baffles, each turning the gas to a different direction) are required, and there may be two or more 'banks' (sets of baffles) with some distance between them.

The main drawback to the horizontal scrubber is that the gas flowing up tends to keep separation mist in the chevron rather than allowing it to drain back into the scrubber. As a result, mist can remain on the surface of the chevron and be re-entrained into the gas. Test have shown that much better performance is obtained if the slats are slanted longitudinally or are set vertically, so that the mist can drain off. This can be accomplished either by placing the unit in a horizontal duct after the scrubber or by using a slanted arrangement in the top of the scrubber, similar to the rafters of a house.

Thus, in very general terms, scrubber characteristics vary with design, which in turn dictates performance.

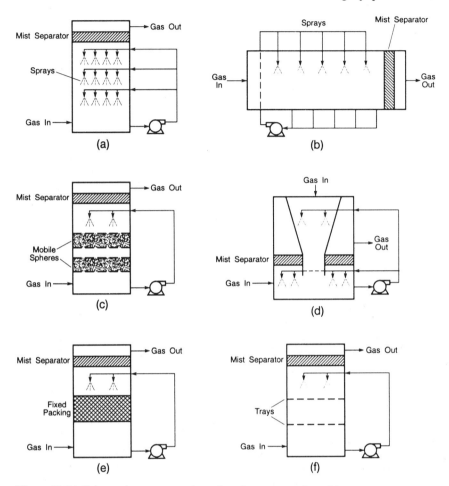

Figure 11.14 Schematic representation of various types of scrubbers

A general comment on the types of scrubbers in operation is warranted here, although there is no attempt to discuss in detail the mechanical and engineering principles involved. The purpose of the following subsection is to give a brief overview of the types of scrubbers available for gas-cleaning operations.

Spray scrubbers

The vertical open spray tower (Figure 11.15) is perhaps the simplest system in use and, perhaps, the most well known in so far as it brings to mind the operation of the bathroom shower in terms of its construction and operation.

The system offers protection against plugging and the ability to adjust to high and/or low gas flows but the mass transfer characteristics are relatively poor. The horizontal spray chamber has the advantage that it can be generally operated at higher velocity than the vertical tower. The higher velocity not

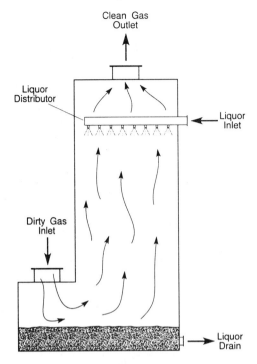

Figure 11.15 Schematic representation of an open spray tower

only reduces scrubber size and cost but also increases mass transfer to the point where it offsets the lack of packing or other internal elements.

Cyclone spray scrubbers (Figure 11.16) use the principle that when dust-laden gases are brought to a high velocity and water droplets impact with the dust particles, the result is the formation of a slurry which is separated from the gas stream through the centrifugal forces produced. Energy input is chiefly from the relatively high entrance gas velocity, supplied by fans or blowers.

Wet scrubbers

Wet scrubbers require relatively high gas velocities, but the liquid is introduced as a fluid at low pressure and the gas stream provides the motive power to break up the liquid into drops. There is the simple application of using a pipe, or using a venturi shape to carry the gas and produce the droplet dispersion effective for the operation. This type of scrubbing may be effectively combined with a cyclonic separator to remove liquid droplets from the effluent gas, as an entrainer.

The venturi scrubber (Figure 11.17) has no internal elements, the velocity is high, and there is no nozzle wear. Turndown, however, requires a variable throat mechanism, and retention time is so short that two venturis in series have been required for the adequate removal of sulphur dioxide. Performance can be improved by the use of additives (e.g. MgO) and sprays in the mist

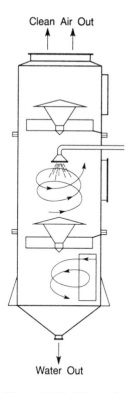

Figure 11.16 Schematic representation of a cyclone scrubber

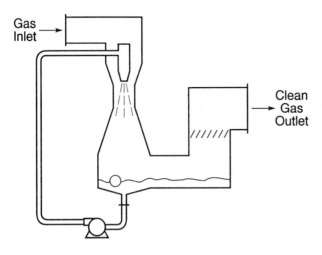

Figure 11.17 Illustration of a water-induced venturi absorber

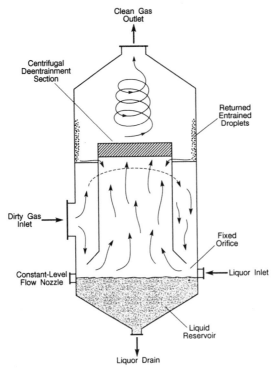

Figure 11.18 Illustration of a liquor impingement scrubber

knockout section, making a single venturi adequate unless a very high removal efficiency is required.

Water-jet scrubbers (Figure 11.18), in contrast to the operation of spray towers, use a high-velocity stream of water, after which the wash solution or wash slurry (if dust removal is also an objective) is discharged in the turbulent state into a collector for removal from the system; the liquid is recirculated. This operation could be combined with a final entrainment device for the effluent gas, but one preferably of low-pressure-drop characteristics.

Mechanical scrubbers

Mechanical scrubbers (Figure 11.19) include a variety of arrangements, often arranged so that liquid enters the inlet of a fan and is dispersed while the gas is being accelerated to a required velocity. Some types employ rotating screens to pass through and pick up the liquid from a pool at the base of the rotator.

Fibrous-bed scrubbers

Fibrous-bed scrubbers (Figure 11.20) are used for the separation of submicron-size particles by means of a thin fibre filter (0.195 in, 5 mm, thick) which is elutriated during the passage of gas and particles through the fibre. Unlike typical filtration, the particles are washed from the downstream side of felted cloth, impaction being produced without clogging the fibrous cloth.

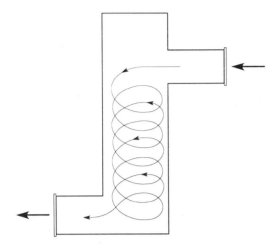

Figure 11.19 Illustration of a mechanical scrubber

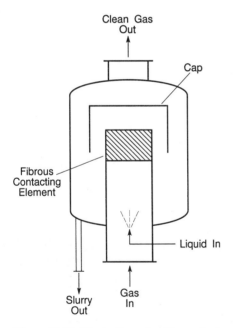

Figure 11.20 Illustration of a fibrous-bed scrubber

Packed-bed scrubbers

Packed-bed scrubbers (Figure 11.21), of which there are various types, may generally be regarded as versions of packed towers as used for gas cleaning, but for dust-laden gases these must be arranged for solids disposal. This may be done by adding sprays to the conventional wash given to the packed bed, for example by placing sprays beneath the packing support plate. The packing

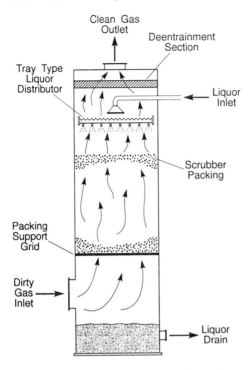

Figure 11.21 Schematic representation of a countercurrent flow packed scrubber

beds may sometimes be kept in a partially flooded state. A floating bed of spheres (Figure 11.22), to facilitate the washing out of collected solids assisted by the motion of the packing when the gas passes through, might be used.

Adsorbers

Adsorbers work on a different principle to absorbers: the usual method of adsorption involves the sorption of material on to the surface (or within the pore system) of a solid as opposed to the 'dissolution' of material into a liquid (absorption) (Kovach, 1979; Juntgen *et al.*, 1981) (see also Chapter 9).

Adsorbers for many gas-cleaning systems are typically vertical vessels but for high gas flow systems it may be more preferable to support the vessel in a horizontal position (Figure 11.23), the adsorbent being held on supports horizontally in the adsorber (Zenz, 1979).

The most critical part of the adsorber is the support holding the adsorbent bed. The design has to result in both weight-supporting rigidity and sufficient open area to prevent the development of large pressure drops. The direct adsorbent bed support can consist of perforated metal, expanded metal, or perforated plastics.

In some cases where the entry of high boilers into the adsorber cannot be prevented, it is advantageous to divide the adsorbent bed into two sections. The inlet section should be sized to remove completely the high boilers and

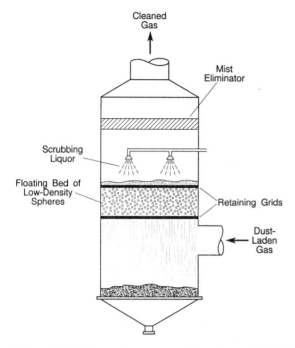

Figure 11.22 Schematic representation of a mobile-bed scrubber

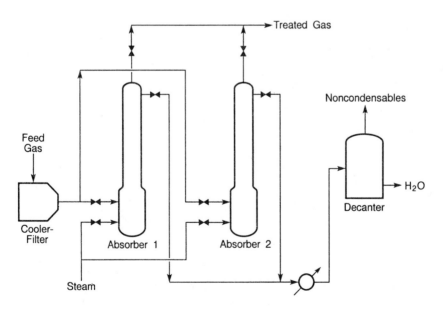

Figure 11.23 Illustration of a fixed-bed adsorber system

prevent their migration into the main adsorber section. This manner of operation permits the periodic removal and reactivation of only this narrow section instead of the entire bed. The 'sacrificial' (or 'guard') bed (see Speight, 1981) should not be placed in a non-regenerable section of the system if the adsorbate can decompose easily in the adsorbed state.

When adsorbent media sampling is planned, ports should be placed to permit the removal of adsorbent at different positions along the direction of the carrier gas flow. The rate of 'poisoning' or high-boiler loading of the adsorbent media can then be monitored in a quantitative manner.

Adsorbent loading and unloading can be well accomplished by vacuum techniques, rather than pressure loading. The use of vacuum will tend to remove dust from the adsorbent material, which is desirable whether loading or unloading is used. Carbon-type adsorbents can also be safely loaded in water–slurry form. This type of loading will result in sufficient retention of water in the adsorbent to prevent high-temperature spots developing as a result of the heat of adsorption of the first cycle.

Venturis

The venturi system consts of a cylindrical inlet, convergent cone, throat and divergent cone (Figure 11.24). The purpose of the divergent cone is to reduce the overall pressure loss of the system. The system (Figure 11.25) is often used when substantial quantities of particulates are present in the gas stream or when the scrubbing liquid has large concentrations of suspended solids.

As a point of reference and illustration of use, when high gas velocities are necessary for operation, venturi systems are used as, for example, in the wet-approach type of scrubbers. The initial mixing of gas and liquid is achieved in a venturi-shaped housing where the liquid enters tangentially to meet the gas which is directed towards the throat of the venturi-shaped housing (installed in a vertical position) (Figure 11.17). Slurry, gas and the liquid load will leave the venturi mixer to enter a cyclone for gross separation.

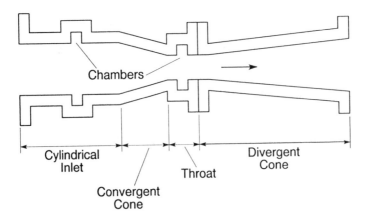

Figure 11.24 Illustration of a venturi tube

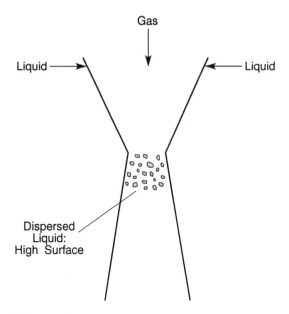

Gas

Liquid ⟶

⟵ Liquid

Dispersed
Liquid:
High Surface

Figure 11.25 Schematic representation of a venturi operation

References

Avallone, E.A. and Baumeister, T. (eds) (1987) *Marks' Handbook for Mechanical Engineers*, 9th edn, McGraw-Hill, New York

Fair, J.R., Steinmeyer, D.E., Penney, W.R. and Crocker, B.B. (1984) In *Perry's Chemical Engineers' Handbook* (ed. R.H. Green and J.A. Maloney), McGraw-Hill, New York, Section 18

Green, R.H. and Maloney, J.A. (eds) (1984) *Perry's Chemical Engineers' Handbook*, McGraw-Hill, New York

Juntgen, H., Klein, J., Knoblauch, K., Schroter, H.-J. and Schulze, J. (1981) In *Chemistry of Coal Utilization. Second Supplementary Volume* (ed. M.A. Elliott), Wiley, New York, Chapter 30

Kean, J.A., Turner, H.M. and Price, B.C. (1991) *Hydrocarbon Processing*, **70**, (4), 47

Kovach, J.K. (1979) In *Handbook of Separation Techniques for Chemical Engineers* (ed. P.A. Schweitzer), McGraw-Hill, New York, Section 3.1

Maas, J.H. (1979) In *Handbook of Separation Techniques for Chemical Engineers* (ed. P.A. Schweitzer), McGraw-Hill, New York, Section 6.1.

Mantell, C.L. (1951) *Adsorption*, McGraw-Hill, New York

Schweitzer, P.A. (ed.) (1979) *Handbook of Separation Techniques for Chemical Engineers*, McGraw-Hill, New York

Sherwood, T.K. (1937) *Absorption and Extraction*, McGraw-Hill, New York

Slack, A.V. (1981) In *Chemistry of Coal Utilization. Second Supplementary volume* (ed. M.A. Elliott), Wiley, New York, Chapter 22

Speight, J.G. (1981) *The Desulfurization of Heavy Oils and Residua*, Marcel Dekker, New York

Zenz, F.A. (1979) In *Handbook of Separation Techniques for Chemical Engineers* (ed. P.A. Schweitzer), McGraw-Hill, New York, Section 3.2

12

Process descriptions and flowsheets

Inrtroduction

In any text about gas processing, it is appropriate to give not only some indication of the types of processes available but also a listing of the different processes (by name). However, it is not the intent to reproduce all of the processes here, a virtually impossible task where so many of the processes might be derived from a single concept, but to give selected examples of specific processes.

In terms of process history, processes for the removal of gaseous contaminants from factory emissions have varied and have evolved along many different paths. Using the production of fuel gases from coal as the example, there has been some form of gas clean-up for almost as long as the plants have been in operation – not perhaps with the health of the various workers and those who lived nearby in mind but certainly with the quality of the product in mind! And now the concern has come full circle to include not only the quality of the product but also the quality of the emissions.

There are many processes for the removal of contaminant materials from natural gas (Kohl and Riesenfeld, 1979; Newman, 1985) and these processes are basically members of the clean-up types described and classified in Chapters 9 and 10. The construction and operation of gas-processing plants in the refining industry alone is a worldwide phenomenon (*Oil and Gas Journal*, 1991). Indeed, there are so many process variations that each process has its own particular niche in natural gas processing.

The inclusion of a process in this section is based upon several factors yet omission of a process from this section is not to be construed as a negative comment about the suitability of that process.

The first factor considered for the inclusion of a process is whether or not the concept achieved commercialization and the success realized. Another factor is the promise that the concept has, or may have, shown in pre-commercialization testing. Another consideration is concept novelty even though commercialization has not, or may not have, been realized. In this way, it is anticipated that a comprehensive overview of the process types available for gas cleaning will be made available to the reader.

Process descriptions and flowsheets

Activated Carbon Catalyst process

Carbon moistened with acid can catalyse the oxidation of sulphur dioxide to sulphur trioxide:

$$2SO_2 + O_2 = 2SO_3$$

The reaction between the adsorbed sulphur dioxide and oxygen in the carbon is much slower, but detectable, when dry and is considerably more rapid in the presence of (liquid) water. This phenomenon has been developed into a process (the Sulfacid process) for removing sulphur dioxide from acid plant tail gas (see also Hitachi Wet process).

In this process, tail gas from an absorption tower is passed through a bed of activated carbon over which weak sulphuric acid is circulated (Figure 12.1). The acid which is formed on the activated carbon dissolves in the circulating acid. Some acid is bled away from the system as product, and water is added as make-up. The process is configured to have one adsorber on-stream whilst the second adsorber is undergoing regeneration.

Adip process

The Adip process (Figure 12.2) uses diisopropanolamine as the chemical agent to capture gases such as hydrogen sulphide (H_2S) and carbonyl sulphide (COS) without the degradation that is inclined to occur when ethanolamine is used to treat gas streams that contain carbonyl sulphide (Ouwerkerk, 1978; Benson, 1981; *Hydrocarbon Processing*, 1990).

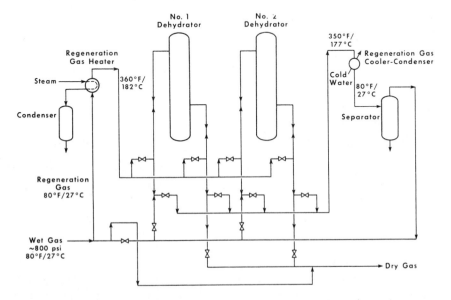

Figure 12.1 Activated Carbon Catalyst process

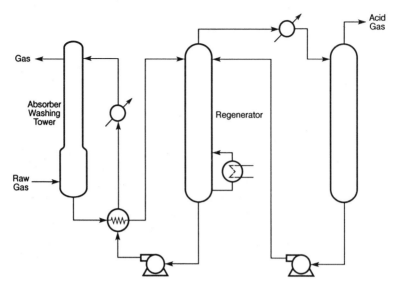

Figure 12.2 Adip process

Alkalized Alumina process

This process (Figure 12.3) uses 'alkalized' (sodium-treated) alumina $(Na_2O.Al_2O_3)$ to remove sulphur dioxide from flue gas.

In this process sulphur dioxide is oxidized by the adsorbent to form the sulfate (Field *et al.*, 1967). Alkalized alumina, prepared by heating an aluminium ore such as Dawsonite $(NaAl(CO_3)(OH)_2)$ at 650 °C (1200 °F), can contain about 20% wt/wt weight-free alumina (Al_2O_3). The spent adsorbent is reactivated by reheating it at 650 °C (1200 °F) during which time it is contacted with a reducing gas such as hydrogen, or producer gas (Katell, 1966). Sulphur, which is held on the adsorbent as sulphate, is reduced to hydrogen sulphide and enters the effluent gas stream (Field *et al.*, 1967):

$$Na_2SO_4 + 4H_2 = 3H_2O + H_2S + Na_2O$$
$$Na_2SO_4 + 4CO + H_2O = 4CO_2 + H_2S + Na_2O$$

These reactions take place readily at 650 °C (1200 °F) and the spent adsorbent is reformed to its original state.

The effluent gas stream from the adsorbent regenerator can then be processed in a Claus sulphur plant where elemental sulphur is recovered from hydrogen sulphide:

$$2H_2S + O_2 = 2H_2O + 2S$$

Alkazid process

The Alkazid process (Figure 12.4) has three different process variations with differing degrees of efficiency (Figure 12.5). This process is not obviously a carbonate process but is included here because of the similarities to the

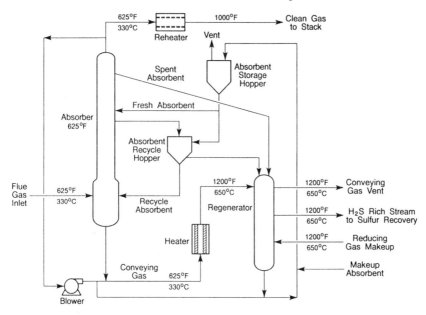

Figure 12.3 Alkalized Alumina process

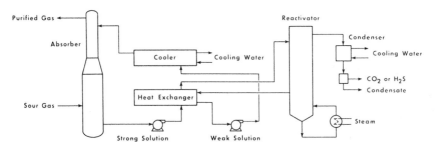

Figure 12.4 Alkazid process

carbonate processes by virtue of the use of the salts of alkali metals and organic radicals.

The Alkazid DIK process uses the potassium salt of diethyl glycine or dimethyl glycine for the selective removal of hydrogen sulphide from gases that contain both hydrogen sulphide and carbon dioxide. Alkazid M uses sodium alanine and is effective in removing both hydrogen sulphide and carbon dioxide. The third process variation, Alkazid S, uses a sodium phenolate mixture to remove contaminants such as carbon disulphide, mercaptans and hydrogen cyanide.

Amine processes

Probably the most widely used processes for removing acid gases such as hydrogen sulphide and/or carbon dioxide from gas streams are those which

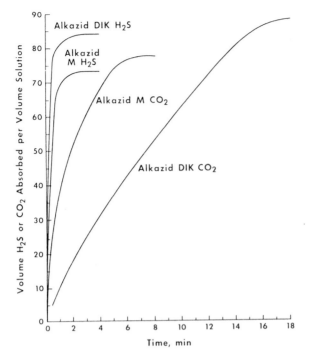

Figure 12.5 Characteristics of the various Alkazid processes

depend upon the reactivity of the acid gases with a variety of amines:

$$2RNH_2 + H_2S = (RNH_3)_2S$$
$$(RNH_3)_2S + H_2S = 2RNH_3HS$$
$$2RNH_2 + CO_2H_2O = (RNH_3)_2CO_3$$
$$(RNH_3)_2CO_3 + H_2O + CO_2 = 2RNH_3HCO_3$$
$$2RNH_2 + CO_2 = RNHCOONH_3R$$

The alkanolamine processes are the most prominent and widely used processes for the removal of hydrogen sulphide and carbon dioxide. Some of the commonly used alkanolamines (Table 12.1) are ethanolamine (also called monoethanolamine; MEA), diethanolamine (DEA), triethanolamine (TEA), hydroxyethanolamine (usually called diglycolamine, DGA), diisopropano-lamine (DIPA) and methyldiethanolamine (MDEA).

The basic process configuration is quite simple (Figure 12.6) but there are many variations that have been designed for specific improvements in the process operation (Sigmund, 1981; Polasek and Bullin, 1985; *Hydrocarbon Processing*, 1990). The many variations of the typical process configuration include (a) the location of the filtering system; (b) using a packed column instead of a bubble cap; (c) valve-type traps in the contactor and the stripper; or (d) the use of a side-stream reclaimer.

In a typical process (Figure 12.7), sour natural gas is sent upwards through the contactor tower, countercurrent to the flow of ethanolamine, and the rich solution from the bottom of the contactor is sent to a flash tank where

Table 12.1 Characteristics of the various alkanolamines used in gas-treating processes

Name		Chem. formula	Mol. wt	Vapour press. at 100°F (mm Hg)	Rel. capacity (%)
Ethanolamine (monoethanolamine)	MEA	$HOC_2H_4NH_2$	61	1.05	100
Diethanolamine	DEA	$(HOC_2H_4)_2NH$	105	0.058	58
Triethanolamine	TEA	$(HOC_2H_4)_3N$	149	0.0063	41
Hydroxyethanolamine	DGA	$H(OC_2H_4)_2NH_2$	105	0.160	58
Diisopropanolamine	DIPA	$(HOC_3H_6)_2NH$	133	0.010	46
Methyl diethanolamine	MDEA	$(HOC_2H_4)_2NCH_3$	119	0.0061	51

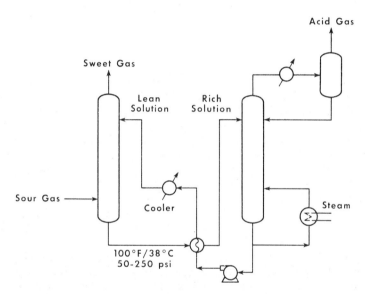

Figure 12.6 Basic configuration for an alkanolamine process

absorbed low-boiling hydrocarbons in solution are vented. The flash tank also serves as a sediment accumulator and provisions must be made for sediment removal.

After the enriched solution has been heated in a heat exchange system by the lean solution, the rich solution enters the top of the stripper where it is stripped by steam generated by the reboiler. Outcoming steam from the top of the stripper is then condensed to recover the ethanolamine. Any acid gases released at this stage are sent to the flare stack and the liquid that is accumulated by the reflux is sent to the regenerative system.

Lean ethanolamine that has accumulated at the bottom of the stripper is continuously recirculated through the reboiler. It is possible to remove up to 90% of the acid gases within the first three trays at the bottom of the absorber. The reactions are exothermic and a rise in temperature must be anticipated in this region of the absorber.

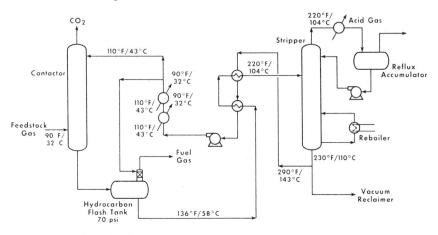

Figure 12.7 Alkanolamine process

Ethanolamine is by far the most widely and extensively used of the amines since it is the strongest base and reacts readily with acid gas constituents. It will remove both hydrogen sulphide and carbon dioxide and is generally considered non-selective between the two. It has a greater potential carrying capacity for acid gases, is chemically stable, and can easily be separated from acid gas constituents by steam stripping. However, ethanolamine reacts irreversibly with carbonyl sulphide and carbon disulphide, which can result in loss of solution and a buildup of solid products in the circulating solution.

In the Girbotol process (page 292), an aqueous solution of ethanolamine is reacted with the gas stream in an absorber, under pressure and at room temperature. The solution is then heated and fed to a reactivation column where the alkanolamine–carbon dioxide complex is dissociated by stripping with steam at 90 to 120°C (195 to 250°F) and near atmospheric pressure. The carbon dioxide and steam leave the regenerator at the top of the column and the regenerated solution leaves at the bottom, is cooled, and is pumped back to the absorber.

The general considerations for selecting amines in sweetening plants have changed over the years (Kohl and Riesenfeld, 1979; Newman, 1985) since each amine has an 'accepted' range of process conditions and parameters associated with it.

Ethanolamine (monoethanolamine, MEA) has the highest acid gas removal capacity, and the lowest molecular weight among the alkanolamines (Dingman and Moore, 1968; Maddox, 1982). Therefore, it offers the highest removal capacity on a unit weight or unit volume basis, from which lower solution circulation rates in a sweetening plant are implied. It is also chemically stable but will undergo some, but often minimal, degradation.

Ethanolamine is generally used as a 15–20% wt/wt solution in water (Table 12.2). The acid gas loading is usually limited to 0.3 to 0.4 moles acid gas per mole of amine although loadings as high as 0.7 to 0.9 mole/mole have been used, depending upon the resistance of the equipment to corrosion. Ethanolamine itself is not considered to be particularly corrosive but its degradation products are very corrosive.

Table 12.2 General operating conditions for the alkanolamine processes

Amine	MEA	DEA	DGA	MDEA
Solution strength (wt %)	15–20	25–35	40–60	30–50
Acid gas loading (mole/mole)	0.3–0.4	0.3–0.4	0.3–0.4	Unlimited
ΔH_r^* for H_2S (Btu/lb)	550	511	674	522
ΔH_r^* for CO_2 (Btu/lb)	825	653	850	600
Ability to absorb H_2S preferentially		Under some		Under most
	No	conditions	No	conditions

* Btu/lb and for total loadings below 0.5 mole acid gas/mole amine.

Diethanolamine is most commonly used in the 25–35% wt/wt range and, because of the nature of the degradation products, corrosion of the equipment is also a concern, although the degradation products of diethanolamine are much less corrosive than those of ethanolamine. Diethanolamine does not appear to lose any of its activity due to the presence of carbonyl sulphide in gas streams (Figure 12.8). As in the SNPA process (page 316), corrosion inhibitors can be employed (Kohl and Riesenfeld, 1979).

Since diethanolamine is a secondary alkanolamine, it has a reduced affinity (relative to ethanolamine) for hydrogen sulphide and carbon dioxide. As a

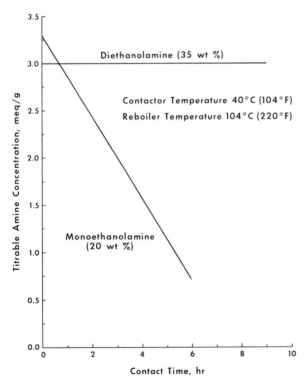

Figure 12.8 Illustration of the effect of carbonyl sulphide on the process life of ethanolamine and diethanolamine (Source: Kohl and Riesenfeld, 1979)

result, diethanolamine is not able to produce pipeline specification gas for all treated streams, especially for some low-pressure gas streams. In general, as the gas pressure is lowered, the stripping steam must be increased or a split-flow design must be used. In some cases, even these measures will not be sufficient and another solvent must be used.

Under some conditions, such as low pressure and controlled (low) liquid residence time, diethanolamine will act selectively towards hydrogen sulphide, allowing a significant fraction of the carbon dioxide to remain in the gas stream. The heat of reaction for diethanolamine and carbon dioxide is approximately 25% lower than the corresponding affinity for carbon dioxide by ethanolamine.

Because of the higher molecular weight of the diethanolamine, relative to monoethanolamine, the higher concentration serves as well as a 15% ethanolamine solution. However, diethanolamine does not react as rapidly with carbonyl sulphide and with carbon disulphide as does monoethanolamine. As a result, diethanolamine can frequently be used for gases (such as refinery gases) which contain appreciable amounts of these constituents.

Triethanolamine has been almost totally replaced in gas stream treating service by ethanolamine and diethanolamine, mostly because of the lower reactivity of the triethanolamine. This results in a relatively lower removal of hydrogen sulphide. Diglycolamine has the same reactivity as diethanolamine which, because of its relatively lower vapour pressure, has resulted in increased commercial use in recent years.

Diisopropanolamine is also used to treat gas streams to desired (pipeline specification) levels and is used in the Sulfinol process (see below) as well as in the Adip process (see above) which employs relatively concentrated solutions of the alkanolamine (diisopropanolamine) solvent (Ouwerkerk, 1978; *Hydrocarbon Processing*, 1990). This solvent can remove carbonyl sulphide and is selective for the removal of hydrogen sulphide in preference to carbon dioxide.

Processes based on diglycolamine occur worldwide (Moore *et al.*, 1985) and the alkanolamine is generally used as 40 to 60% wt/wt solutions in water. The reduced corrosion problems with diglycolamine allow solution loadings equivalent (mole basis) to ethanolamine in most applications even with these high weight percentages. Diglycolamine has a tendency to react preferentially with carbon dioxide at the expense of hydrogen sulphide. It also has a higher pH than ethanolamine and has some definite advantages over the other amines in that higher diglycolamine concentrations in the solution result in lower freezing points. In addition, diglycolamine is not as likely to react irreversibly with carbonyl sulphide, carbon disulphide, sulphur dioxide and sulphur trioxide. In fact, diglycolamine has a high affinity for carbonyl sulphide as well as for methyl and ethyl mercaptan (CH_3SH and CH_3CH_2SH). One of the primary disadvantages of diglycolamine is the heats of reaction (Table 12.2).

Methyldiethanolamine is most commonly used in solution in the 30–50% wt/wt range and, due to considerably reduced corrosion problems, acid gas loadings as high as 0.7 to 0.8 mole/mole are practical. Since methyldiethanolamine is a tertiary amine, it has much less affinity for hydrogen sulphide and for carbon dioxide than diethanolamine and, as for diethanolamine, may be unable to reduce impurities to the desired levels for some low-pressure gas streams.

However, methyldiethanolamine does have several advantages over primary and secondary amines, including lower vapour pressure, lower heats of reaction, higher resistance to degradation, fewer corrosion problems, and (most important) an increased selectivity to hydrogen sulphide in the presence of carbon dioxide (Kohl and Riesenfeld, 1979; *Hydrocarbon Processing*, 1990) as well as an increased performance when activators are employed (Meissner and Heffner, 1990; *Hydrocarbon Processing*, 1990). In addition, due to its lower heat of reaction, methyldiethanomaline can be used in pressure-swing plants for the removal of carbon dioxide. In such plants, the rich amine is merely flashed at, or close to, atmospheric pressure and little or no heat is added for stripping.

Mixtures of glycol and amine have been used for simultaneous dehydration and desulphurization; glycol itself and higher-molecular-weight glycols are also used for dehydration (Dehydrate process, Drizo process) (Smith and Skiff, 1990; *Hydrocarbon Processing*, 1990). Generally, a solution containing 10–30% wt/wt of the alkanolamine (ethanolamine), 45–85% wt/wt triethylene glycol and up to 25% wt/wt water is used.

For high acid gas content of gas streams, a modification might be employed (Kohl and Buckingham, 1960) in which a semiregenerated amine solution is removed from a midpoint of the stripper and pumped through heat exchange and cooling before being introduced to a midpoint of the absorber.

The Amisol process, which is based upon the use of amines such as diisopropylamine and diethylamine with methanol (Figure 12.9) (Kriebel, 1985), has been known since the 1960s and is classified among the chemical/ physical gas purification processes. The chemical component (the amine) and the physical solvent (methanol) are claimed to be highly efficient for gas clean-up because of the good mass transfer between the gas and the liquid which ensures that the number of trays in the absorption column can be kept low.

More recently, hindered amines (Goldstein *et al.*, 1985; Savage *et al.*, 1985; Chludzinski and Wiechert, 1986) have been proposed for gas-treating applications under the general banner of the Flexsorb processes (page 289).

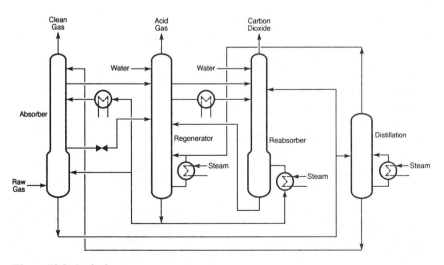

Figure 12.9 Amisol process

Ammonia process

Scrubbing acid plant tail gas with ammonia is an older, or more mature, form of gas treatment.

The process (Showa Denko process; Figure 12.10) involves the reaction of ammonia with sulphur dioxide and water to form ammonium sulphite

$$SO_2 + 2NH_3 + H_2O = (NH_4)_2SO_3$$

with subsequent conversion to ammonium bisulphite

$$(NH_4)_2SO_3 + SO_2 + H_2O = 2NH_4HSO_3$$

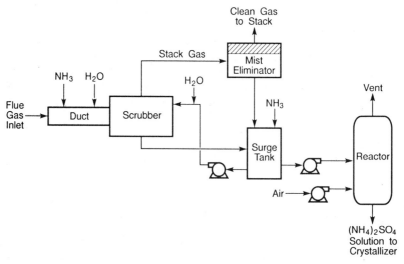

Figure 12.10 Showa Denko Ammonia process

As normally practised, the process (Figure 12.11) is not, in the strictest sense, an absorption–regeneration process. The sulphur dioxide is regenerated by the addition of a strong acid (sulphuric, phosphoric or nitric) and/or by high-temperature decomposition of the ammonium sulphite–bisulphite solution to nitrogen and sulphur dioxide.

Most of the earlier plants used a once-through, single-stage scrubbing to reduce the sulphur dioxide concentration to levels of a 'few hundred' parts per million. Current plants generally use a two-stage scrubbing operation, and sometimes a third washing stage to recover ammonia. Whereas all the earlier plants used sulphuric acid to regenerate the sulphur dioxide (which was then recycled to the drying tower), nitric acid and phosphoric acid are also used.

The use of ammonia originally lay in the fact that ammonium sulphate was a common fertilizer and, hence, a useful product. There are aso cases where other products, such as sodium or potassium sulphite or sulphate, may also be useful by-products and offer a partial answer to pollution issues.

There are other process configurations (Figure 12.12) for the treatment of gas streams other than those containing mainly sulphur dioxide.

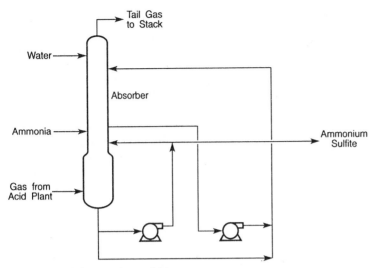

Figure 12.11 Ammonia scrubbing process

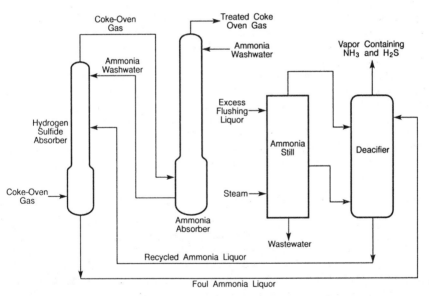

Figure 12.12 Process configuration for ammonia scrubbing of coke oven gas streams

Aquaclaus process

This wet absorption system is capable of producing a treated gas which contains less than 100 ppm of sulphur dioxide (*Hydrocarbon Processing*, 1973).

In this process (Figure 12.13), the Claus tail gas is first incinerated to convert all sulphur-bearing compounds, such as hydrogen sulphide, carbonyl

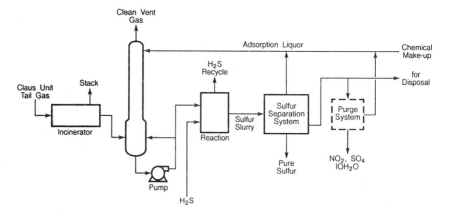

Figure 12.13 Aquaclaus process

sulphide and carbon disulphide, to sulphur dioxide. Then the stream is cooled in a waste heat boiler and/or a direct-contact cooler, and is fed to an absorption tower. The sulphur dioxide is absorbed by the Aquaclaus solution (aqueous sodium phosphate) which is then contacted with fresh hydrogen sulphide feed, from the front of the Claus plant, in a reactor vessel to form elemental sulphur by the classic Claus reaction occurring in an aqueous phase. The solution is heated and liquid sulphur is withdrawn. The Aquaclaus solution is cooled and recirculated to the absorber after sulphur is separated (*Chemical Engineering Progress*, 1973; *Hydrocarbon Processing*, 1973).

Several undesirable side reactions can occur in the absorber and reactor. Compounds such as polythionates, thiosulphates and sulphates can be formed. Some of the thiosulphates are destroyed in the sulphur-melting step. The sodium phosphate solution serves as a buffer to shift the reactions towards the formation of bisulphites, instead of the undesirable compounds.

Aquasorption (Water Wash) process

This process is effective for natural gas that has a high acid gas content (including a high hydrogen sulphide to carbon dioxide ratio) and that is also under high pressure. In this particular type of process, sour natural gas is passed, in an upward direction, through a contactor in which the gas flows countercurrent to the water (Figure 12.14). The partially sweetened gas is then passed on for further treatment (e.g. to an amine unit).

The rich water solution from the bottom of the tower is sent to a pressurized flash tank for the recovery of dissolved hydrocarbons. The water is then repressurized before sending it to a lower-pressure flash tank where all of the acid gas is removed and water obtained for recycling.

Asarco process

This process was originally used for the recovery of sulphur dioxide from smelter gases and used dimethylaniline $(C_6H_5N(CH_3)_2)$ or xylidine $((CH_3)_2.C_6H_3.NH_2)$ as the absorbent. The process uses a water wash (see the

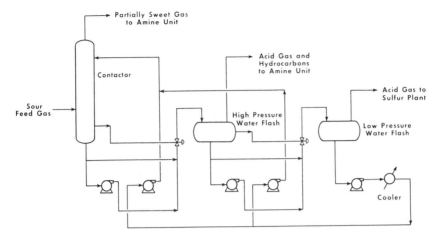

Figure 12.14 Aquasorption (Water Wash) process

Ammonia process and the Aquasorption process for general process configurations) to scrub absorbate from the overhead gas from the stripper after the gas is cooled. The sulphur dioxide stream is dried by a wash with concentrated (about 98 wt %) sulphuric acid.

Basic Aluminium Sulphate process

This process is based upon the absorption of sulphur dioxide in a solution of aluminium hydroxide–sulphate and regeneration of the absorbent by heating. The absorbent $[Al_2(SO_4)_3]$ solution is prepared by treating ammonium sulphate $[(NH_4)_2SO_4]$ solution with limestone in a mixer and separating the gypsum $(CaSO_4)$ precipitate.

A series of absorption towers are used to reduce the sulphur dioxide by countercurrent passage of the solution and gas with cooling between stages to remove the heat of absorption. A stripper vessel equipped with reboilers is used to regenerate the solution by heating. The sulphur dioxide is cooled and dried to produce almost pure sulphur dioxide gas. The stripped solution is passed through heat exchangers and recycled back to the absorber. Phosphoric acid is added to inhibit the crystallization of an alumina–sulphur dioxide complex $(Al_2O_3.3SO_2.5H_2O)$.

Battersea process

This process uses a very dilute aqueous solution of alkaline salts to remove sulphur dioxide by asorption and oxidation.

The Battersea process was one of the first to be applied commercially for the removal of sulphur dioxide from power plant stack gase. An oxidation catalyst (manganese sulphate, $MnSO_4$) was added to the solution from the bottom of the absorber. Oxidation may be desirable to make the waste water more suitable for disposal.

Beavon process

The Beavon sulphur recovery process (Figure 12.15) is used for the conversion of sulphur compounds in tail gas streams (from a Claus plant) to hydrogen sulphide (see also the Parsons Beavon process below) (*Hydrocarbon Processing*, 1990).

The process employs three distinct steps: (a) hydrogenation–conversion of sulphur compounds to hydrogen sulphide in a catalytic converter; (b) cooling of the converter effluent gases; and (c) conversion of the hydrogen sulphide in the tail gas from the cooler to elemental sulphur by use of the Stretford process.

In the Beavon process, the tail gas from a Claus unit is first mixed with 'fuel gas and air' combustion products and then fed to the catalytic converter, which contains a cobalt molybdate catalyst. The hydrogen required for the reduction of sulphur compounds to hydrogen sulphide is supplied by partial combustion of the fuel gas in an in-line burner, which simultaneously raises the tail gas stream temperature to the level required for the hydrogenation reactions.

The converter effluent gases are then cooled with water in a direct-contact condenser, where most of the water vapour contained in the gas is condensed, and at the same time the tail gas is cooled. The purge water produced from the condenser is of good quality, with only a small amount of dissolved hydrogen sulphide. With the removal of hydrogen sulphide in a small 'sour water stripper', it is suitable for cooling tower make-up water. A recirculation loop of water with a shell-and-tube cooler is provided to maintain a flow of water through the direct-contactor condenser from top to bottom. A small bleed or purge stream of water goes to further processing.

The cooled gas then enters the Stretford absorber, where the hydrogen sulphide is removed almost quantitatively:

$$H_2S + Na_2CO_3 = NaHS + NaHCO_3$$

The (bisulphide) hydrosulphide is oxidized to sulphur by sodium vanadate in

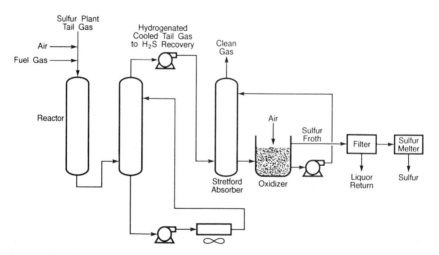

Figure 12.15 Beavon process

solution:

$$NaHS + 2NaVO_3 = S + Na_2V_2O_5 + NaOH$$

Subsequently, the vanadium is oxidized to the pentavalent state by blowing with air, with anthraquinone disulphonic acid (ADA) or sodium anthraquinone disulphonate acting as an oxidation catalyst:

$$Na_2V_2O_5 + O_2 = 2NaVO_3$$

The Stretford process (page 318) can be considered to occur by five steps: (a) hydrogen sulphide is absorbed by the alkali; (b) the sodium vanadate is reduced by a sulphur shift; (c) elemental sulphur is liberated by dissolved oxygen in the solution; (d) the sodium vanadate is oxidized in the presence of the sodium anthraquinone disulphonate; and (e) the alkaline solution is oxygenated by air blowing, which also floats the sulphur out of solution.

The spent absorbent solution flows from the absorber to the regenerator (oxidizer), where air is sparged into the tower. The sulphur is collected as a froth at the top of the oxidizer tower. The sulphur froth is skimmed off the solution and sent to a filter or centrifuge for removal of the filtrate from the sulphur cake. The clear (regenerated) absorbent solution and the filtrate from the filter are recycled back to the absorber.

The Takahax process (page 325) is essentially the same as the Stretford process, except for the chemicals used. This process uses an absorbent solution of sodium carbonate, 1,4-naphthaquinone and 2-sulphonate sodium (Goar, 1971a,b).

Bender process

The main purpose of the Bender process is the sweetening of hydrocarbon gases such as natural gas streams, as well as propane and butane gases, by removal of mercaptans through conversion to disulphides in the presence of a fixed-bed catalyst, a small amount of sulphur, and air (oxygen) (*Petroleum Refiner*, 1955; *Hydrocarbon Processing*, 1990):

$$2R\text{-}SH + O_2 = R\text{-}S.S\text{-}R + H_2O$$

In the process (Figure 12.16), the gas mixture is passed countercurrently over a bed of solid catalyst to a small amount of alkali (as a catalyst activator) whereby the chemical conversion of mercaptans (thiols) to disulphides occurs.

Benfield process

The Benfield process(es) (Benson and Parrish, 1975; *Hydrocarbon Processing*, 1990) (see also Carbonate process(es)) has gained general acceptance for the removal of acid gas from gas streams. However, it is not one process but actually a collection of processes with each one tailored for a particular niche (Bartoo, 1985).

The Benfield process involves the use of a catalyst or an activator (Eickmeyer, 1971) and uses a hot 25–35% aqueous solution of potassium carbonate. The difference in relative absorption rates (about 4) between hydrogen sulphide and carbon dioxide in the solvent is sufficient to allow for a selective absorption design.

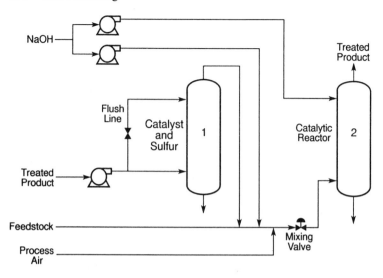

Figure 12.16 Bender process

The differing process configurations under the Benfield name (Figures 12.17 and 12.18) employ conventional packed or trayed towers for the countercurrent contact of liquid and gas and are configured for varying degrees of gas purification. In summary, the Benfield process is versatile and has a broad range of applications with more than 500 units currently in operation.

Carbonate processes

The use of potassium carbonate as a means of gas stream clean-up is a well-known and established procedure (Kohl and Riesenfeld, 1979; Newman, 1985).

Although originally developed for the removal of carbon dioxide from gas streams, the process also serves to remove hydrogen sulphide from gas streams:

$$K_2CO_3 + CO_2 + H_2O = 2KHCO_3$$
$$K_2CO_3 + H_2S = KHS + KHCO_3$$

An elevated temperature is necessary to ensure that the potassium carbonate and the reaction products (potassium bicarbonate and potassium bisulphide) remain in solution. The process requires a relatively high partial pressure of carbon dioxide and is not generally applicable to gas streams that contain only hydrogen sulphide. There are also limitations to the extent of removal of the carbon dioxide and hydrogen sulphide; it is not always possible to clean the natural gas to the desired low levels of acid gas content and, hence, to meet pipeline specifications.

The configuration of the conventional Hot Carbonate process (Figure 12.19) is relatively straightforward but as with the alkanolamine processes there have been many modifications to the process as it evolved (Kohl and Riesenfeld, 1979; Newman, 1985). The Hot Carbonate process (also referred

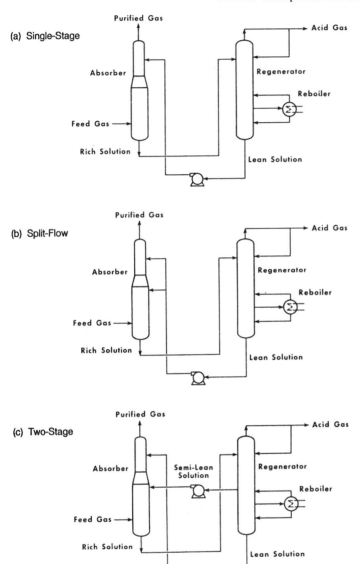

Figure 12.17 Benfield process configurations

to as the Hot Pot process; see also Koppers Vacuum Carbonate process, page 299) uses an aqueous solution of potassium carbonate (cf. Limestone Injection processes below) and a highly concentrated solution is often used to improve process performance.

The process is referred to as the 'hot' process because both the absorber and the regenerator operate at elevated temperatures, usually in the range 110–115°C (230–240°F). This temperature range serves to increase the solubility of potassium carbonate (and thus increase the carrying capacity of

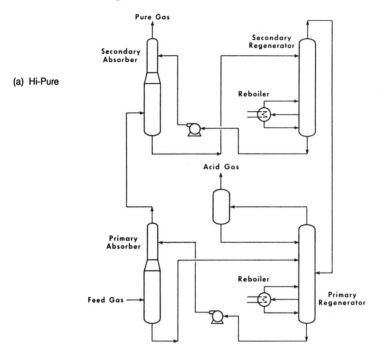

(a) Hi-Pure

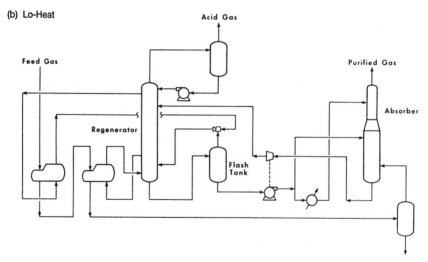

(b) Lo-Heat

Figure 12.18 Benfield process configurations

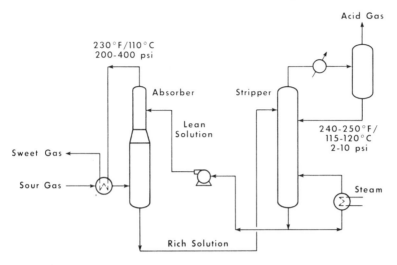

Figure 12.19 Hot Carbonate process

the solution for acid gases) and to prevent the crystallization of the bicarbonate salts.

In addition, the process offers the advantage that any carbonyl sulphide and carbon disulphide in the gas stream will be removed by hydrolysis to afford products (hydrogen sulphide and carbon dioxide) which then proceed to react with the potassium carbonate:

$$COS + H_2O = H_2S + CO_2$$
$$CS_2 + 2H_2O = CO_2 + 2H_2S$$
$$K_2CO_3 + H_2S = KHS + KHCO_3$$
$$K_2CO_3 + CO_2 + H_2O = 2KHCO_3$$

The Molten Carbonate (M-C) process (Figure 12.20) is a closed-cycle process in which sulphur dioxide and sulphur trioxide are absorbed from flue gas on to an absorbent after which the spent adsorbent can be regenerated and recirculated. The absorbent used is a eutectic mixture of lithium, sodium and potassium carbonates (32 wt % Li_2CO_3, 33 wt % Na_2CO_3 and 35 wt % K_2CO_3) which has a melting point of 397°C (747°F) and is a clear, mobile liquid in the molten state. The hot flue gases from the electrostatic precipitator (fly ash and precipitates removal) flow to a scrubber, where the absorbent contacts the gas and reacts chemically with the sulphur oxides, forming sulphites and sulphates which remain dissolved in the unreacted melt.

The salt mixture is then circulated to a filter (for fly ash removal), through heating steps, and into the reducer. In the reducing step, hot carbon monoxide and hydrogen are added to release some carbon dioxide and water from the melt. The reduced melt then flows through a heat exchanger and on to the regenerator, where hydrogen sulphide is released. The regenerated melt then flows back to the scrubber to complete the cycle.

The hydrogen sulphide from the regenerator may be converted to elemental sulphur (in a Claus sulphur plant) or to sulphuric acid (in a contact acid plant).

In the scrubber, the temperature of the melt and the incoming stack gases

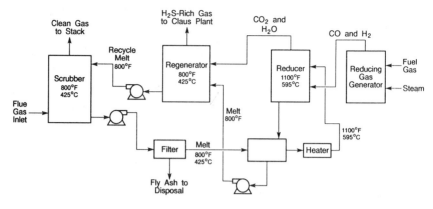

Figure 12.20 Molten Carbonate process

must be kept above the freezing point of the eutectic mixture, a temperature of about 425 °C (800 °F) or higher in actual practice (Oldenkamp and Margolin, 1969).

The regeneration part of the Molten Carbonate process consists of two steps. In the first step, alkali metal sulphites and sulphates are reduced to sulphides using a mixture of hydrogen and carbon monoxide at about 595 °C (1100 °F). The second step involves contact of the melt with carbon dioxide and water vapour at 425 °C (800 °F) which converts the sulphides to hydrogen sulphide gas after which the melt flows back to the scrubber. The carbon dioxide and water vapour required for the second step is generated in the first step (reduction).

The process chemistry is often represented quite simply:

$$M_2CO_3 + SO_2 = M_2SO_3 + CO_2$$
$$M_2CO_3 + SO_3 = M_2SO_4 + CO_2$$

and any sulphites formed may be oxidized to sulphates by oxygen present in the flue gas:

$$2M_2SO_3 + O_2 = 2M_2SO_4$$

In order to increase the rate of the reduction reactions, the spent melt is heated to about 595 °C (1100 °F) and disproportionation of the sulphites occurs:

$$4M_2SO_3 = 3M_2SO_4 + M_2S$$

Because of this, the actual reduction step is carried out by use of carbonates and sulphides:

$$M_2SO_4 + 4H_2 = M_2S + 4H_2O$$
$$M_2SO_4 + 4CO = M_2S + 4CO_2$$

In the regeneration step (second operation), the sulphides are converted back to carbonates and hydrogen sulphide is produced:

$$M_2S + CO_2 + H_2O = M_2CO_3 + H_2S$$

This reaction proceeds more favourably at a lower temperature than that of the reduction step; thus the 'reduced' melt is cooled to 425 °C (800 °F) before the carbon dioxide and water are added. The second regeneration step reaction proceeds rapidly at 425 °C (800 °F) and equilibrium is approached closely.

The hydrogen sulphide stream produced from the regenerator may be fed to a Claus sulphur plant for conversion to sulphur or to a contact acid plant for production of sulphuric acid. Either of these processes may require tail gas treatment of the effluent gases to obtain the ultimate removal of sulphur dioxide.

For gas streams with low acid gas concentrations, there are various process modifications available.

In one particular modification, a portion of the regenerated stream is separated from the main flow and cooled before being fed to the top tray of the absorber. The remainder of the solution is fed without cooling to a lower point in the absorber. Cooling the solution introduced on the top tray lowers the equilibrium partial pressure of the acid gases with the solution and makes possible a lower total acid gas content in the sweetened gas. When using this variation, however, care must be taken to avoid cooling the solution to the point where potassium carbonate or bicarbonate would precipitate from the cooled solution.

Another modification of the Hot Carbonate process is a two-stage process in which a semi-lean solution is removed from an intermediate point in the stripper and passed to an intermediate point in the absorber. Lean stripped solution is removed from the bottom of the stripper and is introduced on the top tray of the absorber.

A further modification of the Hot Carbonate process (Figure 12.21) is a combination of diethanolamine treatment and diethanolamine-activated potassium carbonate treatment. More recently, the use of sterically hindered amines such as 2-amino-2-methyl-1-propanol [$CH_3.(NH_2)C(CH_3).CH_2OH$] as process promoters provides performance advantages, including a 30% higher carbon dioxide equilibrium capacity (Savage et al., 1985; Say et al., 1985; Goldstein et al., 1985; Chludzinski and Wiechert, 1986). These processes fall under the name of the Flexsorb process(es) (page 289).

The choice of the particular carbonate process flow scheme to be used in a given instance is governed by the degree of acid gas removal required and the ratio of hydrogen sulphide to carbon dioxide in the sour gas.

The slow rate of reaction of carbonate solution with acid gases has led to extensive searches to develop catalysts that would speed up the rate of reaction. As a result, several versions of the Hot Potassium Carbonate process have evolved: (a) the Benfield process; (b) the Catacarb process; and (c) the Giammarco–Vetrocoke process. All three are suitable for the removal of carbon dioxide from gas streams as well as the removal of hydrogen sulphide.

The Benfield process (see above) (Eickmeyer, 1971; Benson and Parrish, 1975; Hydrocarbon Processing, 1990), which uses a hot 25–35% aqueous solution of potassium carbonate, has gained general acceptance for acid gas removal from gas streams. However, it is not one process but actually a collection of processes with each one tailored for a particular niche (Bartoo, 1985).

The Catacarb process (page 277) also employs a solution of potassium carbonate for the removal of hydrogen sulphide and carbon dioxide from gas

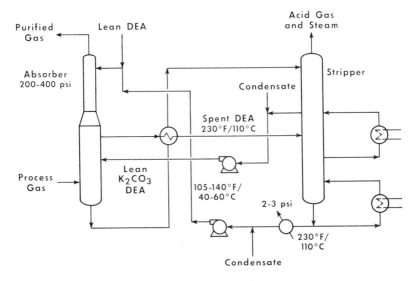

Figure 12.21 Combined Carbonate/Alkanolamine process

streams. Several catalysts and corrosion inhibitors are used in the process but the choice depends upon the composition of the gas to be treated (Gangriwala and Chao, 1985; *Hydrocarbon Processing*, 1990). This process, like the Benfield process, has found general application in gas-treating operations.

The Giammarco–Vetrocoke process (page 291) uses arsenic as an activator for the carbonate solution:

$$KH_2AsO_3 + 3H_2S = KH_2AsS_3 + 3H_2O$$
$$KH_2AsS_3 + 3KH_2AsO_4 = 3KH_2AsO_3S + KH_2AsO_3$$
$$3KH_2AsO_3S = 3KH_2AsO_3 + 3S$$
$$6KH_2AsO_3 + 3O_2 = 6KH_2AsO_4$$

The Giammarco–Vetrocoke process has found many applications; for example, there is one version for the removal of carbon dioxide and another for the removal of hydrogen sulphide, and yet another version for the removal of both of these gases. As for most conventional carbonate processes, impurities such as carbon disulphide, mercaptans and carbonyl sulphide have no detrimental effects on the solution.

Cataban process

In this process, hydrogen sulphide is converted directly to sulphur by an oxidation–reduction reaction involving a chelated iron salt. The hydrogen sulphide is oxidized to sulphur, while the ferric chelate is reduced to ferrous chelate. The iron chelate is regenerated (ferrous back to ferric) by air blowing. Sulphur may be removed from solution by filtration or centrifugation.

Catacarb process

The Catacarb process also employs a solution of potassium carbonate for the removal of hydrogen sulphide and carbon dioxide from gas streams. Several catalysts and corrosion inhibitors are used in the process but the choice depends upon the composition of the gas to be treated (Gangriwala and Chao, 1985; *Hydrocarbon Processing*, 1990). The process is versatile and, like the Benfield process (page 269), has found general application in gas-treating operations.

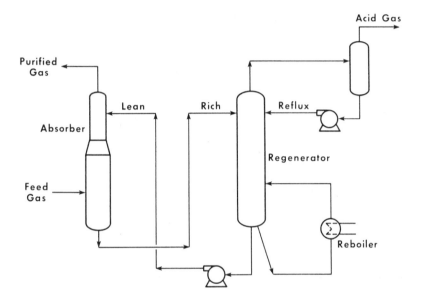

Figure 12.22 Catacarb (single-stage) process

The simplest version of the process is a single-stage unit (Figure 12.22) which is used when high purity is not required in the treated gas. A two-stage design (Figure 12.23) is most efficient in terms of a higher-purity product, and other optional designs (Figure 12.24) are used to reduce the heat requirements significantly.

Catasulf process

This process (Figure 12.25) (*Hydrocarbon Processing*, 1990) involves the passage of a gas stream containing hydrogen sulphide, with a stoichiometric amount of air or oxygen, through an isothermal reactor, packed with a catalyst, in which the majority (up to 94%) of the sulphur compounds are converted to elemental sulphur. The gas stream can be further desulphurized by a second-stage catalytic oxidation in another reactor (adiabatic). The second-stage reactor effects a near quantitative (up to 98%) yield of all of the sulphur compounds originally in the gas stream.

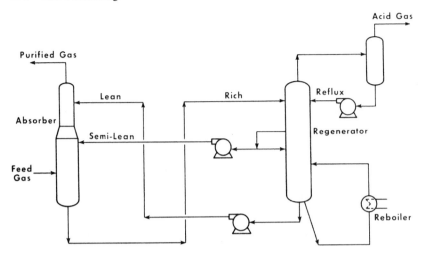

Figure 12.23 Catacarb (two-stage) process

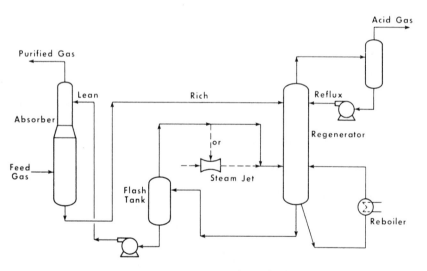

Figure 12.24 Catacarb process (low-heat configuration)

Cat-Ox process (Catalytic Oxidation process)

This process uses an oxidant (usually vanadium pentoxide) to convert sulphur dioxide in flue gases to sulphur trioxide (Stites *et al.*, 1969):

$$SO_2 + [O]_{catalyst} = SO_3$$

In the process (Figure 12.26), flue gas leaving a precipitator is heated to about 455°C (850°F) for conversion of the sulphur dioxide to sulphur trioxide in the catalytic converter by passage through a heat exchanger and then

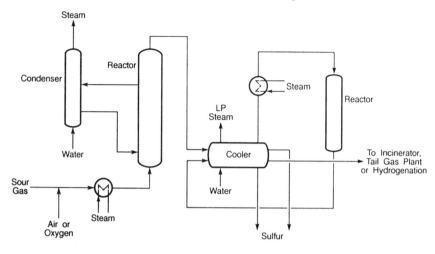

Figure 12.25 Catasulf process

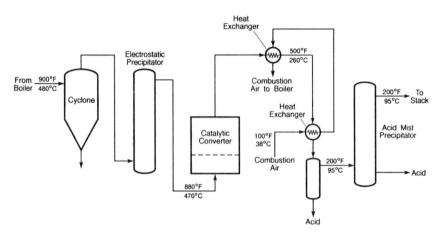

Figure 12.26 Catalytic Oxidation process

brought up to temperature by the direct addition of hot gas from the combustion of fuel oil or natural gas in a reheat furnace. Flue gas from the converter, that is relatively rich in sulphur trioxide, is cooled in a gas–gas heat exchanger (the same exchanger used for preheating the feed gas) by giving up heat to the incoming gas.

Fly ash in the gas not removed by the precipitator collects on the catalyst and must be removed, and periodic catalyst cleaning is required. The converter is designed so that catalyst beds can be cleaned while in service. A vibrating screen is used to remove the fly ash from the catalyst. The clean catalyst is conveyed back to the bed in the converter.

The cooled effluent gas from the converter, at about 95–105°C (200–225°F) and rich in sulphur trioxide, is contacted in a packed-bed absorption tower which operates in conjunction with an external shell-and-tube heat exchanger. the sulphur dioxide is condensed and absorbed in cool sulphuric acid solution in the absorber. Hot sulphuric acid leaves the bottom of the absorber, is cooled in the liquid heat exchanger, and is recycled to the top of the absorber. A side stream of cool sulphuric acid goes to acid storage.

Exit gas leaves the top of the absorber at about 105 °C (225 °F) and passes to a vessel containing mist eliminators which remove the fine sulphuric acid 'mist' from the cleaned flue gas. The acid removed from the gas drains from the bottom of the eliminator vessel to the liquid line going to acid storage.

In the Integrated Cat-Ox process (Figure 12.27), hot flue gas at about 455 °C (850 °F) is passed from the power boiler to the electrostatic precipitator and thence to the catalytic converter, where the majority of the sulphur dioxide is oxidized to sulphur trioxide. The flue gas from the converter is cooled in an economizer and a boiler combustion air heater. The economizer is a finned-tube type of exchanger used to preheat boiler feed water. Both the economizer and air heater operate in a cleaned-gas atmosphere and do not incur erosion from fly ash abrasion. Both of these exchangers are operated above the dewpoint of sulphuric acid which eliminates, or at least reduces, corrosion problems.

In a Reheat Cat-Ox process, flue gas discharged from the induced draught fan bypasses the stack and is directed to an electrostatic precipitator (part of the Cat-Ox unit), which is designed to remove the majority (over 99%) of the fly ash.

Other versions of the oxidation concept, in which contaminants are removed from gas streams by catalytic oxidation, also exist.

A particularly interesting concept involves fluidized-bed combustion in the presence of a metal oxide (Econ–Abator process) (page 288) (Long and Parr, 1980; *Hydrocarbon Processing*, 1990). Efficient oxidation (95–98%) is achieved and mild fluidization of the bed prevents catalyst fouling and allows for the continual addition and removal of the catalyst.

The Sulfint process (page 322) also employs the catalytic oxidation concept

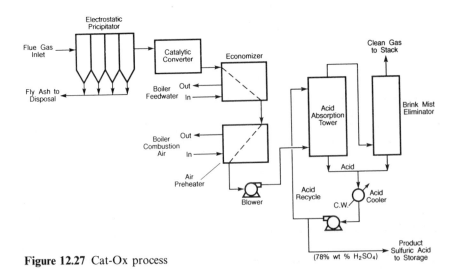

Figure 12.27 Cat-Ox process

to remove hydrogen sulphide from gas streams by conversion to elemental sulphur (Mackinger *et al.*, 1982; *Hydrocarbon Processing*, 1990).

The catalytic conversion of hydrogen sulphide to sulphur is also the basis of the Sulfreen process (page 322) (Davis, 1985; *Hydrocarbon Processing*, 1990) and the Sulfolin process (page 322) (*Hydrocarbon Processing*, 1990).

ChemicoB–Basic process

The ChemicoB–Basic process (Figure 12.28) for the removal of sulphur dioxide from flue gases involves the use of magnesium oxide. The predominant reaction is the formation of magnesium sulphite (Shah, 1971):

$$MgO + SO_2 + 6H_2O = MgSO_3.6H_2O$$

but side reactions do occur:

$$MgSO_3 + SO_2 + H_2O = Mg(HSO_3)_2$$
$$Mg(HSO_3)_2 + MgO = 2MgSO_3 + H_2O$$
$$MgO + SO_2 = MgSO_3$$
$$2MgSO_3 + O_2 + H_2O = 2MgSO_4.7H_2O$$

The magnesium oxide slurry is added as make-up to the venturi scrubber (see page 252) and clean flue gas leaves the scrubber through a demister. The magnesium oxide slurry is recirculated through the scrubber. A bleed stream from the slurry circulation loop flows to a centrifuge, where the magnesium sulphite and sulphate crystals and unreacted magnesium oxide are separated from the mother liquor. The cake produced from the centrifuge contains a minimum of surface moisture content. The mother liquor is recycled to the scrubber.

The wet cake from the centrifuge contains magnesium oxide, magnesium sulphite hexahydrate, magnesium sulphate heptahydrate and surface moist-

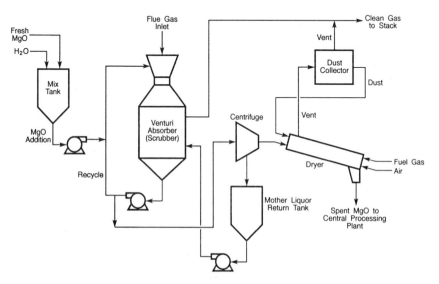

Figure 12.28 ChemicoB–Basic process

ure. This cake is dried in a direct-fired drier to remove both the surface moisture and the water of hydration:

$$MgSO_3.6H_2O = MgSO_3 + 6H_2O$$
$$MgSO_4.7H_2O = MgSO_4 + 7H_2O$$

The hot flue gases from the drier may be used to reheat the clean flue gases from the scrubber; the anhydrous, dry crystals from the drier are then calcined in a direct-fired calciner (kiln) to regenerate the magnesium oxide and release the sulphur dioxide:

$$MgSO_3 = MgO + SO_2$$
$$2MgSO_4 + C = 2MgO + 2SO_2 + CO_2$$

The carbon originates from the fuel gas added to the calciner (kiln). The regenerated magnesium oxide is slurried with water and returned to the scrubber system to complete the cycle.

Chemsweet process

The Chemsweet process (Figure 12.29) uses a suspension of zinc oxide (see also Zinc Oxide process, page 330); the gas stream is contacted by a fine suspension of zinc oxide particles in zinc acetate solution whereby hydrogen sulphide, mercaptans and carbonyl sulphide react instantaneously to produce insoluble zinc sulphide and zinc mercaptide (Manning, 1979; *Hydrocarbon Processing*, 1990):

$$H_2S + ZnO = H_2O + ZnS$$
$$2RSH + ZnO = (RS)_2Zn + H_2O$$

Claus process

The Claus process is the principal method by which hydrogen sulphide is removed from gas streams by conversion to sulphur, or it is the principal method by which sulphur is recovered from hydrogen sulphide (Goar, 1968; Kohl and Riesenfeld, 1979; Dowling *et al.*, 1990; *Hydrocarbon Processing*, 1990).

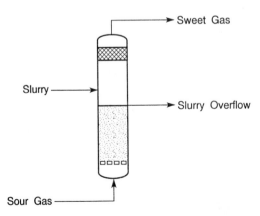

Figure 12.29 Chemsweet process

The process, and variations thereof (Figure 12.30), involves two stages: (a) a thermal stage where part of the hydrogen sulphide is converted to sulphur and sulphur dioxide; and (b) a catalytic stage where the remaining hydrogen sulphide is reacted with sulphur dioxide in the presence of a catalyst to form additional sulphur:

$$2H_2S + 3O_2 = 2H_2O + 2SO_2$$
$$2H_2S + SO_2 = 2H_2O + 3S$$

or

$$2H_2S + O_2 = 2H_2O + 2S$$

In one form of the process, a stoichiometric amount of air sufficient to convert one-third of the hydrogen sulphide to sulphur dioxide is fed to the

(a) Once-Through

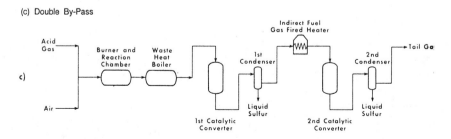

* Sufficient Air is Added to Burn 1/3 of Total H$_2$S to SO$_2$ and All Hydrocarbon to CO$_2$.

(b) Split-Stream

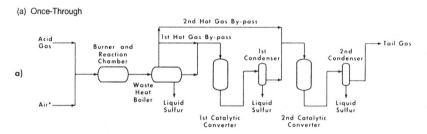

* Sufficient Air is Added to Burn All H$_2$S to SO$_2$ and All Hydrocarbon to CO$_2$ in 1/3 of Acid Gas.

(c) Double By-Pass

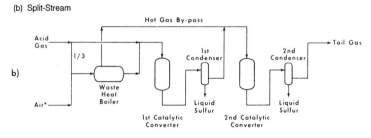

Figure 12.30 Claus process configurations

reaction furnace where the hydrogen sulphide is converted to sulphur at temperatures between 1100 and 1600°C (2010 and 2910°F). The cooled mixture, consisting of two-thirds hydrogen sulphide and one-third sulphur dioxide, is reheated to temperatures in excess of 400°C (750°F), above the sulphur dewpoint, and fed into the catalytic reaction stage where additional sulphur is produced in the presence of an aluminium oxide catalyst. An increased conversion is achieved by passing the gas stream through a series of catalytic reactors at successively lower temperatures. Before the gas enters each reactor it is cooled to condense out any sulphur formed in the preceding reactor and then heated again above the new sulphur dewpoint. The temperature in the last reactor is usually at, or higher than, 200°C (390°F).

The tail gas leaving the last catalytic reactor after the sulphur has been condensed contains small amounts of sulphur compounds and clean-up of this tail gas is required. A number of methods for doing this have been developed and include extension of the Claus reaction, in the liquid phase, and the hydrogenation of the sulphur compounds in the gas stream to hydrogen sulphide with subsequent removal by an amine absorption process or by subsequent conversion to sulphur by the Stretford process (page 318).

Tail gas from a Claus unit usually contains carbon dioxide, hydrogen sulphide, sulphur dioxide, carbon disulphide, carbonyl sulphide and water vapour in concentrations that may be too high to meet emission regulations. In order to comply with emission regulations, several processes have been developed for tail gas treatment of the effluent gas from conventional Claus plants (Table 12.3). These processes are commonly called add-on processes, since they are added to existing Claus plants. There are also sulphur recovery processes other than the modified Claus type which include the Giammarco–Vetrocoke, IFP, Stretford, and Takahax processes (pages 291, 295, 318 and 325).

Cominco process

This process is based on absorption of sulphur dioxide in an aqueous solution of ammonium sulphite $[(NH_4)_2SO_3]$ (see also Sulphite processes, page 324).

The absorbed sulphur dioxide is liberated by the addition of sulphuric acid to the solution, forming ammonium sulphate $[(NH_4)_2SO_4]$ as a by-product. An impure sulphur dioxide stream, containing some air, is produced from the top of the stripper. This stream is usually sent to a sulphuric acid plant. Ammonia is continuously added to the absorber, and sulphuric acid and air are added to the stripper. A 40% wt/wt solution of ammonium sulphate is produced from the bottom of the stripper.

Cryoplus process

This is a cryogenic process (Figure 12.31) (see Refrigeration (cryogenic) processes, page 313) for the removal of non-hydrocarbon contaminants from hydrocarbon gas streams.

Cyclic Lime process

In this process, the flue gas is contacted with a slurry of calcium sulphate $(CaSO_4)$ in water. To keep the solution alkaline, lime (CaO) or chalk $(CaCO_3)$ is added continuously to react with the sulphur dioxide as it is absorbed.

Table 12.3 General summary of selected processes available for clean-up of Claus plant tail gas streams

Process	Required feed sulphur content	Process performance	Sulphur removal mechanism
Low-efficiency processes:			
Sulfreen	Claus tail gas; sulphur compounds 1–3%	Raises overall sulphur recovery to 99%; 2000–3000 ppmv sulphur in treated gas; no COS/CS_2 conversion	Extended Claus reaction below sulphur dewpoint over alumina catalyst
SNPA	Claus tail gas	About 500 ppmv sulphur in treated gas; all sulphur compounds are handled	Independent conversion by catalytic oxidation of sulphur compounds to SO_3, followed by absorption to produce 94% to 98% H_2SO_4
IFP	Claus tail gas or other H_2S-containing feeds, up to 20%	<1500 ppmv sulphur in treated gas; no COS/CS_2 conversion	Extended Claus reaction in liquid phase, with polyethylene glycol as liquid carrier for catalyst
Beavon Mark II	H_2S content of <5% or Claus tail gas	Up to 99% overall sulphur recovery; some COS and CS_2 conversion	Independent catalytic conversion of sulphur compounds to H_2S, followed by successive oxidation and Claus reaction over proprietary Parsons/Union Oil catalyst
High-efficiency processes:			
Beavon	Claus tail gas	99.8% overall sulphur recovery, <300 ppmv sulphur compounds in treated gas, generally <10 ppmv H_2S; some COS/CS_2 conversion	Independent catalytic conversion of sulphur compounds to H_2S, followed by direct oxidation of H_2S to sulphur in Stretford unit
SCOT	Claus tail gas	99.8% overall sulphur recovery if COS/CS_2 content is not excessive; <300 ppmv sulphur compounds in treated gas; H_2S is about 150–200 ppmv	Recycle process, with catalytic conversion of sulphur compounds to H_2S, followed by absorption with diisopropanol amine; H_2S and CO_2 from amine regenerator recycled to Claus plant
Trencor	Claus tail gas	Same as SCOT	Recycle process, same as SCOT, except amine used is MDEA
Wellman–Lord	Claus tail gas	<200 ppmv sulphur compounds in treated gas; all sulphur compounds are handled	Recycle process, with oxidation of all sulphur species to SO_2, followed by absorption in sodium sulphite solution; evaporation of solution released SO_2 which is recycled to Claus plant; up to 10% of entering sulphur may be purged as sulphate solution

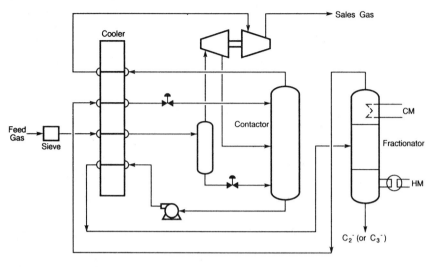

Figure 12.31 Cryoplus process

Calcium sulphite $(CaSO_3)$ is formed first and then oxidized to calcium sulphate in solution by oxygen absorbed from the flue gas. Both calcium sulphite and calcium sulphate are sparingly soluble in water; consequently, these materials precipitate to form a sludge.

DeNox processes

The thermal DeNox process (Figure 12.32) involves a selective non-catalytic nitrogen oxide reduction process (Hurst, 1985; *Hydrocarbon Processing*, 1990). The gas-phase homogeneous reaction between nitrogen oxide(s) and ammonia occurs at 870–1205 °C (1600–2200 °F) using either air or steam as a carrier gas:

$$NO_x + NH_3 = N_2 + H_2O$$

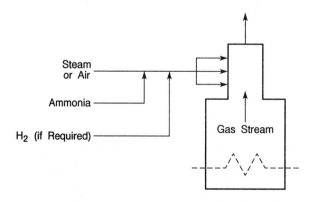

Figure 12.32 Thermal DeNox process

Hydrogen may also be used to bring the temperature down to about 705°C (1300°F).

The Noxout process (Figure 12.33) is also a selective non-catalytic process for the removal of nitrogen oxide(s) from gas streams which uses chemical sprays to convert the nitrogen oxide(s) to nitrogen, carbon dioxide and water (*Hydrocarbon Processing*, 1990).

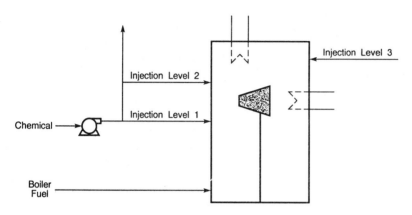

Figure 12.33 Noxout process

Dehydrate (Drizo) process

The Dehydrate process (Figure 12.34) and Drizo process (Figure 12.35) use glycol as a means of removing water from gas streams. Diethylene, triethylene and tetraethylene glycols are suitable absorbents, with triethylene glycol being preferred (Hicks and Senules, 1991).

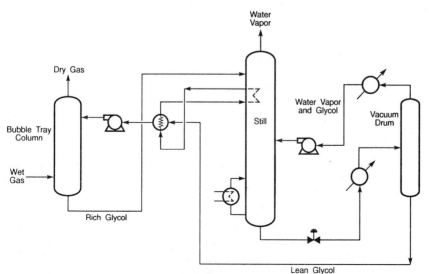

Figure 12.34 Dehydrate process

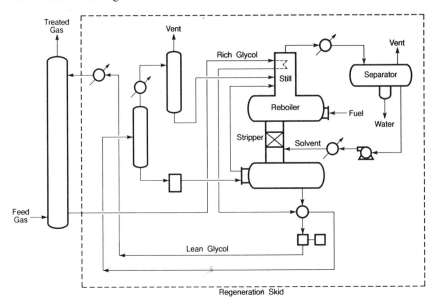

Figure 12.35 Drizo process

Econ–Abator process

The Econ–Abator process (Figure 12.36) is a catalytic oxidation process (page 278) which involves fluidized-bed combustion in the presence of a metal oxide (Long and Parr, 1980; *Hydrocarbon Processing*, 1990). Efficient oxidation (95–98%) is achieved and mild fluidization of the bed prevents catalyst fouling and allows for continual addition and removal of the catalyst.

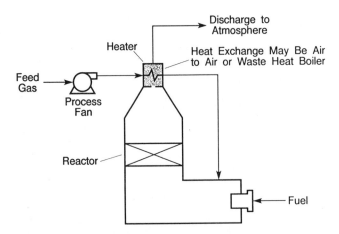

Figure 12.36 Econ–Abator process

Econamine process

The Econamine process (Figure 12.37) is similar in character to an al-kanolamine (i.e. ethanolamine, page 257) process (Huval and van de Venne, 1981; *Hydrocarbon Processing*, 1990). However, the solution employed may contain up to 60% wt/wt diglycolamine (DGA) giving it the ability to dissolve more hydrocarbon material, relative to monoethanolamine solutions, allow-ing for a lower circulation rate. But, as a precaution, an intermediate pressure flash tank between the bottom of the absorber and the heat exchanger is preferable.

Diglycolamine also reacts with carbonyl sulphide and/or carbon disulph-ide. The diglycolamine can be recovered through a reclaiming cycle that is carried out in the range 190–195°C (370–380°F).

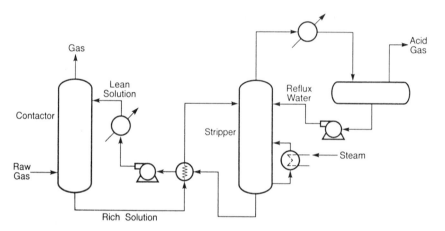

Figure 12.37 Econamine process

Flexsorb process

The Flexsorb process (Figure 12.38) uses hindered amines (Goldstein *et al.*, 1985; Savage *et al.*, 1985; Chludzinski and Wiechert, 1986) for gas-treating applications (page 263).

Fluor process

The Fluor solvent process (Figure 12.39) uses propylene carbonate (b.p. 240°C, 464°F) for the absorption of carbon dioxide and hydrogen sulphide from gas streams (Kohl and Buckingham, 1960; Benson, 1981).

Formate process

This process uses potassium formate solution to reduce sulphur dioxide to thiosulphate and then to hydrosulphide (Buckingham and Homan, 1971):

$$SO_2 + [reduction] = -S_2O_3^{2-}$$
$$-S_2O_3^{2-} + [reduction] = -HS^-$$

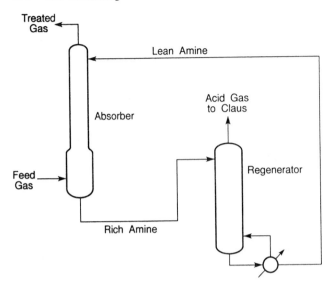

Figure 12.38 Flexsorb process

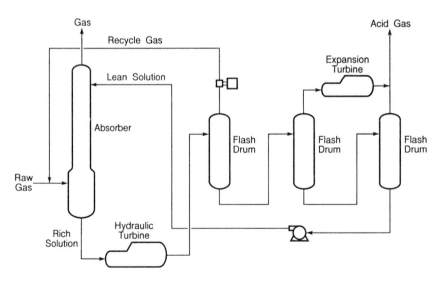

Figure 12.39 Fluor process

Hydrogen sulphide is stripped from the solution and converted to sulphur in a Claus sulphur plant. Formate, consumed in the reduction of sulphur dioxide, is produced from carbon monoxide which is generated using carbon dioxide recovered from the flue gas. The regenerated potassium formate is recycled to the absorption step. Additional carbon dioxide is also required for the stripping stage.

Giammarco–Vetrocoke process

The Giammarco–Vetrocoke process uses arsenic-containing aqueous solutions to absorb hydrogen sulphide from gas streams (*Hydrocarbon Processing*, 1990). After absorption of the hydrogen sulphide, the solution is treated by an (aerial) oxidation step to convert the hydrogen sulphide to sulphur:

$$KH_2AsO_3 + 3H_2S = KH_2AsS_3 + 3H_2O$$
$$KH_2AsS_3 + 3KH_2AsO_4 = 3KH_2AsO_3S + KH_2AsO_3$$
$$3KH_2AsO_3S = 3KH_2AsO_3 + 3S$$
$$6KH_2AsO_3 + 3O_2 = 6KH_2AsO_4$$

or

$$2H_2S + O_2 = 2H_2O + 2S$$

The oxidation step also aids in the regeneration of the absorbing solution.

The first step in the reaction sequence is that hydrogen sulphide is absorbed by reaction with potassium (or sodium) arsenite. The compound that results from this reaction is converted to potassium monothioarsenate by reaction with potassium arsenate. The combination of the arsenite and arsenate is very rapid, and essentially complete absorption of the hydrogen sulphide is obtained. The partial pressure of hydrogen sulphide over the monothioarsenate is decomposed to elemental sulphur and potassium arsenite by lowering the pH of the solution.

In the process (Figure 12.40 and 12.41), the gas stream enters the bottom of the absorber and flows in countercurrent contact with regenerated solution which is introduced at the top of the vessel. A hydrogen-sulphide-free stream leaves the top of the absorber but with little, if any, of the carbon dioxide removed since the process is highly selective for hydrogen sulphide. Rich solution leaves the bottom of the absorber and flows to the digester, which is nothing more than a small surge vessel. The digester allows time for the relatively slow reaction between arsenate and arsenite to go to completion.

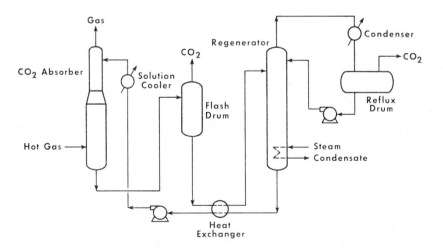

Figure 12.40 Giammarco–Vetrocoke process (carbon dioxide removal, steam regeneration)

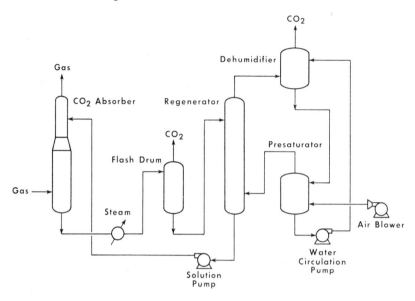

Figure 12.41 Giammarco–Vetrocoke process (carbon dioxide removal, air regeneration)

From the digester the rich solution flows through a cooler and then to an acidification drum, where the pH is adjusted by blowing carbon dioxide through the solution. The slurry of sulphur and solution from the bottom of the acidifier goes to the sulphur filter. The filtrate flows to the top of the regeneration tower or oxidizer, where it is regenerated by contact with air before being pumped back to the top of the absorber.

The Giammarco–Vetrocoke process has different applications. For example, there is one version for the removal of carbon dioxide, another version for the removal of hydrogen sulphide and yet another process version for the removal of both of these gases. As for most conventional carbonate processes, impurities such as carbon disulphide, mercaptans and carbonyl sulphide have no adverse effects on the solution (Jenett, 1962). The process modification that is specifically designed for the removal of hydrogen sulphide produces elemental sulphur of high purity as the by-product and the process also has the ability to reduce the hydrogen sulphide content to less than 1 ppm. The process is reputed to be unable to handle gas streams with hydrogen sulphide concentrations greater than 1.5%.

Girbotol process

In the Girbotol process (see Amine processes, page 257 *et seq.*) a water solution of ethanolamine is reacted with the acid gas in an absorber vessel under pressure and at room temperature. The solution is then heated and fed to a reactivation column where the ethanolamine–carbon dioxide complex is dissociated by stripping with steam at 90 to 120°C (195 to 250°F) and near atmospheric pressure. The carbon dioxide and steam leave the regenerator at the top of the column and the regenerated solution leaves at the bottom, is cooled, and is pumped back to the absorber.

Hitachi Wet process

Another version of the carbon adsorption method (see Activated Carbon Catalyst process, page 255) for the removal of sulphur dioxide involves the use of carbon adsorbent contained in six reactors (towers) which are operated in a cyclic manner. A single tower goes through a cycle of absorption (30 h), washing (10 h), and drying (20 h).

Uncooled flue gas from a 'dust collector' or air heater flows to the tower on 'adsorption'. Water, and no gas, flows to the tower on 'washing'. In the tower on 'drying', the cleaned stack gas passes first through the wet tower and then through a dry tower that removes any acid mist. Minimal flue gas cooling is a main objective of the process and part of the gas is bypassed around the wet tower during the drying cycle (Slack, 1967).

The carbon beds are washed in stages with washes of decreasing concentration; the dilute product acid contains only 10–15% wt/wt sulphuric acid.

Holmes–Stretford process

The Holmes–Stretford process is a modification of the Stretford process (page 318) and is used to remove hydrogen sulphide from gas streams by converting it to elemental sulphur. The process is selective for hydrogen sulphide and can reduce its concentration to as low as 1 ppm. However, the carbon dioxide concentration remains virtually unchanged during application of this process (Ouwerkerk, 1978).

The process (Figure 12.42) uses an aqueous solution containing sodium carbonate and bicarbonate (in the ratio of approximately 1:3), resulting in a pH of about 8.5 to 9.5, and the sodium salts of 2,6- and 2,7-anthraquinone disulphonic acid (Table 12.4).

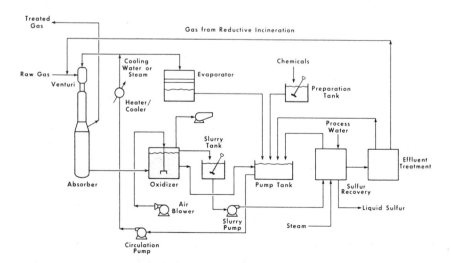

Figure 12.42 Holmes–Stretford process

Table 12.4 Organic derivatives used to enhance the chemistry of the Holmes–Stretford process

Anthraquinone-2,6-disulphonic acid

Anthraquinone-2,7-disulphonic acid

Several possible additives have been tested to increase the solution capacity for hydrogen sulphide and the rate of conversion of hydrosulphide to elemental sulphur. Alkaline vanadates have been found to be excellent additives for reducing hydrosulphide to sulphur, with a simultaneous valence change of vanadium from five to four. In the presence of the anthraquinone disulphonic acid, the vanadate solution can be regenerated to a five-valence state.

The overall chemistry of the Holmes–Stretford process is the atmospheric oxidation of hydrogen sulphide to elemental sulphur:

$$2H_2S + O_2 = 2H_2O + 2S$$

but, in reality, the chemistry is much more complex. The postulated reaction mechanism involves several steps:

(a) absorption of the hydrogen sulphide by the alkali:

$$Na_2CO_3 + H_2S = NaHS + NaHCO_3$$

(b) reduction of the anthraquinone disulphonic acid by the addition of hydrosulphide to a carbonyl group and the liberation of elemental sulphur from reduced anthraquinone disulphonic acid (ADA is anthraquinone disulphonic acid and ADA-H$_2$ is reduced anthraquinone disulphonic acid):

$$2ADA + 2NaHS + H_2O = 2\underset{\text{reduced}}{ADA\text{-}H_2} + 2NaOH + 2S$$

(c) reoxidation of the reduced anthraquinone disulphonic acid by air:

$$2ADA\text{-}H_2 + O_2 = 2ADA + H_2O$$

(d) reoxygenation of the alkaline solution, which also provides oxygen for the conversion of the reduced anthraquinone disulphonic acid to anthraquinone disulphonic acid (above).

According to the chemistry outlined above, it might be assumed that the chemicals could be used indefinitely with only minimal replenishments for losses that occur in the absorber or within the sulphur recovery unit. However, side reactions produce dissolved solids, e.g.

$$2NaHS + 2O_2 = Na_2S_2O_3 + H_2O$$

which increase in concentration until some of the solution must be discarded. The formation of sodium thiosulphate is probably due to some lack of reaction between the sodium bisulphide and the vanadate (because of insufficient time in the absorber) and, thus, the bisulphide is carried to the oxidizer where a reaction with oxygen occurs. These by-products can also be formed in the contactor if the natural gas contains oxygen; high temperature and high pH also promote the formation of thiosulphate.

The absorption rate of the hydrogen sulphide in solution to produce the sodium bisulphide and the sodium bicarbonate is greatly aided by a high pH, whereas the conversion of the sodium bisulphide to elemental sulphur is adversely affected by pH values above 9.5 – hence the preference for a pH range of 8.5 to 9.5.

IFP process

In the Institut Français du Petrole (IFP) process (Figure 12.43), Claus unit tail gas at about 125°C (260°F) enters a vertical, packed-tower reactor vessel where countercurrent contact occurs between the gas and a catalyst–solvent liquid solution (Davis, 1972) at a temperature of 125–140°C (260–280°F). The treated tail gas leaves the top of the reactor at about 125°C (260°F). Liquid sulphur is withdrawn from the base of the reactor.

The basic reaction of the IFP process is the same as in the Claus unit:

$$2H_2S + SO_2 \text{ (+liquid catalyst)} = 3S + 2H_2O$$

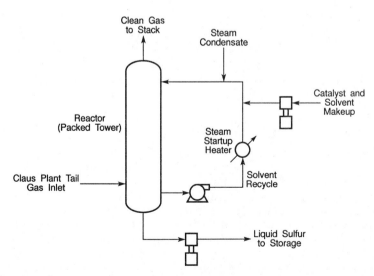

Figure 12.43 IFP process

The hydrogen sulphide and the sulphur dioxide dissolve in the solvent (at times a glycol-type solvent has been used) and react in the presence of the catalyst to form elemental sulphur which is only slightly soluble in the solution.

The Townsend process (page 326) is similar to the IFP process, in that it uses an organic solvent to allow hydrogen sulphide and sulphur dioxide to react (Claus reaction) to form elemental sulphur. The reactor is operated at a temperature above the melting point of sulphur, so that liquid sulphur is produced from the bottom. This process may be applied directly to the treatment of Claus plant gas.

Ionics process

This process uses an aqueous caustic solution (NaOH) to remove sulphur dioxide from flue gases (Remirez, 1968). The resulting sodium bisulphite solution is stripped to yield pure, dry sulphur dioxide which goes to a sulphuric acid plant. The sodium sulphate solution from the stripper is sent to an electrolytic cell. Here, use of a membrane enables the cell to produce caustic soda, sodium bisulphate, dilute sulphuric acid, oxygen and hydrogen.

In the process, the hot flue gases from the electrostatic precipitator (at 95–175°C, 200–350°F) enter the absorber, where the gases are quenched with water (at 55–65°C, 130–150°F) in the bottom of the tower. Some removal of particulate matter is also achieved. The quench water flows to a settling tank from which clear water is recirculated to the lower section of the absorber. Solids are withdrawn from the bottom of the settling tank to a sludge pond.

There are two recirculating absorption stages. Caustic solution is flowed to the upper stage and a solution of sodium sulphite is passed from the upper stage to the lower stage in which a part of the sulphite is converted to bisulphite. Side reactions involving oxygen and nitrogen oxides convert some of the sulphite and bisulphite to sulphates:

$$-SO_3 = -SO_4$$
$$-HSO_3 = -SO_4$$

These reactions may be catalysed by dissolved and suspended fly ash in the solution. The resulting solution containing the sulphite and bisulphite is reacted with dilute sulphuric acid to form sodium sulphate. This reaction releases sulphur dioxide which is stripped from the solution and processed for removal or passed to an acid plant to form sulphuric acid.

Iron Oxide (Iron Sponge, Dry Box) process

One chemically reactive process which does exhibit a high selectivity for hydrogen sulphide is the Iron Oxide (Iron Sponge or Dry Box) process. This process was one of the first to be developed for removing sulphur compounds from gas streams. It was introduced in England about the middle of the nineteenth century and is still widely used in many areas of special application.

In this process (Figure 12.44), the sour gas is passed through a bed of wood chips that have been impregnated with hydrated ferric oxide. Dispersing the iron oxide in this way provides a large surface area-to-weight ratio and maximizes contact between the gas stream and the iron oxide thereby increasing the affinity of the oxide for hydrogen sulphide, whereupon the

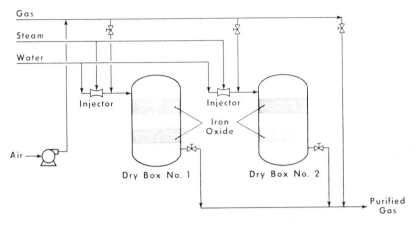

Figure 12.44 Iron Oxide process

hydrogen sulphide is converted to ferric sulphide:

$$Fe_2O_3 + 3H_2S = Fe_2S_3 + 3H_2O$$

and the iron oxide is regenerated from the iron sulphide by passing oxygen/air over the bed:

$$2Fe_2S_3 + 3O_2 = 2Fe_2O_3 + 6S$$

This reaction–regeneration cycle can be repeated several times. Eventually the sulphur formed in the reaction will cover the majority of the surface of the oxide particles. This causes the oxide to lose activity and may also cause excessive pressure drop or channelling in the bed.

The process operates in a batch-type reaction–regeneration cycle and offers the advantages of simplicity and excellent selectivity for the removal of hydrogen sulphide. However, bed regeneration can be difficult and expensive. In addition, sulphur will eventually cover most of the surface of the ferric oxide particles and further regeneration becomes impossible.

A continuous regeneration process has also been developed (Figure 12.45) where small amounts of oxygen or air are added along with the sour gas at the inlet. This latter process gives an improved performance, generating a higher removal efficiency as well as better regeneration (Kohl and Riesenfeld, 1979).

A logical extension of the Iron Sponge process involved the utilization of a liquid suspension of iron oxide for the removal of hydrogen sulphide from gas streams. And there are several such processes.

The chemistry of these processes is based on the reaction of hydrogen sulphide with an alkaline compound, usually sodium carbonate or ammonium carbonate, followed by reaction of the hydrosulphide with iron oxide. The resulting ferric sulphide is regenerated by blowing air through the solution:

$$H_2S + Na_2CO_3 = NaHS + Na_2CO_3$$
$$Fe_2O_3 + 3NaHS = Fe_2S_3 + 3NaOH$$
$$2Fe_2S_3 + 3O_2 = 2Fe_2O_3 + 6S$$

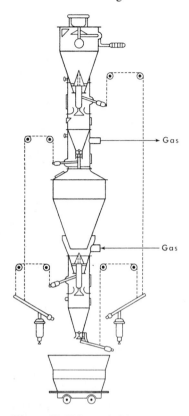

Figure 12.45 Iron Oxide (continuous) process

In natural gas sweetening these are the major reactions that occur with the carbon dioxide going essentially unreacted. In manufactured or synthetic gases the other sulphur compounds or hydrogen cyanide may also react.

The alpha and gamma forms of ferric oxide are satisfactory for gas-sweetening purposes; both forms react readily with hydrogen sulphide and can be easily oxidized to sulphur and ferric oxide. For both, the reactions are best carried out at ambient temperatures in an alkaline atmosphere. Temperatures above 50°C (120°F) and neutral or acid conditions drive the water of crystallization from the ferric sulphide and make it more difficult to regenerate. In the operating plant, provisions must be made to inject water into the gas stream. A solution of sodium carbonate can be used to adjust the pH to the 6.0–8.5 range where the reactions proceed most readily.

The Ferrox process is similar in principle to the Iron Oxide process in so far as it employs a solution containing 0.5% hydrated ferric oxide and 3% sodium carbonate. The gas is contacted with this solution, which can be regenerated by blowing with air. Removal of hydrogen sulphide is reputed to be high and even better when a double-stage unit is used. The sulphur solids produced in the process contain large amounts of ferric oxide and sodium carbonate and chemical entrainment in the sulphur can also occur.

Another variation, the Gluud process, is similar in principle to the Ferrox process. The primary difference is that a solution of ammonium carbonate is used to suspend the iron oxide. The process employs tall regeneration towers that require less regeneration air but at higher pressure.

The Manchester process is another modification of the basic Ferrox process. The process uses several absorbing towers in a series with fresh solution being fed to each absorption stage. Because of this, the Manchester process is capable of treating a gas stream to lower hydrogen sulphide requirements.

Katasulf process

In the Katasulf process (Figure 12.46) (see also the Catasulf process, page 277), the hydrogen sulphide in the natural gas is reacted with oxygen to form water and sulphur dioxide:

$$2H_2S + 3O_2 = 2H_2O + 2SO_2$$

The reaction is exothermic and the heat generated is used to heat the inlet gases to reaction temperature (400°C, 750°F). Catalysts used in the process are activated carbon, bauxite, and alloys of iron or nickel or copper with tungsten or vanadium or chromium. Any one of the first three metals reacts with the hydrogen sulphide to form the metal sulphide whilst any one of the last three metals acts as an oxygen carrier.

Koppers Vacuum Carbonate process

In this process, the sour gas enters the bases of a two-stage absorber where countercurrent contact takes place (in each stage) with a sodium carbonate solution (see also Carbonate process(es), page 270). The fat reacted solution from each absorber is pumped to a steam stripper after which it is cooled and returned to the absorbers.

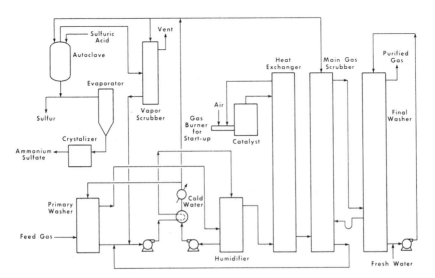

Figure 12.46 Katasulf process

Lignite Ash process

This process advocates the use of ash from lignite combustion as the adsorbent (Slack, 1967). Hydration of the ash converts the lime constituent to calcium hydroxide. Flue gas is contacted by the adsorbent in a series of countercurrent reactors (Maurin and Jonakin, 1970) where the sulphur dioxide is converted to calcium sulphite and calcium sulphate:

$$CaO + H_2O = Ca(OH)_2$$
$$Ca(OH)_2 + SO_2 = CaSO_3 + H_2O$$
$$2Ca(OH)_2 + 2SO_2 + O_2 = 2CaSO_4 + 2H_2O$$

An electrostatic precipitator is used to remove particulate matter from the flue gas and the spent ash is heated to yield sulphur dioxide. The regenerated absorbent is recycled to the reactors.

Limestone (Dolomite) Injection processes

The use of limestone or lime as an absorbent for sulphur oxides has been investigated and carried out for several decades (Figure 12.47) (Barron, 1971) (see also Carbonate processes above).

The concept of direct injection of an alkaline reagent, such as limestone ($CaCO_3$), dolomite ($CaCO_3.MgCO_3$), or derivatives thereof such as lime (CaO), into a utility furnace to reduce sulphur dioxide emissions evolved from early work and studies are continuing with the goal of increasing the effectiveness of the limestone injection procedure (Squires, 1967).

The simplest of all the proposed schemes for removing sulphur dioxide from flue gases is the injection of pulverized limestone directly into the furnace of a utility boiler (Figure 12.48) (cf. Carbonate processes, page 270). However, this technique removes only 25–35% of the sulphur oxides in the furnace (Maurin and Jonakin, 1970) and there are reports (Falkenberry and Slack, 1969) that the dry limestone injection process cannot be expected to remove more than about one-half of the sulphur oxides, even if twice the stoichiometric quantity of limestone is used. However, the process can be applied to either coal-burning or oil-burning power plant boilers.

At the high temperatures of the furnace (above 815°C, 1500°F), the limestone and/or dolomite additive is calcined to yield the more reactive calcium oxide and/or magnesium oxide

$$CaCO_3 = CaO + CO_2$$
$$MgCO_3 = MgO + CO_2$$

which react with the sulphur oxide(s) in flue gas to form calcium or magnesium sulphites and sulphates. The adsorbent removes all of the sulphur trioxide from the flue gas and the large majority of the nitrogen oxides:

$$CaO + SO_2 = CaSO_3$$
$$MgO + SO_2 = MgSO_3$$
$$CaO + SO_3 = CaSO_4$$
$$MgO + SO_3 = MgSO_4$$
$$2CaSO_3 + O_2 = 2CaSO_4$$
$$2MgSO_3 + O_2 = 2MgSO_4$$

In the wet scrubber, calcium and magnesium oxides that have not yet reacted with sulphur dioxide and sulphur trioxide in the furnace react with

(a)

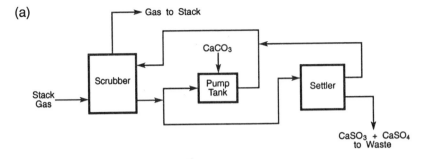

(b)

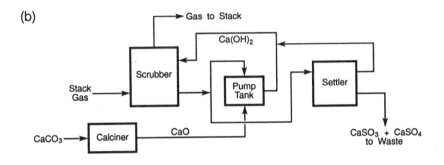

(c)

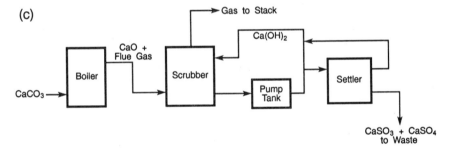

Figure 12.47 Illustration of the use of limestone scrubbing for stack gas clean-up

recycled wash water to form hydroxides:

$$4CaO + 4SO_2 + O_2 = 2CaSO_3 + 2CaSO_4$$
$$4MgO + 4SO_2 + O_2 = 2MgSO_3 + 2MgSO_4$$

In addition, the wet-scrubbing action serves to remove fly ash and other particulate matter from the stack gas.

The calcium sulphate, calcium sulphite and magnesium sulphite are only slightly soluble in water and precipitate. However, magnesium sulphate, which has an appreciable solubility in water, becomes highly concentrated in the solution since the water is recycled. The precipitated solids are carried by the wash water to a clarifier where they settle with the fly ash to permit removal.

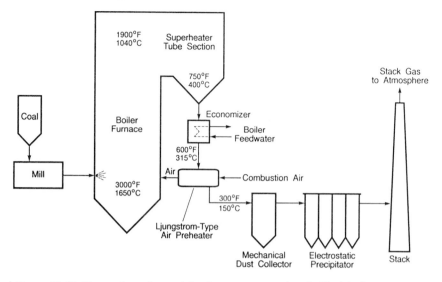

Figure 12.48 Illustration of a coal-fired power generation (utility) boiler

Lo-Cat process

In this process (Figure 12.49), sour gas is contacted with a chemical solution in which the hydrogen sulphide is converted by a chelated iron compound to produce elemental sulphur. The iron is reduced to a lower valency state after which it is regenerated with air. The absorber vessels can be vessels full of liquid, venturi scrubbers, packed columns, spray chambers, static mixers or mobile beds (Hardison, 1985a,b; *Hydrocarbon Processing*, 1990a,b).

Application of heat is not required because of the exothermic nature of the reaction and pressures are low (below 1500 psi, 10.3×10^3 kPa). The solution is regenerated by contact with air (or oxygen).

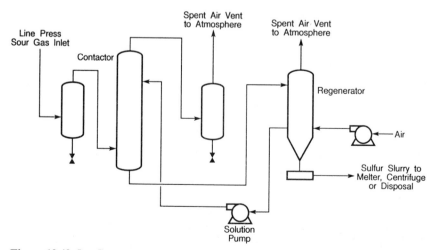

Figure 12.49 Lo-Cat process

Chelated iron compounds are also the heart of the SulFerox process (page 320) which is designed for the removal of hydrogen sulphide from process gases (*Oil and Gas Journal*, 1987; *Hydrocarbon Processing*, 1990).

Manganese Dioxide (DAP–Mn) process

This process (Figure 12.50) uses dry, activated manganese dioxide (MnO_2) in a dilute fluidized-bed reactor to contact the flue gas and remove sulphur dioxide by forming manganese sulphate ($MnSO_4$) (Uno *et al.*, 1970). The reaction occurs readily in the temperature range 100–180 °C (212–356 °F). The absorbent is used in the form of particles of about 40 μm size which are also suitable for removal by a downstream electrostatic precipitator. Any soot deposited on the adsorbent can be separated by flotation, using a small amount of kerosene as an additive. The sulphur dioxide is recovered as a by-product of ammonium sulphate.

The process can be described as four steps: (a) adsorption of sulphur dioxide; (b) regeneration of the adsorbent by the addition of ammonia and air; (c) soot removal and separation from the 'lean' adsorbent; and (e) crystallization and separation of ammonium sulphate:

$$MnO_2.H_2O + SO_2 = MnSO_4 + H_2O$$
$$MnO_2.H_2O + SO_3 = MnSO_4 + O_2 + H_2O$$
$$MnSO_4 + 2NH_4OH = Mn(OH)_2 + (NH_4)_2SO_4$$
$$2Mn(OH)_2 + O_2 = 2MnO_2.H_2O$$

The sulphur dioxide and sulphur trioxide react with the adsorbent while it is being transported through the fluidized adsorber by the total flue gas stream to form manganese sulphate.

The effluent gas stream passes through cyclones and an electrostatic precipitator, where about 90% of the unreacted adsorbent and manganese sulphate are collected. 'Lean' adsorbent is returned to the disintegrator for mixing with the bypassed gas stream. The remaining solids are removed by

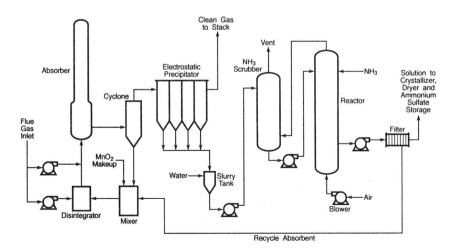

Figure 12.50 Manganese Dioxide (DAP–Mn) process

cyclones and precipitator and are then formed into a slurry containing about 70% water which is passed through an ammonia recovery tower to the regenerator (oxidizer). Air and ammonia are added to the regenerator.

The aqueous solution in the tower contains excess ammonium hydroxide which reacts with the manganese sulphate to form manganese hydroxide and the oxygen in the air oxidizes the manganese hydroxide to manganese oxides. The resultant fresh adsorbent is filtered from the ammonium sulphate solution and returned to the disintegrator. The filtrate is passed to the crystallizer where ammonium sulphate crystallizes from the solution and is separated by filtration (or centrifugation) and dried.

A variation of the DAP–Mn process has been proposed for the desulphurization of fuel oil; dimanganese trioxide (Mn_2O_3) is the active adsorbent (Yulish, 1971).

Membrane processes

The ability of polymeric membranes to separate gases was initially applied on an industrial scale to the recovery of hydrogen and the application has been extended to other services, such as the removal of carbon dioxide and hydrogen sulphide from gas streams (Russell, 1985; Stookey et al., 1985; Chiu, 1990).

Polymeric membranes separate gases by selective permeation of gas species in contact with the membrane; the gases dissolve in the polymer and are transported across the membrane barrier under the partial pressure gradient. The pressure gradient is accomplished by feeding high-pressure gas to the outside of the hollow fibre membrane while the permeate side is operated at substantially lower pressure. Typically, water vapour, carbon dioxide and hydrogen sulphide are 'fast' species in comparison to methane, nitrogen and other hydrocarbon gases, hence their utility in the removal of acid gas from natural gases (Stookey et al., 1985).

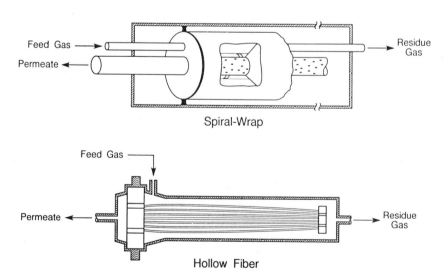

Figure 12.51 Schematic representation of membrane separators

Some membrane separators utilize bundles of hollow fibres which are enclosed in pressure vessels (Figure 12.51). Feed enters from one end and the acid gas components, which then permeate to the bores of the hollow fibres, pass through an internal tube and discharge through an exit in the pressure vessel.

Several separators may be connected in series and parallel arrangements to meet specific requirements and may also include a two-stage system (Figure 12.52). In such an arrangement, the first stage might produce the desired gas product and all acid gases removed by the system are concentrated in the second stage and leave in its permeate. A single membrane stage (Figure 12.53) is usually sufficient for bulk removal applications. In some cases it may be desirable to split the first stage into two sections. This enables some of the permeate to be recycled to reduce hydrocarbon losses with the permeate stream. This is particularly true when the feed gas is at low pressures and must be compressed (Figure 12.54).

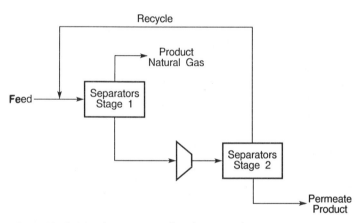

Figure 12.52 Membrane separation (two-stage) process

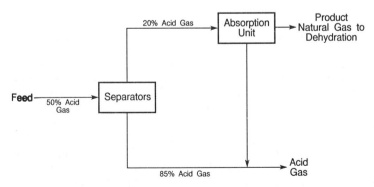

Figure 12.53 Bulk acid gas removal by membrane separation

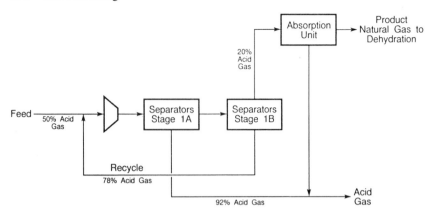

Figure 12.54 Bulk acid gas removal by membrane separation (with recycle)

Mixed Metal Oxide process

The adsorbent used in this process is a mixture of manganese dioxide and magnesium hydroxide, which is first deposited on a carrier, for example coke (Slack, 1967) (see also the Manganese Dioxide (DAP–Mn) process, page 303).

The adsorbent reacts with the sulphur dioxide present to form manganese sulphate and magnesium sulphate. The adsorbent is removed from the carrier by shaking and screening, after which the carrier is recycled. The adsorbent is regenerated by mixing with crushed coal followed by heating to 900°C (1650°F) to reduce the sulphur compounds to sulphur, hydrogen sulphide and, on occasion, carbonyl sulphide.

The sulphur-rich gas stream is subsequently burned to produce sulphur dioxide which may be used as feedstock for a sulphuric acid plant. The adsorbent, converted to the oxide form by calcining, is quenched in water to hydrate the magnesium contacted with the carrier to complete the cycle.

Molecular Sieve process

Molecular sieves are synthetic forms of crystalline sodium–calcium alumina silicates which are porous in structure and have a very large surface area. The pores are uniform throughout the material and several grades of molecular sieves are available with each grade corresponding to a very narrow range of pore sizes.

Molecular sieves remove contaminants from natural gas through a combination of size selectivity ('sieving') and the physical adsorption process. Because of the narrow pore sizes, molecular sieves can discriminate among the adsorbates on the basis of molecular size. The sieves also possess highly localized polar charges on their surface that act as adsorption sites for polar materials. Therefore, small molecules that are polar (or which are polarizable) and which could conceivably pass through the pores of the sieve are also removed from the gas stream by the sieve.

Molecular sieves are highly selective for the removal of hydrogen sulphide (as well as other sulphur compounds) from natural gas and offer a continuously high absorption efficiency. They are also an effective means of removing water and thus offer a process for the simultaneous dehydration and desulphurization of natural gas. Gas that has an excessively high water content may, however, require upstream dehydration (Rushton and Hays, 1961).

The Molecular Sieve process (Figure 12.55) is similar to the Iron Oxide process (page 296) (Conviser, 1965) but the molecular sieves are susceptible to poisoning by chemicals such as glycols and require thorough gas-cleaning methods prior to the adsorption step. The sieve can be offered some degree of protection by the use of 'guard beds' in which a less expensive catalyst is placed in the gas stream, prior to contact of the gas with the sieve, to protect the more expensive catalyst from poisoning. This concept is analogous to the use of guard beds or attrition catalysts in the petroleum industry (Speight, 1981).

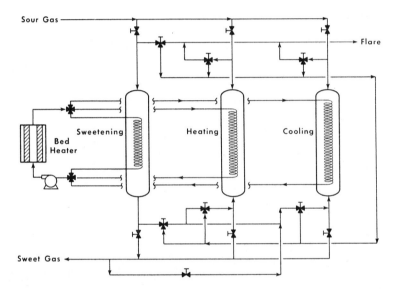

Figure 12.55 Molecular Sieve process

The sieve 'bed' is regenerated by passing part of the heated clean gas over the bed, and as the temperature of the bed increases, it releases the adsorbed hydrogen sulphide into the regeneration gas stream; up to 2% of the gas treated can be lost in the regeneration process (Rushton and Hays, 1961). A portion of the natural gas may also be lost by the adsorption of hydrocarbon components by the sieve.

The molecular sieve adsorption concept can also be employed in conjunction with other process concepts as, for example, in the process for purifying gas streams that are rich in carbon dioxide (Figure 12.56). On the other hand, the PuraSiv S process (Figure 12.57) is designed for the removal of sulphur dioxide from sulphuric acid plants.

Caution is advised in the use of some molecular sieves for the clean-up of mixed gas streams since they have been known to catalyse the formation of

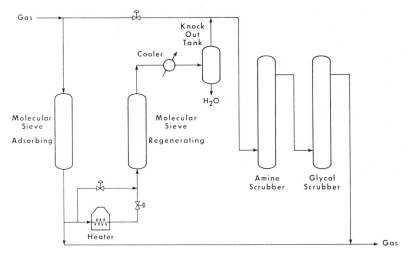

Figure 12.56 Molecular sieve concept used with other gas-cleaning processes

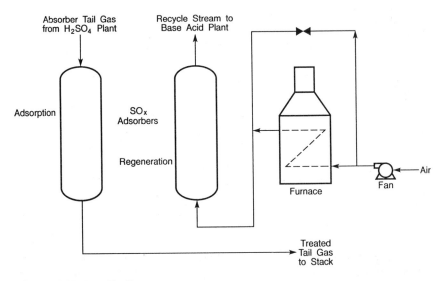

Figure 12.57 PuraSiv S process

carbonyl sulphide: $$H_2S + CO_2 = COS + H_2O$$

New molecular sieves have been developed to retard the formation of carbonyl sulphide. If further clean-up of the gas stream is desired, caution is advised in the use of alkanolamines since the presence of carbonyl sulphide in the gas stream can bring about irreversible reactions that lead to alkanolamine loss in the process.

Noxout process

The Noxout process (see the DeNox processes, page 286) is a selective non-catalytic process for the removal of nitrogen oxide(s) from gas streams which uses chemical sprays to convert the nitrogen oxide(s) to nitrogen, carbon dioxide and water (*Hydrocarbon Processing*, 1990).

Oil Absorption process

Until the early 1970s, most hydrocarbon recovery plants used the concept of oil absorption, although very few oil absorption units are included in newly constructed plants. Nevertheless, there are many older plants which still use the oil absorption principle and the concept needs to be described here with some degree of detail.

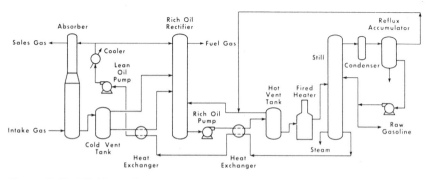

Figure 12.58 Oil Absorption process

The Oil Absorption process (Figure 12.58) uses the principle of countercurrent contact of the lean (or stripped) oil with incoming wet gas (Figure 12.59) (see also Chapter 11) with the temperature and pressure conditions programmed to maximize the dissolution of the liquefiable components in the oil.

The plant may also be of a dual nature in so far as refrigeration may also be used to obtain lower temperatures. The remainder of the plant consists of: (a) separation of light ends from the oil; (b) separation of absorbed materials from the oil; (c) removal of light ends from the raw product; and (d) separation of the raw product into various finished products.

The removal of any light ends may also be necessary and can be achieved using one or more additional steps. For example, methane and, in some plants, ethane can be removed in the rich-oil rectifier by pressure reduction and heating. Following rich-oil rectification, the absorbed material is removed from the oil in a stripper or in a still.

If a heavy absorption oil is employed, stripping is usually achieved by preheating the oil followed by countercurrent contact with steam. Most processes use a compromise between these two operations and some plants use two stills in series: (a) a high-pressure still to condense the light ends; and (b) a low-pressure still to ensure good stripping of the heavier gasoline fractions. If the oil is not stripped efficiently, the lighter components remaining in the oil will be vaporized in the absorber and lost in the residue gas stream.

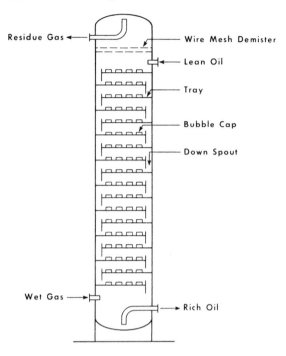

Residue Gas ——— Wire Mesh Demister

——— Lean Oil

——— Tray

——— Bubble Cap

——— Down Spout

Wet Gas ———

——→ Rich Oil

Figure 12.59 Oil absorption tower

The absorption processes offer reasonable selectivity for the removal of acid gas. In addition, the solvent used is generally recovered in good yield by flashing the rich solvent in flash tanks at successively lower pressures which requires little, or no, heat. Most solvents currently in use have a relatively high solubility for the higher-molecular-weight hydrocarbons, particularly the unsaturated and aromatic components which, because of chemical interactions, may be responsible for yielding a product that is contaminated with sulphur. Thus, for sour gases containing these particular hydrocarbons, care must be taken during the regeneration step to prevent their entry into the acid gas stream that is to be sent to a sulphur recovery unit.

Parsons Beavon process

This process employs three distinct steps: (a) hydrogenation–conversion of sulphur compounds to hydrogen sulphide in a catalytic converter; (b) cooling of the converter effluent gases; and (c) conversion of the hydrogen sulphide in the tail gas from the cooler to elemental sulphur by use of the Stretford process (see also the Beavon process above).

In the process, the tail gas from a Claus unit is first mixed with 'fuel gas and air' combustion products and then fed to the catalytic converter, which contains a cobalt molybdate catalyst. The hydrogen required for the reduction of sulphur compounds to hydrogen sulphide is supplied by partial combustion of the fuel gas in an in-line burner, which simultaneously raises the tail gas stream temperature to the level required for the hydrogenation reactions.

The converter effluent gases are then cooled with water in a direct-contact condenser, where most of the water vapour contained in the gas is condensed, and at the same time the tail gas is cooled. The purge water produced from the condenser is of good quality, with only a small amount of dissolved hydrogen sulphide. With the removal of hydrogen sulphide in a small 'sour water stripper', it is suitable for cooling tower make-up water. A recirculation loop of water with a shell-and-tube cooler is provided to maintain a flow of water through the direct-contactor condenser from top to bottom. A small bleed or purge stream of water goes to further processing.

The cooled gas then enters the Stretford absorber, where the hydrogen sulphide is removed almost quantitatively:

$$H_2S + Na_2CO_3 = NaHS + NaHCO_3$$

The hydrosulphide is oxidized to sulphur by sodium vanadate in solution:

$$NaHS + 2NaVO_3 = S + Na_2V_2O_5 + NaOH$$

Subsequently, the vanadium is oxidized to the pentavalent state by blowing with air, with anthraquinone disulphonic acid (ADA) or sodium anthraquinone disulphonate acting as an oxidation catalyst:

$$Na_2V_2O_5 + O_2 = 2NaVO_3$$

The Stretford process (page 318) can be considered to occur by five steps: (a) hydrogen sulphide is absorbed by the alkali; (b) the sodium vanadate is reduced by a sulphur shift; (c) elemental sulphur is liberated by dissolved oxygen in the solution; (d) the sodium vanadate is oxidized in the presence of the sodium anthraquinone disulphonate; and (e) the alkaline solution is oxygenated by air blowing, which also floats the sulphur out of solution.

The spent absorbent solution flows from the absorber to the regenerator (oxidizer), where air is sparged into the tower. Tiny particles of sulphur are collected as a froth at the top of the oxidizer tower. The sulphur froth is skimmed off the solution and sent to a filter or centrifuge for removal of the filtrate from the sulphur cake. The clear (regenerated) absorbent solution and the filtrate from the filter are recycled back to the absorber.

The Takahax process (page 325) is essentially the same as the Stretford process, except for the chemicals used. This process uses an absorbent solution of sodium carbonate, 1,4-naphthaquinone, and 2-sulphonate sodium (Goar, 1971a,b).

Pintsch–Bamag/Bergbau Adsorption process

This process depends on the ability of coke to absorb sulphur dioxide from flue gases after which the sulphur dioxide is converted to sulphuric acid. The coke, made from bituminous coal, is given a high catalytic activity by coating it with oxides and applying other catalysts. The pelletized coke is abrasion resistant enough to stand up to multipass operation (*Chemical Engineering*, 1969).

Pritchard Cleanair process

The Pritchard Cleanair process (Figure 12.60) (*Oil and Gas Journal*, 1970) is quite similar to the Parsons Beavon process (page 310) and also uses the Stretford process (page 318) as the final treatment step.

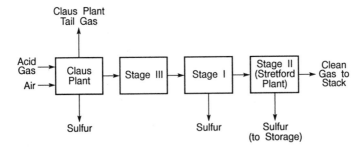

Figure 12.60 Pritchard Cleanair process

The process uses three processing steps: (a) tail gas from a Claus unit passes through a section where the carbonyl sulphide and carbon disulphide levels in the tail gas are reduced: (b) the tail gas then flows to another section which converts essentially all of the residual sulphur dioxide in the tail gas to sulphur, with some conversion of hydrogen sulphide to sulphur. The remaining gas then flows to another section, which is the Stretford plant. Essentially all of the hydrogen sulphide is converted to sulphur.

Purisol process

The Purisol process (Figure 12.61) is a physical solvent process which uses N-methyl-2-pyrrolidone to remove carbon dioxide from gas streams.

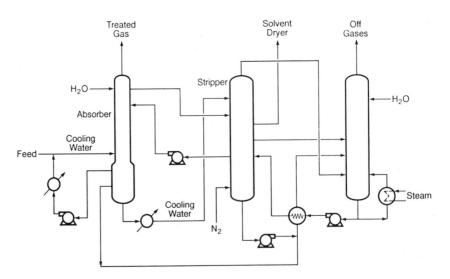

Figure 12.61 Purisol process

Rectisol process

The Rectisol process (Figure 12.62) (Ranke and Mohr, 1985; *Hydrocarbon Processing*, 1990) is a gas absorption, physical solvent, process that was designed primarily to clean the gases from a coal gasifier. The process uses organic solvents, such as methanol, at temperatures between -60 and $-1°C$ (-80 and $30°F$) and involves countercurrent contact of the gas with the solvent in a trayed absorption column. The spent solvent is regenerated by stripping or by reboiling and the lean solvent is then recycled to the top of the absorption column.

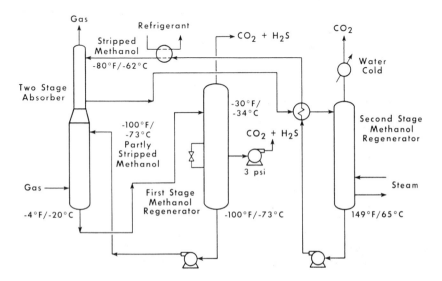

Figure 12.62 Rectisol process

The process depends upon the solubility of hydrogen sulphide and carbon dioxide in cold methanol whilst the desired product gases such as hydrogen, carbon monoxide and methane have little or no solubility in the methanol under process conditions, generally $-40°C$ ($-40°F$) or lower. The considerably higher solubility of hydrogen sulphide, relative to carbon dioxide, in cold methanol also offers the potential for the partial separation of hydrogen sulphide from gas streams containing both gases. However, high solvent losses can be anticipated because of the high vapour pressure of the methanol, even at the low process temperatures.

Refrigeration (cryogenic) processes

Refrigeration (cryogenic) processes can be used to upgrade gas streams; the processes are essentially a means of separating the non-hydrocarbon contaminants from the hydrocarbon constituents of the gas (e.g. the Cryoplus process, page 284) (Paradowski and Castel, 1987; *Hydrocarbon Processing*, 1990).

Refrigeration processes generally use mechanical or compression-type refrigeration to reduce the temperature whereupon the basic separation (phase separation of the crystallized or solid product) occurs. The most common refrigerants in current use are propane and ammonia but it is also advisable to inject ethylene glycol into the system at points where icing, or the formation of gas hydrates, can occur.

Reinluft Dry process

This process uses a slowly moving bed of activated char formed from a 'semicoke' carbonized under vacuum at 595°C (1100°F) (Katell, 1966).

In this dry adsorption method (Figure 12.63), flue gas with a low dust content passes through a vertical, two-stage countercurrent adsorber vessel. The flue gas, at a temperature above its dewpoint, enters the first stage where the coke adsorbs the sulphur dioxide and sulphuric acid mist (if present). The flue gas is then drawn off (at 145°C, 290°F), cooled to 95°C (200°F) and then fed to the second stage. The gas is cooled to enhance the oxidation of sulphur dioxide to sulphur trioxide to be converted to sulphuric acid. The cleaned flue gas leaves the reactor at about 101°C (215°F).

The activated char, with its adsorbed sulphuric acid, drops to the regeneration (desorber) section of the reactor where the char temperature is raised to 370°C (700°F). The sulphuric acid dissociates into sulphur dioxide and water and the sulphur trioxide reacts with the carbon to form sulphur dioxide and carbon dioxide. The product gas leaves the top of the regenerator section at 150°C (300°F), is heated to 370°C (700°F) and then returned to the base of the regenerator.

The regenerated char leaves the base of the regeneration section of the reactor, is screened to remove fines, mixed with make-up char (consumed in the process) and then conveyed back to the top of the adsorber (reactor).

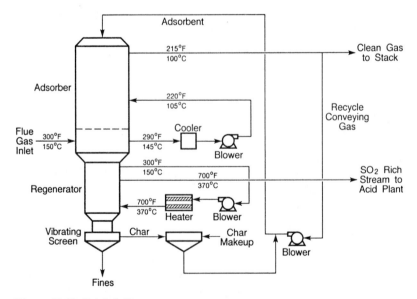

Figure 12.63 Reinluft Dry process

Selexol process

The Selexol process, a physical solvent process, uses the dimethyl ether of polyethylene glycol as a solvent since the solubilities of hydrogen sulphide, carbon dioxide and other acid gas components in this solvent are directly proportional to the partial pressures of these components. Different Selexol-based processes have been designed and used successfully for a wide range of ratios of hydrogen sulphide to carbon dioxide (Hegwer and Harris, 1970; Sweny, 1980; Johnson and Homme, 1983; *Hydrocarbon Processing*, 1990).

This process uses an organic solvent (the dimethyl ether of polyethylene glycol) and does not require the cold temperatures of the Rectisol process (see above); the absorption of gas stream 'acid' contaminants can be achieved at temperatures of the order of 20°C (68°F). The 'glycol' solvent is about ten times more selective for hydrogen sulphide then for carbon dioxide and, as with the Rectisol process, partial separation of the two gases can be accomplished.

The basic process configuration (Figure 12.64) has a somewhat different configuration to that employed (Figure 12.65) when there are low ratios of

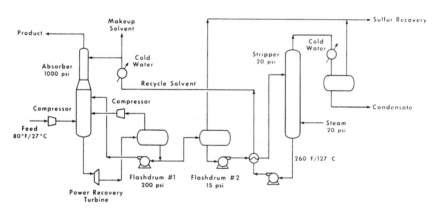

Figure 12.64 Selexol process

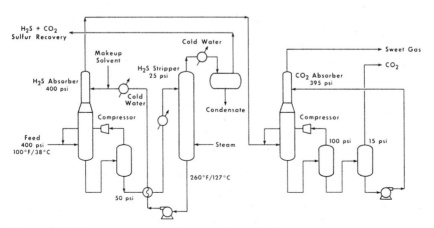

Figure 12.65 Selexol process (optional configuration)

hydrogen sulphide to carbon dioxide in the gas streams. As an example of the latter (i.e. a gas stream having a low ratio of hydrogen sulphide to carbon dioxide) the sour natural gas is dehydrated, cooled and sent to the absorber where it is contacted (countercurrent) with the Selexol solvent. Rich Selexol from the bottom of the absorber is sent via a surge tank, to remove entrained gas that is recycled back into the absorber, to a high-pressure flash unit where most of the absorbed methane and part of the carbon dioxide are released. There is also a second flash unit where most of the vapour released is carbon dioxide which is vented. Finally, the Selexol is sent to the low-pressure flash unit where hydrogen sulphide and any remaining carbon dioxide are flashed off as the vapour stream and vented to the atmosphere.

Shell Claus Off-Gas Treating (SCOT) process

The Shell Claus Off-Gas Treating (SCOT) process (cf. the Beavon process, page 268 and the SFGD process below) (*Chemical Week*, (1972) consists of two basic sections. In the first section, all sulphur compounds in the Claus tail gas stream are reduced to hydrogen sulphide over a cobalt–molybdenum catalyst at 300°C (570°F) in the presence of a reducing gas such as hydrogen and carbon monoxide (*Hydrocarbon Processing*, 1990). The exit gas is cooled and passed on to the second section where it is contacted, in a absorber, with an alkanolamine solution for the selective absorption of hydrogen sulphide. The hydrogen sulphide can be stripped from the amine solution and recycled back to the front of the Claus plant (*Hydrocarbon Processing*, 1972).

Shell Flue Gas Desulphurization (SFGD) process

This process employs a solid, fixed bed of copper on alumina as a sulphur dioxide accepter at 370–425°C (700–800°F) in a parallel passage reactor (Figure 12.66) which does not require the removal of particulate matter (to prevent bed plugging) (*Petrochemical News*, 1971). Desorption of the sulphur dioxide, at about 400°C (750°F), is accompanied by the addition of a reducing gas such as hydrogen or a hydrogen–carbon monoxide mixture. The sulphur dioxide may be used to produce sulphur, sulphuric acid or other by-products.

In the version of the process which may be applied to Claus plant tail gas treatment, the tail gas is first incinerated to oxidize all sulphur compounds to sulphur dioxide. The gas is cooled before being passed to an adsorption reactor where the removal of sulphur dioxide occurs. The sulphur dioxide produced upon regeneration is recycled to the front end of the Claus plant (Barry, 1972).

SNPA process

The SNPA (Société Nationale des Petroles d'Aquitaine) catalytic oxidation process is a wet contact process for treating Claus unit tail gases (Kohl and Riesenfeld, 1979; Polasek and Bullin, 1985).

In the process (Figure 12.67), Claus unit tail gases are first incinerated to convert sulphur to sulphur dioxide after which the gases are cooled in a waste heat boiler to 420°C (790°F). They are then passed through a converter containing a vanadium oxide–base catalyst where sulphur dioxide is converted to sulphur trioxide. The converted effluent gases are cooled in a boiler

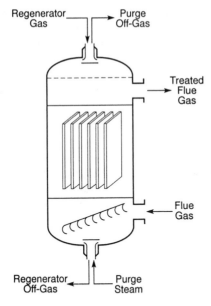

Figure 12.66 Parallel passage reactor for the Shell Flue Gas Desulphurization (SFGD) process

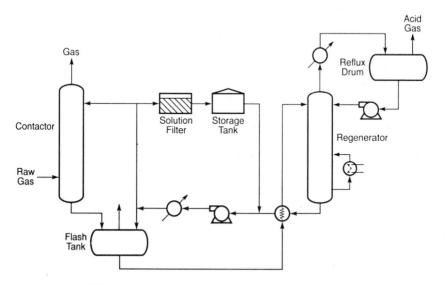

Figure 12.67 SNPA process

feed-water economizer to 300°C (570°F) and then pass to an acid concentrator and on to the absorber, in which the sulphur trioxide is absorbed to form 80 wt % sulphuric acid. The acid is then sent to the concentrator in which heat from the incoming gases evaporates part of the water to produce a 94 wt % acid.

The SNPA Sulfreen process (page 322) converts hydrogen sulphide and sulphur dioxide to sulphur at low temperatures (125–150°C, 260–300°F):

$$2H_2S + SO_2 = 3S + 2H_2O$$

Activated charcoal is used as the adsorbent and catalyst in the reactors.

The contaminants in the tail gas stream are converted by passage through several on-line parallel reactors whilst the beds in other reactors are undergoing regeneration. During the regeneration of a bed, the sulphur is vaporized by heating with a flow of inert gas circulated by a blower. The inert gas is cooled in a standard sulphur condenser where liquid sulphur is condensed and removed. When desorption is completed, the bed must be cooled before going back on-stream. To accomplish this, a portion of the cool inert gas is taken from the blower discharge and flowed through the bed to be cooled.

Stretford process

Because of deficiencies in the use of sodium carbonate/ferric oxide systems, improved liquid contacting systems were sought, especially in the area of organic materials that could be used as oxygen carriers. Such materials would tend to cut down on holding time by speeding up the reaction. Thus, the Stretford process was born! And development of the process continues (Ryder and Smith, 1963; Wilson and Newell, 1985; *Hydrocarbon Processing*, 1990) (see also the Holmes–Stretford process above).

The process uses anthraquinone disulphonic acid (ADA) as the organic oxygen carrier and utilizes alkaline solutions of sodium carbonate. The 2,7-isomer of the disulphonic acid is preferred to the 2,6-isomer because of the greater solubility of the 2,7-isomer in water. One of the products is finely divided sulphur and the process is capable of treating gases to very low hydrogen sulphide concentrations.

The overall chemistry of the process is essentially the oxidation of hydrogen sulphide to sulphur:

$$2H_2S + O_2 = 2S + 2H_2O$$

However, as might be anticipated, there being nothing so simple in life, the actual chemistry of the process is more complex and can be considered to take place in several steps (Miller and Robuck, 1972):

$$H_2S + Na_2CO_3 = NaHS + NaHCO_3$$
$$4NaVO_3 + 2NaHS + H_2O = Na_2V_4O_9 + 4NaOH + 2S$$
$$Na_2V_4O_9 + 2NaOH + H_2O + 2ADA = 4NaVO_3 + 2ADA(reduced)$$

In the oxidizer, or regenerator, the reduced anthraquinone disulphonic acid is reoxidized by blowing with air. The precipitated sulphur is permitted to overflow as a froth.

In the process (Figure 12.68), sour gas enters the bottom of the absorber and flows upwards in countercurrent contact with the Stretford solution where

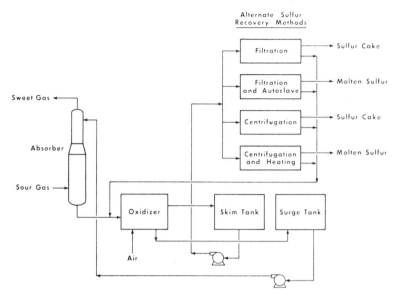

Figure 12.68 Stretford process

hydrogen sulphide is selectively absorbed. After contact with the sour gas the Stretford solution must be held or retained for several minutes in order for the reactions to pass to completion and all of the hydrogen sulphide to be converted to sulphur. The solution holdup can be either in the bottom of the absorber or in a separate vessel.

From the holding tank the Stretford solution flows to the oxidizer, where air is blown upwards through the solution to reoxidize the quinone. In the oxidizer sulphur rises to the top of the solution and is taken off for filtration. Liquid from the oxidizer and from the filter together with make-up water is mixed in a pump tank for recirculation to the absorber. Sulphur from the filter is fed to an autoclave for melting and higher purification.

The rate of thiosulphate formation depends on the partial pressure of oxygen in the sour gas stream, and on the pH and temperature of the circulating liquor. Formation of thiosulphate increases with temperature, and efforts are made to hold the contacting temperature below 40°C (100°F).

Sulfacid process

In the Sulfacid process (Figure 12.69) (Ellwood, 1969) flue gases from the fly ash removal step are contacted with weak sulphuric acid in a packed tower as a prelude to conversion of the sulphur dioxide to sulphur trioxide in the reactors (see Activated Carbon Catalyst process). Weak acid produced from the process is concentrated by this step and the incoming gas is cooled to the reactor temperature of 60–70°C (140–160°F). In the packed tower, the acid scavenges impurities from the gas.

If the content of fly ash or dust is excessive or if large volumes of flue gas are to be processed, a venturi scrubber may be used to contact the incoming gas.

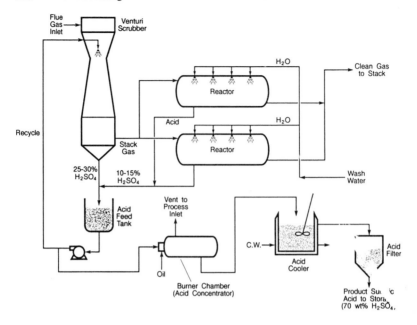

Figure 12.69 Sulfacid process

In this case, dilute acid is strengthened to only 25–30% wt/wt sulphuric acid because of the temperature drop in the venturi scrubber (page 252). A 70% wt/wt sulphuric acid may be obtained by installing a submerged flame burner in an evaporator vessel to drive off excess water. The concentrated acid may then be cooled and passed through an acid-resistant vacuum filter for the removal of soot and/or ash.

After cooling and cleaning, the incoming flue gas flows to the reactor(s) which contains beds of the catalyst (mainly of porous carbon) where the sulphur dioxide is adsorbed. Oxygen and water (sprayed through nozzles located above the beds) are then injected to promote direct conversion of the sulphur dioxide to sulphuric acid which is continuously washed out of the pore systems of the carbon by the water spray. A dilute acid of a strength of 10–15% wt/wt sulphuric acid is produced from the beds. This dilute acid is returned to the scrubber for concentration of the acid and cooling of the feed gas. The concentration of the dilute acid from the reactor(s) can be adjusted somewhat by the amount of water sprayed into the reactors. Normally, 25–30% wt/wt sulphuric acid from the scrubber bottoms is concentrated to 70% wt/wt strength by burner–evaporator, cooling, and filtration methods.

SulFerox process

Chelated iron compounds are also the heart of the SulFerox process (Figure 12.70) (see also the Lo-Cat process, page 302) which is designed for the removal of hydrogen sulphide from process gases (*Oil and Gas Journal*, 1987; *Hydrocarbon Processing*, 1990).

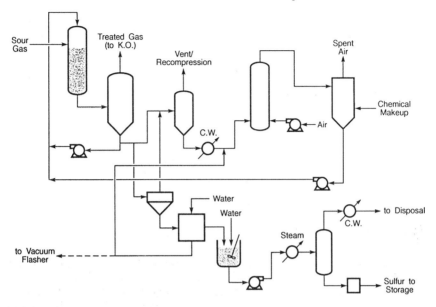

Figure 12.70 SulFerox process

Sulfiban process

In this process, hydrogen sulphide is removed from gas streams by countercurrent scrubbing with ethanolamine (MEA) solution in either a packed or spray
contactor (see Chapter 11). The pressure can be a little higher than
atmospheric pressure. The rich solution is pumped from the contactor to the
stripping column where contact can also be made with a carbon filter to
remove degradation products. The lean solution is cooled before being
recycled to the contactor.

Sulfidine process

This was the first commercially successful process of the type using aromatic
amines and was used in Germany during the 1930s. The process uses as an
absorbent a mixture of xylidine and water (approximately 1:1 by weight).
Contact takes place in a packed absorber, where the xylidine reacts with
sulphur dioxide to form xylidine sulphite. The liquid absorbent is stripped of
the sulphur dioxide by heating.

Flue gases are cleaned in electrostatic precipitators prior to processing.
Vapours from the stripping column are cooled and then scrubbed in a water
wash column to reduce the xylidine content; essentially pure sulphur dioxide is
produced from the water wash column.

Sulfinol process

The Sulfinol process (Figure 12.71) is a physical solvent process which
involves the use of a solvent (tetrahydrothiophene 1,1-dioxide or sulfolane)

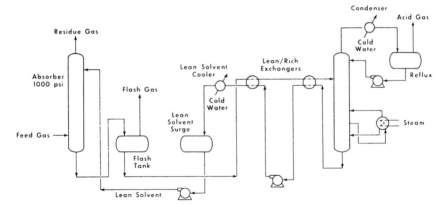

Figure 12.71 Sulfinol process

and a chemical agent (usually diisopropanolamine, but other alkanolamines are also used) (*Hydrocarbon Processing*, 1990; Taylor *et al.*, 1991) in the treating solution. This offers the distinct advantage of a continuing increase in solution carrying capacity with acid gas partial pressure. At the same time it offers the advantage of the chemically reacting solution: extremely low acid gas concentrations in the sweetened gas.

The Adip process (page 255) also uses diisopropanolamine as the chemical agent to capture gases such as hydrogen sulphide (H_2S) and carbonyl sulphide (COS) without the degradation that is inclined to occur when mono-ethanolamine is used to treat gas streams that contain carbonyl sulphide (Benson, 1981).

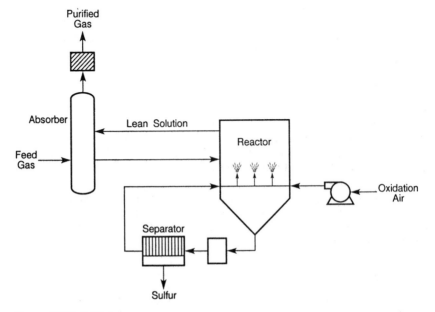

Figure 12.72 Sulfint process

Sulfint process

The Sulfint process (Figure 12.72) employs the catalytic oxidation concept (see Cat-Ox process or Catalytic Oxidation processes above) to remove hydrogen sulphide from gas streams by conversion to elemental sulphur (Mackinger *et al.*, 1982; *Hydrocarbon Processing*, 1990).

Sulfolin process

The catalytic conversion of hydrogen sulphide to sulphur (see Cat-Ox process or Catalytic Oxidation processes, page 278) is the basis of the Sulfolin process (Figure 12.74) (Davis, 1985; *Hydrocarbon Processing*, 1990) and the Sulfreen process (see Figure 12.75) (*Hydrocarbon Processing*, 1990).

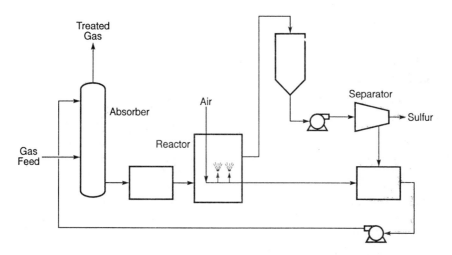

Figure 12.73 Sulfolin process

Sulfreen process

The catalytic conversion of hydrogen sulphide to sulphur (see Cat-Ox process or Catalytic Oxidation processes and SNPA process above) is the basis of the Sulfreen process (Figure 12.75) (Davis, 1985; *Hydrocarbon Processing*, 1990) and the Sulfolin process (see above) (*Hydrocarbon Processing*, 1990).

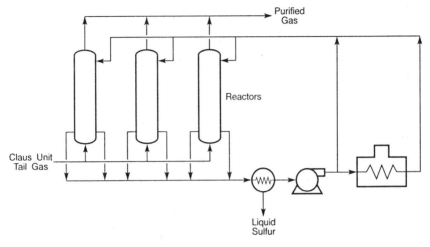

Figure 12.74 Sulfreen process

Sulphite process(es)

The basic chemistry of the sulphite process(es) is quite similar to that of the Wellman–Lord process in which a stack gas rich in sulphur dioxide is contacted with a sodium sulphite solution. The resulting sodium bisulphite can be converted back to sodium sulphite, purified and crystallized to yield a high-purity product. Or the sodium bisulphite can be neutralized with calcium hydroxide and catalytically oxidized with air to form a gypsum product.

This process (Figure 12.73) (Davis, 1971), which is similar to the Hot Potassium Carbonate process (page 270) used for the removal of carbon dioxide, depends on the reaction of sulphur dioxide with sodium or potassium sulphite to form the corresponding bisulphite:

$$SO_2 + Na_2SO_3 + H_2O = 2NaHSO_3$$

The lower solubility of sodium sulphite may cause crystallization to occur in the regeneration stage, thus increasing the proportion of bisulphite in solution and favouring regeneration. The sulphite crystallized is redissolved and returned to the absorber system. The rich solution from the absorber is injected into the circulating stream of the evaporator–crystallizer, where sulphur dioxide is released and sodium sulphite formed. The sodium sulphite is separated from the mother liquor in a centrifuge or settling tank, redissolved and returned to the absorber.

The regenerated sulphur dioxide leaves the evaporator–crystallizer mixed with about ten times its volume of water vapour. Some of this vapour is condensed and returned to the system, and the wet sulphur dioxide gas is fed into the drying tower or other convenient point in the sulphuric acid plant. The process therefore operates to increase the overall yield of acid from the available raw material.

A portion of the active absorbent may react to form sulphate which cannot be regenerated. This results partly from the fact that the tail gas contains sulphuric acid as a mist or vapour and traces of sulphur trioxide and partly

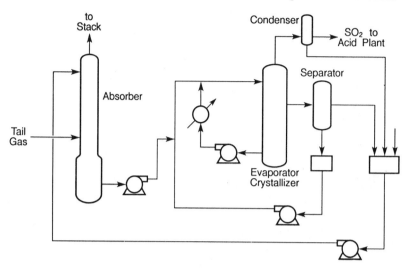

Figure 12.75 Sulphite process

from oxidation of the sodium sulphite or bisulphite by the excess oxygen also present in the tail gas. This oxidation in solution may be prevented by the use of anti-oxidants such as hydroquinone. However, so long as sodium sulphate is produced in the cycle, it must be purged from the system. Disposal of this purge stream remains one of the difficulties of the process.

Takahax process

The Takahax process is similar to the Stretford process (see above) in so far as it employs a quinone derivative (the sodium salt of 1,4-naphthaquinone 2-sulphonic acid; NSS) as the organic oxygen carrier and alkaline solution (sodium carbonate) as well as producing finely divided sulphur. Moreover, the process is capable of treating gases to very low hydrogen sulphide concentrations:

$$Na_2CO_3 + H_2S = NaHS + NaHCO_3$$
$$NaHS + NSS = S + NSS(reduced)$$

In the regenerator, air is bubbled through the solution to provide oxidation of the naphthahydroquinone sulphonate to naphthaquinone sulphonate. The sulphur produced in the process is removed from the system by taking a slip stream off the circulating system and passing it through a filter.

In cases where carbon dioxide is present in the gas stream, the carbonate solution may pick up some of the carbon dioxide but the amount depends upon the pH of the solution. Unless the content of the carbon dioxide in the gas stream is exceedingly high, the sodium bicarbonate is decomposed and the carbon dioxide stripped from the solution by the regeneration air. There is no release of hydrogen sulphide from the regenerator.

As a general process description (Figure 12.76), the sour gas stream enters at the bottom of the absorber and flows upwards in countercurrent contact with

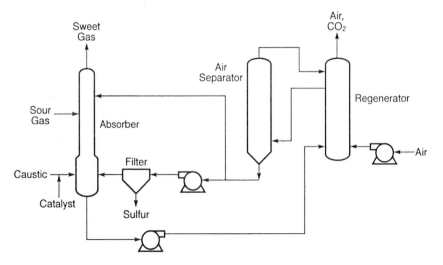

Figure 12.76 Takahax process

fresh solution from the regenerator. Sweetened gas leaves the top of the absorber, and rich solution flows to the bottom of the absorber, where make-up caustic and catalyst are added, as is the filtrate from the sulphur filter.

The solution circulation pump moves solution to the regenerator, where it is blown with air for regeneration. Any carbon dioxide absorbed is rejected with the regeneration air at the top of the regenerator. Solution overflows the regenerator into a separator. The liquid underflow from the separator is divided: part of it flows to the top of the absorber for contacting the sour gas; and the other part flows to the filter pump and through the filter for the removal of sulphur.

If hydrogen cyanide is present in the gas stream, as might be the case with industrial gases, it will be absorbed by the alkali and be oxidized to sodium thiocyanate causing a net consumption of the alkali with the ensuing purging of the solution to limit the concentration of thiocyanate. If sulphur dioxide is present, it will consume caustic and be oxidized to sodium sulphate. This may also build up in concentration in the solution, requiring a purge step. Should ammonia be present in the gas, it may well be beneficial since it will reduce the addition of alkali required to maintain the pH of the solution.

Townsend process

The Townsend process is similar to the IFP process (page 295), in that it uses an organic solvent to allow hydrogen sulphide and sulphur dioxide to react (Claus reaction) to form elemental sulphur. The reactor is operated at a temperature above the melting point of sulphur, so that liquid sulphur is produced from the bottom. This process may be applied directly to the treatment of Claus plant gas.

Trentham Trencor process

The process uses an organic solvent to absorb sulphur dioxide preferentially from the gas. A catalytic oxidation step is first used to convert all sulphur compounds to sulphur dioxide and the gas then flows to the absorber where the sulphur dioxide is absorbed. The sulphur dioxide is released from the solvent in the regenerator and recycled to the front end of the Claus plant. The process flow scheme is similar to an amine gas-sweetening plant scheme.

Water Wash (Aquasorption) process

This process is effective for natural gas that has a high acid gas content (including a high ratio of hydrogen sulphide to carbon dioxide) that is also under high pressure. In this particular type of process, sour natural gas is passed in an upward direction through a contactor in which the gas flows countercurrent to the water (Figure 12.77). The partially sweetened gas is then passed on for further treatment (e.g. to an amine unit).

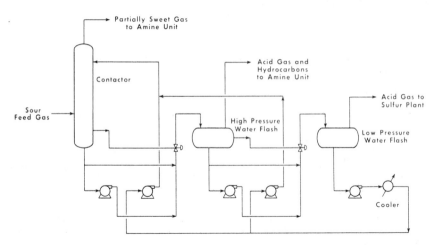

Figure 12.77 Water Wash (Aquasorption) process

The rich water solution from the bottom of the tower is sent to a pressurized flash tank for the recovery of dissolved hydrocarbons. The water is then repressurized before sending it to a lower-pressure flash tank where all of the acid gas is removed and water obtained for recycling.

Wellman–Lord process

In the Wellman–Lord process (Figure 12.78) (Potter and Craig, 1972) the first step may be a wet scrubbing with water to remove fly ash and sulphur trioxide. The cleaned and cooled flue gases then flow to a second gas-scrubbing step, where the sulphur dioxide is removed by contact with an aqueous solution of sodium sulphite (Na_2SO_3):

$$Na_2SO_3 + SO_2 + H_2O = 2NaHSO_3$$

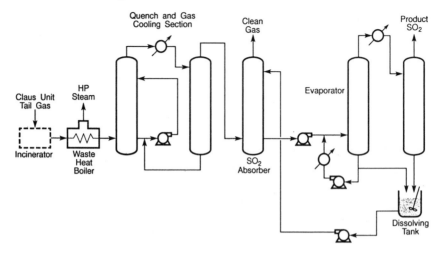

Figure 12.78 Wellman–Lord process

Dehydration of the sodium bisulphite solution brings about the formation of sodium pyrosulphite ($Na_2S_2O_5$):

$$2NaHSO_3 = Na_2S_2O_5 + H_2O$$

When sulphur trioxide is present, the pyrosulphite is formed in the scrubber:

$$2Na_2SO_3 + SO_3 = Na_2SO_4 + Na_2S_2O_5$$

In the presence of oxygen, oxidation of the sodium sulphite occurs:

$$2Na_2SO_3 + O_2 = 2Na_2SO_4$$

The spent solution from the scrubber passes to an evaporator–crystallizer where the water is evaporated from the sodium bisulphite solution. The sulphur dioxide is released and sodium sulphite is formed:

$$2NaHSO_3 = Na_2SO_3 + SO_2 + H_2O$$
$$Na_2S_2O_5 = Na_2SO_3 + SO_2$$

Caustic (sodium hydroxide) is added to the solution recycled from the regeneration step to the scrubber (absorption step). Reaction between the caustic and the sodium bisulphite occurs to form additional sodium sulphite:

$$NaOH + NaHSO_3 = Na_2SO_3 + H_2O$$

Alternatively, the reaction of caustic with sulphur dioxide may yield sodium bisulphite:

$$NaOH + SO_2 = NaHSO_3$$

The Wellman–Lord process for the removal of sulphur dioxide also finds application in the areas of Claus tail gas treatment and boiler flue gas treatment. In the process modification for Claus tail gas treatment, the tail gas is first incinerated, then cooled in a waste heat boiler and by an adiabatic water quench system to 60°C (140°F), and finally fed to the absorber for the removal of sulphur dioxide. The product sulphur dioxide released from the absorbent in the evaporator–crystallizer is cooled and recycled to the front end of the Claus unit.

Westvaco Dry process

This process uses the activated carbon adsorption approach to remove sulphur dioxide from flue gases. The reactor is located between the electrostatic precipitator and the stack (Figure 12.79). In the sulphur dioxide removal step, the activated carbon catalyses the reaction of sulphur dioxide with oxygen in the flue gas to form sulphur trioxide which is hydrolysed by water to form sulphuric acid which remains adsorbed in the carbon pores (see also Activated Carbon Catalyst process, page 255).

The carbon is regenerated by reacting the sulphuric acid on the carbon with hydrogen sulphide to form sulphur in the sulphur generator:

$$H_2SO_4 + 3H_2S = 4S + 4H_2O$$

One-fourth of the sulphur derived from the sulphuric acid is recovered from the carbon by direct vaporization in the sulphur stripper. The remaining sulphur is reacted with hydrogen in the hydrogen sulphide generator to provide hydrogen sulphide for the reduction step:

$$H_2 + S = H_2S$$

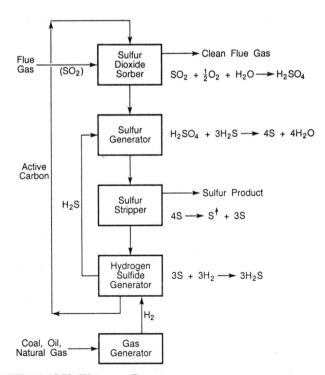

Figure 12.79 Westvaco Dry process

The hydrogen for the hydrogen sulphide generator is supplied by a gas producer or reformer.

The overall chemistry of the process may be represented as

$$2SO_2 + O_2 + 6H_2 = 2S + 6H_2O$$

Thus, sulphur dioxide is removed from the flue gas at the expense of hydrogen consumption, but the process yields elemental sulphur as a useful by-product.

Zinc Oxide process

In this process, flue gases are contacted with a solution of sodium sulphite (Na_2SO_3) and sodium bisulphite ($NaHSO_3$). The absorption of sulphur dioxide causes an increase in the bisulphite content after which the solution leaves the absorber and passes to a clarifier where particulate matter is separated, and finally to a mixer where it is treated with zinc oxide (ZnO). At this point, the original ratio of sulphite to bisulphite is restored and a precipitate of zinc sulphite ($ZnSO_3$) is formed. This zinc sulphite precipitate is removed by settling and filtration, and the filter cake is dried and calcined.

Calcination of the zinc sulphite produces a gas consisting of 30% sulphur dioxide and 70% water vapour. The gas is cooled, dried and compressed to produce a virtually pure liquid sulphur dioxide. Zinc oxide from the calciner is recycled back to the process.

In another variation of the process, called the Chemsweet process (page 282) (Manning, 1979; *Hydrocarbon Processing*, 1990), the gas stream is contacted by a fine suspension of zinc oxide particles in zinc acetate solution whereby hydrogen sulphide, mercaptans and carbonyl sulphide react instantaneously to produce insoluble zinc sulphide and zinc mercaptide:

$$H_2S + ZnO = H_2O + ZnS$$
$$2RSH + ZnO = (RS)_2Zn + H_2O$$

Zinc ferrite ($ZnFe_2O_4$) has also been claimed to be suitable for gas clean-up (Grindley and Steinfeld, 1985). It is produced by heating a mixture of zinc and iron oxides, and tends to combine the high sulphur affinity of zinc oxide with the high regenerability of iron oxide.

References

Barron, A.V. Jr (1971) *Combustion*, p. 44 (October)

Barry, C.B. (1972) *Hydrocarbon Processing*, p. 102 (April)

Bartoo, R.K. (1985) In *Acid and Sour Gas Treating Processes* (ed. S.A. Newman), Gulf Publishing Company, Houston, TX, Chapter 13

Benson, H.E. (1981) In *Chemistry of Coal Utilization* (ed. M.A. Elliott), Wiley, New York, Chapter 25

Benson, H.E. and Parrish, R.W. (1975) *Hydrocarbon Processing*, **53**, (4), 81

Buckingham, P.A. and Homan, H.R. (1971) *Hydrocarbon Processing*, **50**, (8), 121

Chemical Engineering (1969) 28 July, p. 67

Chemical Engineering Progress (1973) **67**, (12), 54

Chemical Week (1972) 4 October, p. 47

Chiu, C.-H. (1990) *Hydrocarbon Processing*, **69**, (1), 69

Chludzinski, G.R. and Wiechert, S. (1986) *Spring Meeting, American Institute of Chemical Engineers, Houston, Texas, April*, Paper 58e

Conviser, S.A. (1965) *Oil and Gas Journal*, **63**, (49), 130

Davis, G.W. (1985) *Oil and Gas Journal*, **83**, (8), 108

Davis, J.C. (1971) *Chemical Engineering*, **78**, (27), 43

Davis, J.C. (1972) *Chemical Engineering*, 15 May, p. 66

Dingman, J.C. and Moore, T.F. (1968) *Hydrocarbon Processing*, **47**, (7), 138

Dowling, N.I., Hyne, J.B. and Brown, D.M. (1990) *Industrial Engineering Chemistry Research*, **29**, 2327

Eickmeyer, A.G. (1971) *Oil and Gas Journal*, **69**, 74

Ellwood, P. (1969) *Chemical Engineering*, 16 June, p. 62

Falkenberry, H.L. and Slack, A.V. (1969) *Chemical Engineering Progress*, **65**, (12), 61

Field, J.H., Kurtzrock, R.C. and McCrea, D.H. (1967) *Chemical Engineering*, 19 June, p. 158

Gangriwala, H.A. and Chao, I-M. (1985) In *Acid and Sour Gas Treating Processes* (ed. S.A. Newman), Gulf, Houston, TX, Chapter 14

Goar, B.G. (1968) *Hydrocarbon Processing*, **47**, (9), 248

Goar, B.G. (1971a) *Oil and Gas Journal*, 12 July, p. 75

Goar, B.G. (1971b) *Oil and Gas Journal*, 12 July, p. 84

Goldstein, A.M., Edelman, A.M. and Ruziska, P.A. (1985) In *Acid and Sour Gas Treating Processes* (ed. S.A. Newman), Gulf Publishing Company, Houston, TX, Chapter 12

Grindley, T. and Steinfeld, G. (1985) In *Acid and Sour Gas Treating Processes* (ed. S.A. Newman), Gulf Publishing Company, Houston, TX, Chapter 16

Hardison, L.C. (1985a) *Hydrocarbon Processing*, **64**, (4), 70

Hardison, L.C. (1985b) In *Acid and Sour Gas Treating Processes* (ed. S.A. Newman), Gulf Publishing Company, Houston, TX, Chapter 24

Hegwer, A.M. and Harris, R.A. (1970) *Hydrocarbon Processing*, **49**, (4), 103

Hicks, R.L. and Senules, E.A. (1991) *Hydrocarbon Processing*, **70**, (4), 55

Hurst, B.E. (1985) *Proc. Joint Symp. on Stationary Combustion NO_x Control, Boston, 9 May*

Huval, M. and van de Venne, H. (1981) *Oil and Gas Journal*, 17 August, p. 91

Hydrocarbon Processing (1972) **51**, (10), 23

Hydrocarbon Processing (1973) **53**, (10), 95

Hydrocarbon Processing (1990) **69**, (4), 69 *et seq.*

Johnson, J. and Homme, A. (1983) *Proc. Gas Processing Symp. Summer National Meeting, American Institute of Chemical Engineers*

Katell, S. (1966) *Chemical Engineering Progress*, **62**, (10), 67

Kohl, A.L. and Buckingham, P.A. (1960) *Oil and Gas Journal*, **58**, (19), 35

Kohl, A.L. and Riesenfeld, F.C. (1979) *Gas Purification*, McGraw-Hill, New York

Kriebel, M. (1985) In *Acid and Sour Gas Treating Processes* (ed. S.A. Newman), Gulf Publishing Company, Houston, TX, Chapter 4

Long, K. and Parr, G. (1980) *Chemical Processing*, September

Mackinger, H., Rossati, F. and Schmidt, G. (1982) *Hydrocarbon Processing*, **61**, (3), 169

Maddox, R.N. (1982) *Gas Conditioning and Processing*, Vol. 4, *Gas and Liquid Sweetening*, Campbell, Norman, OK

Manning, W.P. (1979) *Oil and Gas Journal*, **77**, (42), 122

Maurin, P.G. and Jonakin, R. (1970) *Chemical Engineering*, 27 April, p. 27

Meissner, H. and Heffner, W. (1990) *Proc. 1990 Eur. Conf. on Energy Efficient Production of Fertilizers, Bristol, England*

Miller, S.G. and Robuck, R.D. (1972) *Journal of Petroleum Technology*, p. 545 (May)

Moore, T.F., Dingman, J.C. and Johnson, F.L. Jr (1985) In *Acid and Sour Gas Treating Processes* (ed. S.A. Newman), Gulf Publishing Company, Houston, TX, Chapter 11

Newman, S.A. (1985) *Acid and Sour Gas Treating Processes*, Gulf Publishing Company, Houston, TX, Chapter 14

Oil and Gas Journal (1970) **68**, (30), 131

Oil and Gas Journal (1987) **85**, (21), 54

Oil and Gas Journal (1991) **89**, (29), 54

Oldenkamp, R.D. and Margolin, E.D. (1969) *Chemical Engineering Progress*, **65**, (11), 73

Ouwerkerk, C. (1978) *Hydrocarbon Processing*, **57**, (4), 89

Paradowski, H. and Castel, H. (1987) *Proc. XVIIth Int. Congr. of Refrigeration, Vienna (Wien)*, A, p. 167

Petrochemical News (1971) 28 June, p. 3

Petroleum Refiner (1955) **34**, (9), 182

Polasek, J. and Bullin, J. (1985) In *Acid and Sour Gas Treating Processes* (ed. S.A. Newman), Gulf Publishing Company, Houston, TX, Chapter 7

Potter, B.H. and Craig, T.L. (1972) *71st Annual Convention, American Institute of Chemical Engineers, Dallas, 22 February*, Paper No. 52c

Ranke, G. and Mohr, V.H. (1985) In *Acid and Sour Gas Treating Processes* (ed. S.A. Newman), Gulf Publishing Company, Houston, TX, Chapter 3

Remirez, R. (1968) *Chemical Engineering*, 21 October, p. 54

Rushton, D.W. and Hayes, W. (1961) *Oil and Gas Journal*, **59**, (38), 102

Russell, F.G. (1985) In *Acid and Sour Gas Treating Processes* (ed. S.A. Newman), Gulf Publishing Company, Houston, TX, Chapter 21

Ryder, C. and Smith, A.V. (1963) *Journal of the Institute of Gas Engineers*, **3**, 283

Savage, D.W., Bisio, A., Sartori, G., Say, G.R., Heinzelmann, F.J. and Iyengar, J.N. (1985) In *Acid and Sour Gas Treating Processes* (ed. S.A. Newman), Gulf Publishing Company, Houston, TX, Chapter 15

Say, G.R., Heinzelmann, F.J., Iyengar, J.N., Savage, D.W., Bisio, A. and Sortori, G. 1985) In *Acid and Sour Gas Treating Processes* (ed. S.A. Newman), Gulf Publishing Company, Houston, TX, Chapter 15

Shah, I.S. (1971) *Chemical Engineering Progress*, **67**, (5), 51

Sigmund, P.W. (1981) *Hydrocarbon Processing*, **60**, (5), 118

Slack, A.V. (1967) *Chemical Engineering*, 4 December, p. 188

Smith, R.S. and Skiff, T.B. (1990) *Proc. Lawrence Reid Gas Conditioning Conf., March*

Speight, J.G. (1981) *The Desulfurization of Heavy Oils and Residua*, Marcel Dekker, New York

Squires, A.M. (1967) *Chemical Engineering*, 20 November, 133

Stites, J.G., Horlacher, W.R. Jr, Bachofer, J.L. Jr and Bartman, J.S. (1969) *Chemical Engineering Progress*, **65**, (10), 74

Stookey, D.J., Graham, T.E. and Pope, W.M. (1985) In *Acid and Sour Gas Treating Processes* (ed. S.A. Newman), Gulf Publishing Company, Houston, TX, Chapter 20

Sweny, J.W. (1980) *Proc. 59th Annual Convention, Gas Processors Association, Houston, TX 17–19 March*

Taylor, N.A., Hugill, J.A., van Kessel, M.M. and Verburg, R.P.J. (1991) *Oil and Gas Journal*, **89**, (33), 57

Uno, Y., Fukui, S., Atsukawa, M., Higashi, M., Yamada, H. and Kamei, K. (1970) *Chemical Engineering Progress*, **66**, (1), 61

Wilson, B.M. and Newell, R.D. (1985) In *Acid and Sour Gas Treating Processes* (ed. S.A. Newman), Gulf Publishing Company, Houston, TX, Chapter 23

Yulish, J. (1971) *Chemical Engineering*, 14 June, p. 58

Appendix

Conversion factors for US units and SI units

SI units	Multipliers	US units
°C (diff)	$\times 1 \rightarrow$	K (diff)
°C	$\times (t_C \times \frac{9}{5}) + 32 \rightarrow$	°F
	$\leftarrow (t_F - 32)\frac{5}{9} \times$	
cm	$\times 0.3937 \rightarrow$	in
	$\leftarrow 2.54 \times$	
cm^2	$\times 0.1550 \rightarrow$	in^2
	$\leftarrow 6.452 \times$	
cm^3	$\times 3.382 \times 10^{-2} \rightarrow$	oz (US fluid)
	$\leftarrow 29.57 \times$	
g	$\times 2.205 \times 10^{-3} \rightarrow$	lb
	$\leftarrow 453.6 \times$	
g	$\times 28.35 \rightarrow$	oz (avoirdupois)
	$\leftarrow 0.03527 \times$	
J	$\times 0.2390 \rightarrow$	cal
	$\leftarrow 4.186 \times$	
J	$\times 9.480 \times 10^{-4} \rightarrow$	Btu
	$\leftarrow 1055 \times$	
J	$\times 0.7376 \rightarrow$	ft lb
	$\leftarrow 1.356 \times$	
J/kg K	$\times 2.388 \times 10^{-4} \rightarrow$	Btu/lb °F
	$\leftarrow 4.187 \times 10^3 \times$	
J/m s K	$\times 0.5778 \rightarrow$	Btu/h ft °F
	$\leftarrow 1.731 \times$	
kg	$\times 2.205 \rightarrow$	lb (mass)
	$\leftarrow 0.4536 \times$	
K	$t_K = (t_F + 459.7)\frac{5}{9}$	°F
	$t_F = (t_K - 273.1)\frac{9}{5} + 32$	
kg	$\times 10^{-3} \rightarrow$	t (metric tons)
	$\leftarrow 1000 \times$	
kg	$\times 1.102 \times 10^{-3} \rightarrow$	US short ton
	$\leftarrow 907.2 \times$	
kg	$\times 9.842 \times 10^{-4} \rightarrow$	US long ton
	$\leftarrow 1016 \times$	
kg/m^3	$\times 0.06243 \rightarrow$	lb/ft^3
	$\leftarrow 16.02 \times$	
kJ	$\times 2.777 \times 10^{-4} \rightarrow$	kW h
	$\leftarrow 3.600 \times 10^3 \times$	
km	$\times 0.6214 \rightarrow$	mile (US statute)
	$\leftarrow 1.609 \times$	

SI units	Multipliers	US units
kPa	$\times 9.872 \times 10^{-3} \rightarrow$ $\leftarrow 101.3 \times$	std atm
kW	$\times 1.341 \rightarrow$ $\leftarrow 0.7457 \times$	hp (US)
l	$\times 0.03532 \rightarrow$ $\leftarrow 28.31 \times$	ft^3
l	$\times 0.2642 \rightarrow$ $\leftarrow 3.785 \times$	gal (US)
m	$\times 10^{10} \rightarrow$ $\leftarrow 10^{-10} \times$	Ångstrom
m	$\times 39.37 \rightarrow$ $\leftarrow 0.0254 \times$	in
m	$\times 3.281 \rightarrow$ $\leftarrow 0.3048 \times$	ft
m^2	$\times 10.76 \rightarrow$ $\leftarrow 0.0929 \times$	ft^2
m^3	$\times 35.31 \rightarrow$ $\leftarrow 0.02832 \times$	ft^3
m^3	$\times 264.2 \rightarrow$ $\leftarrow 3.785 \times 10^{-3} \times$	gal (US)
m^3	$\times 6.290 \rightarrow$ $\leftarrow 0.1590 \times$	bbl (US petroleum, 42 gal)
m^3/s	$\times 1585 \rightarrow$ $\leftarrow 6.308 \times 10^{-5}$	gpm
MPa	$\times 0.1 \rightarrow$ $\leftarrow 10 \times$	bars
MJ	$\times 0.3724 \rightarrow$ $\leftarrow 2.685 \times$	hp h
MJ/m^3	$\times 26.84 \rightarrow$ $\leftarrow 3.725 \times 10^{-2} \times$	Btu/ft^3
N	$\times 0.2248 \rightarrow$ $\leftarrow 4.448 \times$	lb (force)
Pa	$\times 75.02 \rightarrow$ $\leftarrow 1.333 \times 10^{-2} \times$	mm of Hg
Pa	$\times 2.961 \times 10^{-4} \rightarrow$ $\leftarrow 3377 \times$	in of Hg
Pa	$\times 1.450 \times 10^{-4} \rightarrow$ $\leftarrow 6895 \times$	lb/in^2
Pa s	$\times 10 \rightarrow$ $\leftarrow 0.100 \times$	poises (abs. viscosity)
W	$\times 1.341 \times 10^{-3} \rightarrow$ $\leftarrow 745.7 \times$	hp (US)

Index